Energy Futures

Daniel Soeder

Energy Futures

The Story of Fossil Fuel, Greenhouse Gas, and Climate Change

 Springer

Daniel Soeder
Soeder Geoscience LLC
Tunnelton, WV, USA

ISBN 978-3-031-15383-9   ISBN 978-3-031-15381-5   (eBook)
https://doi.org/10.1007/978-3-031-15381-5

This Springer imprint is published by the registered company Springer Nature Switzerland AG
The registered company address is: Gewerbestrasse 11, 6330 Cham, Switzerland

To my grandson, Riley James.

Preface

Energy "futures" are commodities that are bought and sold by stockbrokers. This book is not about those kinds of energy futures, but about the real future of energy, and the choices we have to make. Human civilization is currently on an energy path that is not sustainable. We are eventually going to run out of fossil fuels, but long before we burn the last lump of coal or pump the last drop of oil, we will severely impact the climate and degrade the environment. Decisions we make over the next decade will determine our future a century from now.

Climate change is not a hoax, despite those who claim it is. Burning fossil fuel increases the amount of heat-trapping gases in the atmosphere that warm things up and change the climate. The evidence is clear and compelling. Stronger storms, retreating glaciers, more intense and longer-lived droughts, killer heat waves in unusual places like Siberia, blizzards in unusual places like southern Texas, and huge wildfires are all signs of human influence on climate. Given that humans have changed the flow of rivers, modified coastlines, contaminated vast swaths of groundwater, polluted the oceans with plastic, created hazardous air in cities, and radically changed landscapes, how can anyone doubt that our actions are also affecting the climate?

Climate issues are complex, politically charged, and often either innocently or deliberately misunderstood. There are copious amounts of both information and disinformation on social media, blogs, podcasts, and in old-fashioned written documents. Sorting it out can be a challenge. Ordinary people need to understand climate because the issue affects us all. Most existing books and articles on climate change tend to approach it with either overly simplified stories (that are sometimes inaccurate) or dive deeply into complex technical explanations that almost require a PhD to understand. My entry into this field attempts to accurately explain the causes and effects of climate change in a manner understandable to most members of the general public along with providing some suggestions for practical solutions to fix the climate crisis.

Denial and resistance from the fossil fuel industry and their political supporters are intended to delay climate actions and maintain our dependence on fossil energy. Their goal seems to be for society to continue filling industry coffers with profits until civilization collapses. What we are all supposed to do after that is never quite made clear in the narrative.

Some people react to the climate crisis with indifference, while others wring their hands in despair and say we're all doomed. The problem with both of these extremes is that nothing gets done. On the one hand, ignoring the climate crisis won't stop heatwaves or drought-driven wildfires. On the other, I will remind those in despair that every doomsday prediction throughout history has been wrong and this one almost certainly is wrong as well.

The climate can be fixed, and it must be fixed. We know exactly what caused the problem and we know exactly what must be done to address it. The cause was technology, and technology can solve it. First, we must replace fossil fuels with clean, sustainable energy to keep the problem from getting worse. Many of these clean energy technologies either already exist or will become practical quickly with an infusion of funding. A process called geoengineering can then be used to employ existing technology to reduce the concentrations of greenhouse gas in the atmosphere to pre-industrial levels. A determined effort backed by undeniable facts is needed, and research on both clean energy and geoengineering would move faster if governments gave it the funding and the urgency they give to defense programs. We are in big trouble here, and every delay adds challenges and costs.

In the four decades since the issue of human-induced climate change became widely understood, neither the fossil fuel industry nor the "free market" took any action to address it. The only option remaining to make the changes necessary to avoid the worst consequences of the climate crisis is government policy. Thus, it is critical for the concerned citizens of all nations to be informed and knowledgeable about the details of fossil fuels, greenhouse gas, and climate change. We have seen time and again that when enough citizens demand that their government do the right thing, changes are made. Indeed, in the United States, civil rights, gender equality, voting rights, workplace safety rules, environmental laws, affordable medical care, and the lifting of racial and gender marriage restrictions all came about because people demanded them from political leaders. The climate crisis requires the same level of commitment, if not more.

An informed populace can push on government officials to act. Politicians of all stripes and all parties everywhere have a tendency to dodge or change the subject when asked difficult questions. Knowledgeable constituents give them less room to maneuver. There is no doubt that dealing with the climate is challenging and will cause some political pain. Many political leaders seem eager to produce a sound bite about the urgency of the climate crisis but then leave the politically unpopular and hard decisions for someone else. This is unacceptable, and we need leaders who will take a stand today to address climate change.

It is impossible to write a book about climate and government policy without expressing some opinions. I have been careful to identify my opinions as such, and separate them from the straight, factual information. The more important facts are referenced, but I've tried to use these sparingly because I find excessive references distracting when reading a book. Climate science, like most science, has a lot of specialized acronyms and terms. Hopefully, these are kept to a minimum – so readers are not swimming in alphabet soup – and each is defined at first use. There is also a glossary included at the end for quick reference. The only acknowledgement I want to offer is to the readers of this book. Thank you for making the investment in time, money, or effort to learn about the climate crisis. It's important.

Maybe you've heard of me; maybe you haven't. I spent 45 years as an Earth science researcher and educator, and I have explained a lot of things to a lot of people. I think I've gotten pretty good at it, including publishing three other books prior to this one. I've worked on oil and gas, nuclear energy, groundwater, and environmental issues. I know a few things about energy and climate. If you don't trust me, trust the references. I have tried to keep the text accurate, but as they say when there is a flaw in a Persian carpet, "Only Allah is perfect." Since this is the first edition, there will be an opportunity to correct mistakes in future editions, and I would appreciate knowledgeable readers pointing these out.

The climate crisis is the existential threat of our times. Certainly, people have survived some bad things in the past like world wars, pandemics, and economic downturns, but the climate crisis could overshadow all of these. The only equivalent thing I can think of is a full scale nuclear war. Both will leave large swaths of the planet literally uninhabitable.

Unlike many of the other crises, we know this one is coming. If we don't act, we have no one to blame but ourselves. Advance warnings about climate were being forecast by primitive computer models in the 1980s, most of which under-predicted the severity of the effects. Things have actually turned out to be worse, and our current, much more sophisticated computer models show some really bad scenarios in the future if we don't change.

A smart deer in the road recognizes that the headlights of an oncoming car are an obvious, imminent danger and gets out of the way. The not-so-smart deer ends up as roadkill. Which one are we?

Tunnelton, WV, USA   Daniel Soeder

Contents

About the Author

Daniel Soeder has 45 years of experience as a research scientist and geologist working on issues related to energy and the environment. His background includes a decade of research on the geology of natural gas resources at the Gas Technology Institute in Chicago, followed by 18 years with the U.S. Geological Survey (USGS) coordinating hydrologic and geologic fieldwork at the proposed Yucca Mountain high-level nuclear waste repository site in Nevada, and researching coastal hydrology, wetlands, water supply, and water contamination in the Mid-Atlantic. He chaired the Scientific and Technical Advisory Committee (STAC) for the Delaware Estuary Program for 3 years. He transferred from the USGS to the U.S. Department of Energy (DOE) National Energy Technology Laboratory in Morgantown, West Virginia, in 2009 where he spent 8 years performing energy and environmental research on gas shale and other unconventional fossil energy resources. He took an early retirement from the government to direct the Energy Resources Initiative at the South Dakota School of Mines & Technology in Rapid City, South Dakota, from 2017 to 2020, and he currently runs a consulting firm primarily engaged in the geologic sequestration of carbon dioxide. He has authored multiple reports, scientific papers, and three books on shale development and hydraulic fracturing. He has a B.S. degree in geology from Cleveland State University, and an M.S. degree in geology from Bowling Green State University (Ohio). He has three adult children and a grandchild who deserve a better world.

The author at left leading a public tour group on the crest of Yucca Mountain in May 1996. (Photograph by Chuck Savard, U.S. Geological Survey. Used with permission.)

Chapter 1
The Controversy

Keywords Climate change · Climate skeptics · Fossil fuel

****

The hot-button question in the first quarter of the twenty-first century is the following: are humans responsible for changing the Earth's climate or not? Nearly all scientists, including me, are fully convinced that burning the fossil fuels that provide 80–90% of our energy emits combustion products into the atmosphere that are contributing significantly to changes in the climate of Planet Earth. This claim is supported by abundant evidence that is obvious, clear, compelling, aligned with the known laws of physics and chemistry, internally consistent, and reproducible. These are all hallmarks of good science. Estimates for the potential future impacts on climate range from "moderate" to "calamitous."

This alarming notion is vigorously rejected as hyperbole by some politicians and business people who claim that human activities are not affecting the Earth at all. They declare that any change in the climate is a natural event that has nothing to do with the combustion of fossil fuels. Those who deny that humans are affecting the climate prefer the term "climate skeptic" to "climate denier." I would like to point out that in order to be a climate skeptic, you must first deny the validity of substantial amounts of detailed and internally consistent climate science. However, I will use the term skeptic instead of denier if it will avoid offending people. Climate skeptics are not going to be happy with this book in any case.

Some scientists aligned directly or indirectly with the fossil fuel industry are among the climate skeptics, providing a veneer of scientific legitimacy to the climate science denial arguments.[1] Pseudo-technical discussions from these people claiming that carbon dioxide emissions from humans are insignificant or that CO_2 emissions don't actually trap heat only serve to confuse the public. This is exactly the purpose. Confusion produces uncertainty, which prevents climate action, preserves the status quo, and allows the fossil fuel industry to continue selling their products at a profit. We will explore the evidence in detail in Chap. 5, but it is

---

[1] Koonin, Steven E., 2021, Unsettled: What Climate Science Tells Us, What It Doesn't, and Why It Matters: Dallas, TX; BenBella Books, 320 p.

D. Soeder, *Energy Futures*, https://doi.org/10.1007/978-3-031-15381-5_1

important for readers to understand that there are huge amounts of data supporting the notion of human-induced climate change, and very little data that refute it.

Although climate skeptics by and large are closely associated with the very fossil fuels that are the root cause of the problem, they will often present the issue as being unfairly biased in favor of climate change activists. A typical response is to call for a more "balanced" approach to invoke a notion of fairness for allowing equal time arguments on each side, even though one side has way more evidence than the other. Skeptics will also promote a sense of uncertainty about the validity of the data and urge caution about not jumping to conclusions with so much that is unknown. Their arguments often finish off with a claim that "the science is not settled" in response to climate scientists supposedly saying that it is.

If this playbook sounds familiar, it should. This is the same approach used for years in religious arguments against teaching evolution and the true age of the Earth in public schools, by anti-vaxxers against vaccine effectiveness, and others opposed to certain aspects of established science. Even the Flat Earthers have used these tactics in the face of overwhelming evidence that the Earth is, in fact, a sphere. We have photographs to prove it.

The truth is that scientists have never claimed that climate science is settled because that is not the way science works. Scientists acknowledge that there is always some uncertainty in science, and human-induced or "anthropogenic" climate change is no exception. The public often equates uncertainty to a lack of knowledge because there are many instances in everyday life where uncertainty is considered a bad thing. For example, I doubt anyone would want to get on an airplane with pilots who were uncertain if they had enough fuel to reach the intended destination. However, uncertainties in science are expected and part of the process. There is always the possibility that some new discoveries will come along and change your conclusions. So most scientists leave room for argument and rarely speak in absolutes, but that doesn't mean we are clueless.

The uncertainties in climate result from complex atmospheric models and the varying probability of different kinds of responses to a warming atmosphere. There isn't much doubt that anthropogenic climate change is happening, and that things will be getting bad, but it is more of a question along the lines of "just how bad are they going they get?"

The phrase about climate science being "settled" is a misquote from former President Bill Clinton, who actually said, "The science is clear and compelling: We humans are changing the global climate." Clear and compelling is not the same as settled. Nevertheless, a strawman argument ginned up in the 1990s by climate skeptics claimed that arrogant scientists, biased media, and lying politicians were saying the science was settled to cash in on public concerns. Casting the climate scientists as bad guys allowed the climate skeptics to then pursue *ad hominem* attacks against their opponents rather than argue about the scientific merits of the issue. Dr. Michael Mann at Penn State has been the subject of many of these assaults because of his early warnings about the climate crisis. Attack publications with titles like "Mann-Made Climate Change" have relentlessly criticized the professor. Other prominent climate scientists like Dr. James Hansen at NASA and Dr. Katharine Hayhoe at

Texas Tech have faced similar assaults. The best response I've found for these attacks is to defuse them by agreeing with the person, and then return to the issue at hand. For example, if someone says, "You're an idiot," the reply is "That may be true but it's a different discussion. So how do you explain these temperature data?" If your ego can handle this approach, it is very effective.

Former President Clinton was correct in saying that the science is clear and compelling. Several decades worth of environmental data show that combustion products from the human use of fossil fuels are accumulating in our atmosphere and affecting climate. A review of 88,125 climate-related papers published in the mainstream scientific literature since 2012 found that more than 99% of these publications link climate change to humans.[2] In fact, the claims that there are no links between fossil fuel combustion and climate change are so lacking in supporting evidence that it suggests to me the people who continue to deny the reality of anthropogenic climate change are not motivated by honest scientific disagreements but by something else.

Some climate skeptics claim that such a consensus on the science doesn't necessarily validate it. There are certainly past examples like the seventeenth century consensus that the Earth was 6000 years old and had been subjected to a world-wide flood. These beliefs were widely held based on what was in the bible but did not hold up under actual, physical observations. By the eighteenth century, scholars were developing the "scientific method" still in use today, which requires repeatable observations, interpretations based on multiple working hypotheses, empirical testing, and evidence-based conclusions.

The scientific method was invented because many of the things people were accepting on the basis of religious faith or from the thought experiments of ancient philosophers were flat-out wrong. As a case in point, the ancient Greek philosopher Aristotle thought that it was quite reasonable for a heavier object like a rock to fall faster than a lighter object like a stick. Galileo proved him wrong during the Renaissance by dropping two different size balls from the Leaning Tower of Pisa and both hit the ground simultaneously. Could Aristotle have performed the same experiment from a tower 1900 years earlier? Yes, but he preferred logical discourse instead.

Today, long-standing principles like Newton's laws of motion, Einstein's mass-energy equivalence, and Darwin's rules for natural selection have so much compelling evidence that few people question the validity of the underlying theories. In these cases the science receives wide consensus precisely because it has been so thoroughly validated. Nevertheless, there is always an opportunity to prove Newton wrong and establish a new theory. Such an undertaking would require substantial amounts of new evidence that contradicts everything Newton and those who came after him observed to support his theories. As Carl Sagan often said, "extraordinary claims require extraordinary evidence." Skeptics who claim that the human

---

[2]Lynas, M., Houlton, B.Z., and Perry, S., 2021, Greater than 99% consensus on human caused climate change in the peer-reviewed scientific literature, *Environmental Research Letters*, v. 16, no. 11

combustion of fossil fuels has no effect on the climate need to provide compelling evidence for their case. So far they have not done so.

This is not just an academic debate. The latest climate assessment from the United Nations Intergovernmental Panel on Climate Change (IPCC) has declared a "Code Red" emergency for the world on climate. Climate skeptics are stoking levels of false uncertainty that only confuses poorly-informed citizens and gives political leaders an excuse to delay the tough actions that are needed to avert the worst parts of this looming disaster. Climate change is often described as crossing a threshold that we didn't know existed until after it was crossed. I think a better analogy is a boulder rolling down a hill. The longer we wait, the more momentum it gains and the harder it is to stop. Most climate scientists agree that it is too late at this point to avoid climate change completely, but at least we can make it less bad.

Climate science appears inexact to the average person because the atmospheric circulation system that drives climate is incredibly complex. Climate changes occur naturally over geologic time periods and are caused by solar fluctuations, shifts in the Earth's orbital cycles, drifting continents that change ocean circulation, and erupting volcanoes that affect the atmosphere. Anthropogenic climate change is imposed over this natural background variation and can be hard to single out, although in recent decades the human influence has become more obvious. Climate predictions are based largely on the outputs of sophisticated computer models that analyze and correlate input data to chart the probabilities of future climates. Probability outputs by their very nature provide a range of answers rather than a single, precise answer, and this creates some uncertainty. Climate skeptics exploit these uncertainties in an attempt to get the unwary to question the validity of the entire anthropogenic climate change hypothesis.

Climate and weather are often confused. Weather is an event and climate is a trend, and to understand climate you have to look at weather trends over time instead of single events. For example, abnormally hot temperatures in Siberia in the summer of 2020 were followed by a very unusual and devastating snowstorm that reached as far south as the Texas-Mexico border in February 2021. The following summer, hot and dry Siberia became the location of the largest forest fire on the planet. Places like British Columbia experienced record high temperatures while the western U.S. fell into an intense drought. The Greenland ice sheet is undergoing record melting and the highest point on the ice experienced rainfall for the first time on record instead of snow. Are these events weather or climate? Looked at individually, they could be considered extreme but still normal ranges of weather. Observed as a trend, however, I believe they point to significant changes in the Earth's climate system. When combined with other observed phenomena like the loss of polar sea ice, retreating mountain glaciers, devastating floods, massive forest fires, and record-setting hurricanes, the pattern indicates that these events are related and represent more than just "weather."

Probability and risk are important concepts for understanding climate change. Climate predictions are probabilities, and in my experience the public has an abysmal level of ignorance when it comes to understanding probability. People regularly buy million-to-one lottery tickets because "someone has to win, and it could be me."

Thousands of visitors to Las Vegas arrive with high hopes of hitting the jackpot, ignoring the fact that those fancy hotels were not financed by giving away money in the casinos. They are, in fact, money factories for the owners.

The casino operators understand probability quite well and know that with the odds in favor of the house, even though the players might win some of the time, if a person keeps playing long enough the casino will eventually get all their money. This is why casinos are timeless places with no clocks or windows. They never close. Midnight and noon under the bright lights inside a casino look almost exactly the same. Probability works through repetition: the more often a coin is flipped, the closer the results of heads versus tails will be to the theoretical odds of 50/50. Players receive rounds of free drinks to keep them rooted to the gaming tables or slot machines. The longer you play, the closer you come to the house odds.

I used to live in Las Vegas, and when people seated next to me on airplanes discovered that, they would often ask me for advice on how to win at gambling, or gaming as it is called there (clearly ignoring the fact that a successful high-roller would have been sitting in first class, not in a cramped coach seat next to them). My initial advice was always "don't play" and just enjoy the extravagant shows, lavish buffets, free drinks, and discounted hotel rooms paid for by other people's gaming losses. If they insisted on gambling, I told them to "quit when you're ahead." But of course almost no one does. They keep playing, lose their winnings and more on top of that, and the casinos prosper. Las Vegas is indeed a vacation destination for people who can't do math.

As poorly as probability is understood by the public, I think risk is even less well understood. Risk is defined as the probability of an event multiplied by the consequences. A high probability mishap with minimal consequences such as taking a tumble while ice skating is considered a low risk (although having done that, I can report it is a bit hard on the knees). On the other hand, a high probability mishap with severe consequences, such as driving a car at extreme speeds down a wet highway with sharp curves is considered such a high risk that it is reckless. Even actions with lower probabilities of mishap still can be risky if the consequences are severe. Although skydiving has been done safely by many people for many years and the odds of a parachute failure are low, the consequences of such a failure are so catastrophic that it is still considered risky. As comedian Steven Wright has said, "If at first you don't succeed, skydiving is not for you."

Risk can be lowered two ways: by reducing the probability of a mishap occurring in the first place or by reducing the severity of the consequences if it does happen. Flying used to be considered a high risk back in the 1930s when airplanes crashed on a regular basis. The consequences of an airplane crash still remain severe and in fact may have gotten even worse because of the higher speeds of modern jets, but the commercial airlines have taken so many steps to reduce the probability of a crash that flying is now rated as the safest form of travel.

In contrast, even though the probability of getting into an automobile wreck has not dropped much in the past 50 years, the addition of seatbelts, airbags, and energy-absorbing auto body construction have all reduced the consequences of a collision. A friend of mine's daughter was recently involved in a head-on collision that put her in the hospital with relatively minor injuries. After extracting her from the wreck, the police stated flatly that if she hadn't been wearing a seatbelt and the airbags hadn't deployed properly, they would have been taking her to the morgue instead of the hospital. Thus, the white-knuckle flier who drives to the airport without wearing a seatbelt clearly does not understand risk.

The IPCC has predicted[3] that if current fossil fuel emissions continue, there is a one in six chance over the next century that the mean global temperature will increase by less than 2.0 °C (3.6 °F) with moderate consequences for the climate. This is not cheery news, because there is also a one-in-six chance that temperatures will increase by more than 5.4 °C (9.7 °F) and cause serious climate disruptions such as killer heat waves and the likely melting of the polar ice sheets that could raise sea levels as much as 80 m (260 feet). This one-in-six probability of a bad outcome has the same odds as Russian roulette, which most people would not consider a low-risk activity. Keep in mind that risk assessment considers both the probability and the consequences. Like Russian roulette, the true risk from climate change is the one-in-six chance of severely bad consequences. In my opinion, that alone justifies doing something about it.

Economics are another important concept in climate discussions. Climate change to date has been addressed primarily with technology such as wind turbines and solar panels to replace fossil fuel electricity, and electric cars to replace gasoline and diesel engines. Despite making some modest inroads, these technologies face an uphill economic battle because fossil fuels are cheaper. This is partly because of the inherent inefficiencies in renewable power supplies versus the economics of scale in large fossil fuel plants. Part of it also stems from tax breaks and subsidies governments give to the fossil fuel industry. As long as fossil is the cheapest energy source, it will be favored. A tax on carbon or increased subsidies for renewables can help level the playing field.

Few people would agree to willingly pay a higher electric bill just because it is good for the environment. However, if humanity is serious about addressing climate change, fossil has to get more expensive to compete with renewables, and the bottom line is that energy prices will increase. Economists call this the "social cost of carbon" and they have been debating for years[4] about how various national economies might be adversely affected by restricting fossil fuels and requiring higher-cost renewable energy. However, when costs of climate change are considered, including more frequent and intense storms, sea level rise, crop failures from droughts, fatalities from heat waves, wildfire losses, and the refugees resulting from these events, the social cost of carbon actually looks relatively cheap. Not dealing with climate change will be a lot more costly than dealing with it.

---

[3] IPCC, 2018: Global warming of 1.5 °C: An Intergovernmental Panel on Climate Change Special Report (Summary for Policymakers): Geneva, Switzerland, World Meteorological Organization, 32 p., ISBN 978-92-9169-151-7.

[4] Nordhaus, W.D., 2017, Revisiting the social cost of carbon: *Proceedings of the National Academy of Sciences,* v. 114, no. 7, p. 1518–1523; DOI: 10.1073/pnas.1609244114

Another important economic concept is "externalized cost." The combustion products of fossil fuel are vented directly into the atmosphere where they negatively affect the climate. It is the cheapest way of getting rid of them, but the damage being done to the environment is borne by everyone on the planet, not just those using the fossil fuels. This "externalizes" the cost. The release of fossil fuel combustion products into the air could be halted by using carbon capture and storage (CCS) technology that removes carbon dioxide from the flue gases and places it deep underground, away from the atmosphere. This would stop greenhouse gas (GHG) emissions and transfer the externalized cost of carbon back onto the fossil fuel users.

Higher fossil energy prices are not necessarily a bad thing. Renewable energy technology development over the past five decades has been focused on improving efficiency to bring down the cost of renewables to levels that are economically competitive with fossil fuels. Raising the price of fossil fuels is another way to achieve economic competitiveness, while higher energy costs overall will encourage conservation.

****

Some environmental advocates are quick to condemn the oil companies as greedy business entities that relentlessly push their products and willingly sacrifice the environment for profit. Coal producers are cast as corporate behemoths that callously forfeit the lives of workers and tear apart the landscape to extract their product, leaving the mess for others to clean up. As with most stereotypes, there is a kernel of truth to these, but also like most stereotypes, the real story is more complex and nuanced.

I know many people working in the fossil fuel industry, and by and large they are decent human beings who do care about the future. Coal mining and the extraction of oil and gas are well-paying jobs, and people often become locked into careers that they can't easily leave. Most fossil energy workers do recognize that the industry is likely to face some downsizing in the future, and many are casting about for new job opportunities in environmental fields such as CCS that can utilize their skill sets. In any case, individual fossil energy workers are no more responsible for climate change than fast food workers are responsible for the increase in obesity across America or pharmaceutical workers are responsible for the opioid epidemic.

That being said, it is clear that the fossil fuel industry itself has been promoting climate skepticism to continue making profits. Fossil fuel producers claim that they are just meeting a market demand, which is true. The demand for fossil fuel remains high because the industry nurtures it. Gas-guzzling, full-size pickup trucks and behemoth sport-utility vehicles are the most popular automobiles sold in America, encouraged by abundant and relatively cheap gasoline. Cheap natural gas and coal produce 70% of U.S. electricity.

Americans are anxious about energy supplies and tend to panic at the first sign of a gasoline shortage; witness people filling up empty milk jugs with gas (not recommended) when an East Coast oil pipeline was temporarily shut down by hackers in the spring of 2021. Fossil energy has become so tightly woven into our economy that switching to other energy sources will be disruptive and challenging no matter how it is done.

Although fossil energy companies have been aware internally for decades about the potential impacts of their products on climate, they donate heavily to political candidates who deny climate change. The industry-affiliated scientists claiming to be climate skeptics provide cover for the politicians who call climate change a hoax. Even some legislators who agree that climate change is real have not taken any significant action for fear of jeopardizing their own oil industry contributions or the support of their constituencies. Climate science has evolved from a technical debate into a tribal issue through what is called "cultural cognition," where information that may go against someone's personal beliefs is contorted to match their cultural values. The division is symptomatic of the many other divisions that rip apart the fabric of American society.

Passing off climate change as a hoax is becoming increasingly difficult as glaciers melt, storms and droughts become more frequent and more intense, killer heat waves occur in places like Siberia and the Pacific Northwest, wildfires are more common, and temperatures rise. Most people worldwide believe their own eyes that things are changing and agree that humanity must do something very soon to avoid a climate calamity.

The fossil fuel industry has responded by changing tactics from outright denial into what are known as "discourses of climate delay.[5]" These consist of redirecting responsibility for GHG emissions onto consumers instead of the fossil fuel industry. In this narrative, people are at fault for driving cars and heating homes with fossil fuel and the industry is only supplying a product. The discourse states that individuals can do nothing meaningful about climate change and perhaps we'd all be better off just giving up on any idea of mitigating it and continue using fossil fuels to the profit of the industry. Related claims suggest that higher levels of GHG are good for agriculture and warming up the planet is good for humans. The resulting heat waves and sea level rise from melting ice sheets are also presumably good, although not specifically mentioned.

A social justice argument called "wokewashing" is a common strategy of climate skeptics that warns poor and marginalized communities of being the most adversely affected by a transition away from fossil fuels. This states that ending the use of fossil fuels will lead to radical, disruptive change that will be detrimental to society and the economy. It claims that affordable, renewable energy is out of reach for disadvantaged communities or poor nations, and that these communities receive more benefits than problems from cheap fossil energy. This is patently untrue and will be addressed in more detail later.

It is important to recognize that these denials about the existence of the climate crisis are not just an honest disagreement among scientists. An organized effort is attempting to discredit climate science using $64 million in annual funds from 140

[5] Lamb, W.F., Mattioli, G., Levi, S., Roberts, J.T., Capstick, S., Creutzig, F., Minx, J.C., Müller-Hansen, F., Culhane, T., and Steinberger, J.K., 2020, Discourses of climate delay: *Global Sustainability*, v. 3, e17; DOI: https://doi.org/10.1017/sus.2020.13

different conservative foundations, who generally conceal donations through the use of donor-directed philanthropies.[6] The money goes to think tanks and institutes that use tactics developed for the tobacco industry (by some of them, as a matter of fact) to convince the American public that climate change is neither real nor serious. Like the fossil fuel industry, the tobacco industry was fully aware of the hazards of their products but created enough uncertainty and doubt to maintain robust cigarette sales until governments finally addressed secondhand smoke and banned indoor smoking, prompting many people to quit. Thus, anyone who believes that "climate change is a hoax" should also remember that "4 out of 5 doctors smoke [*insert brand name here*] cigarettes."

One of the recipients of this anti-climate funding is the Heartland Institute, an Illinois-based think tank that has moved to the forefront of climate skepticism after downplaying the health hazards of tobacco in the 1990s. It views climate change as a conspiracy by world governments to control people's lives. "The global climate agenda, as promoted by the United Nations, is to overhaul the entire global economy, usher in socialism, and forever transform society as one in which individual liberty and economic freedom are crushed," reads an event description for a recent climate skeptics conference sponsored by the institute. Their message that climate change is nothing more than a construct of an evil liberal agenda is contradicted by wildfires, droughts, heatwaves, melting glaciers, and fierce storms.

Another conservative institute that is at the forefront of climate skepticism is the Washington, D.C. -based Heritage Foundation. Their approach is generally a bit softer than the Heartland Institute, asserting that we don't really understand all the uncertainties of climate and have no hard data to show just how much warming the increase in GHG emissions will actually induce. According to Heritage, government policy decisions are supposedly being based on the IPCC "unrealistic" worst-case scenarios (I will remind readers that the "worst case" scenario in both Russian roulette and climate models can actually be quite bad). Whether or not one chance in six should be considered "unrealistic" is up for debate. The Heritage Foundation claims their climate skepticism serves as an "antibody to flawed assumptions and preconceptions" as if those who support the notion of anthropogenic climate change are afflicted with some kind of disease. In the early 1990s, an organization called the Global Climate Coalition (GCC) representing the oil and coal industries engaged E. Bruce Harrison to build a campaign sowing doubt about the science of climate change. Harrison's previous successes included discrediting research on the toxicity of pesticides for the chemical industry, discounting the hazards of smoking for the tobacco industry, and campaigning against tougher emission standards for the auto industry. His firm was considered one of the best. The tactics he developed for the GCC included claims that the science was unsettled and reducing fossil fuel use would negatively affect American jobs, trade, and prices. Harrison specifically

---

[6] Brulle, R.J., 2014, Institutionalizing delay: foundation funding and the creation of U.S. climate change counter-movement organizations: *Climatic Change*, v. 122, no. 4, p. 681–694

sought spokespeople who were scientists, economists, academics, or other experts because they carried greater credibility than industry representatives.

In 2016, U.S. Senator Sheldon Whitehouse (D-RI) called out this chicanery in an effort to expose "the web of denial" about climate change. In his remarks on the Senate floor, he said his purpose was to "spotlight the bad actors who are polluting our American discourse with phony climate denial." He called their actions a disgrace and said, "our grandchildren will look back at this as a dirty time in America's political history." The press release is available on Senator Whitehouse's web page.

The end result of all this conservative foundation and oil money being used to promote the denial of anthropogenic climate change is divisiveness, confusion, and political dithering in national governments that has stalled out any significant policy initiatives on climate. Little or nothing is actually being done to deal with climate change, except to talk about it. This is exactly the point. The delay in implementing measures to address climate change means that most of our technology remains directly or indirectly powered by fossil fuel. With government climate actions on hold, the fossil fuel industry continues to rake in the profits as they keep on selling their products. It is all about the money and always was.

To be fair, the fossil fuel industry is not the only one that uses campaign contributions and lobbying practices to tie up policies in knots. Stall tactics by the insurance industry have stymied most political efforts to reform health care costs. Insurance providers have even come close several times to rescinding the Affordable Care Act. Likewise, the gun lobby and the National Rifle Association have fought tooth and nail against every proposed limitation on guns from universal background checks to banning assault weapons to restrictions on large magazines. This has resulted in wide open access to guns, including for a lot of people who should not have them, and a rash of tragic shootings. Firearms are now the leading cause of death for children in the United States, surpassing automobile accidents and poisonings.[7]

In my opinion, America has become a republic where corporations and capitalism have taken precedence over the wishes of the majority of citizens. Most people support some kind of gun restrictions. Most people want access to medical care that won't leave them bankrupt. And most people are concerned about the climate crisis and want alternative clean energy. Self-serving corporations driven by greed are pouring tons of money into political campaigns to elect candidates who will look after their best interests and profit margins. What is good for the corporations is not necessarily good for the nation.

Some free market proponents and libertarians claim that government climate policies are unnecessary because industry will develop some miraculous carbon-free energy technology to replace fossil fuel. If nuclear fusion, zero point energy, antimatter, di-lithium crystals, black holes, or some other exotic energy source comes about in the future, it would be terrific, but we can't wait for it. The climate crisis is urgent, and we must work with what is available now.

---

[7] Goldstick, J.E., Cunningham, R.M., and Carter, P.M., 2022, Current Causes of Death in Children and Adolescents in the United States: *New England Journal of Medicine*, v. 386, May 19, p. 1955–1956; DOI: 10.1056/NEJMc2201761

It has also become clear over the past half-century that a breakthrough in technology or economics to replace fossil energy with decarbonized sources will not happen in the investor-driven free market. The financial risks are too high, the unknowns are too great, and venture capitalists have plenty of safer places to put their money. Even near-term solutions to replace coal, oil and natural gas with sustainable, carbon-neutral technologies such as nuclear and geothermal still require some significant development to become commercial and competitive. This may happen in the near future with strong government support, but it almost certainly won't happen in the free market.

I believe that government policy and support is the only tool that can provide the incentives and penalties to force industries, utilities, and consumers to adopt new energy behaviors that can help mitigate climate change. Relying on the fossil fuel industry to end our dependence on fossil fuels is like asking the dairy industry to ban ice cream. Relying on the free market to address climate concerns expects famously skittish investors to drop a familiar and fiscally safe energy technology and adopt something new, unproven, and with unknown risks. Although it is clear that fossil fuels must go if we hope to stop making the climate crisis even worse, the industry and investors are not going to do it voluntarily or out of the goodness of their heart. There is just too much money involved for them to walk away from fossil fuels on their own. It has to be a government policy.

These government policies have been a long time coming as the climate crisis continues to slowly worsen. Former U.S. President Barack Obama has pointed out that "most people who serve in Washington have been trained either as lawyers or as political operatives; professions that tend to place a premium on winning arguments rather than solving problems."[8] This is true not only in Washington but worldwide among most of the elected and appointed government officials who would be responsible for implementing climate policies. It may explain why some political leaders have been so slow to take action against climate change, despite talking up a good game.

The lack of global progress on climate suggests that many politicians are dodging the hard choices that must be made to deal with climate change. Political leaders generally want to keep their jobs, and I think they recognize that subjecting constituents to economically painful climate policies would cost votes or support. A far better strategy for a politician is to kick the can down the road and let some future prime minister or president deal with it. In this view, the ongoing climate "debate" provides cover for timid leaders who insist that the uncertainty in climate science prevents bold actions. Nothing gets done and the fossil fuel industry cheerfully continues to dominate the energy economy.

So if we agree with the IPCC and the 99% of mainstream scientists who believe that climate change is a real thing that must be taken seriously and acted upon immediately, how do we fix it? The IPCC recommends doing two things: (1) stop burning fossil fuels and making it worse; and (2) reduce the existing CO_2

---

[8] Obama, Barack, 2006, The Audacity of Hope: New York, Crown/Three Rivers Press, 362 p.

concentrations in the atmosphere to pre-industrial levels. These are simple to say, but in fact are complicated and challenging to carry out.

For example, the amount of electricity generated by fossil fuel combustion in the United States is approximately 273 billion watts, or 273 gigawatts. It would take at least 140,000 two megawatt wind turbines to replace this. We currently have around 70,000 such wind turbines. The IPCC says that 700 billion tons (700 gigatons) of carbon dioxide must be removed from the atmosphere by the year 2100. If we start in 2030, that is ten gigatons per year until the end of the century. The situation is not hopeless, but it is still daunting.

Technologies are being developed to replace fossil fuel with geothermal or nuclear heat in existing electrical generating plants. New types of carbon-neutral biofuels can replace petroleum and natural gas. The technology for wind and solar power, and that for electric vehicles keeps improving. Both biological and engineering solutions are being explored for removing and sequestering atmospheric CO_2. Given the potential consequences from the climate crisis, readers might be surprised to learn that governments so far have not thrown huge research budgets at these technologies in crash programs to fully develop and implement them. Although the funding for atmospheric carbon dioxide removal (CDR) has recently increased in the United States, the research on engineered geothermal and new nuclear technology is minimally funded and proceeding slowly.

****

The scientific principles behind human-induced climate change are not particularly difficult to understand, which makes the success of climate skeptics even more remarkable. These are explored in more detail in later chapters, but in brief, burning fossil fuels such as coal, petroleum, and natural gas adds CO_2 into the air as a combustion product. CO_2 absorbs infrared (heat) radiation and warms up the atmosphere.

The Earth has natural, trace amounts of carbon dioxide in the atmosphere that warm the planet and are used by plants during photosynthesis. Fossil fuel carbon was trapped underground and isolated from the atmosphere for millions of years. Releasing it into the air as a fossil energy combustion product is putting additional carbon dioxide into the atmosphere on top of what was already there. This has increased the concentration of atmospheric CO_2 far above historic levels, resulting in more heat absorption and a warmer atmosphere.

Mathematical climate models show that a warmer atmosphere leads to unstable climates. The models predict that melting polar ice caps will alter ocean currents and raise sea levels, which is already happening. Basic physics states that warmer air holds more water vapor than cold air; thus a warmer atmosphere will result in more intense droughts along with more intense storms, which we are also seeing.

There is a difference between doubt and distrust. Doubt can be overcome by facts and evidence. Distrust cannot. Some of those denying climate change are asserting that the very science itself cannot be trusted and it's all a hoax, despite the fact that the heat-trapping properties of CO_2 were discovered and documented back in 1824. Others say scientists are self-serving and untrustworthy, and that scientific claims about an onrushing climate calamity are only being used to create panic so scientists can profit off large sums of research grant funding. If this really was just a strategy for obtaining grant money, I would have to call it an abject failure because the research has been woefully underfunded for decades, especially with respect to non-fossil energy sources. A few climate skeptics have called CCS research a "welfare program for geologists" and several have even said that if the climate scientists are wrong, humanity will have cleaned up the planet for nothing (!).

Trying to reason with climate skeptics who distrust science is often fruitless and frustrating. As author Lee McIntyre[9] has stated, "Whatever evidence is presented to debunk these claims is explained as part of a conspiracy: it was faked, biased, or at least incomplete, and the real truth is being covered up. No amount of evidence can ever convince a hardcore science denier because they distrust the people who are gathering the evidence." According to McIntyre, the only way to reach these people is to talk to them calmly and respectfully—to put ourselves out there and meet them face-to-face. This can be especially challenging when your conspiracy-loving uncle is only interested in trying to get you riled up during Thanksgiving dinner. People refuse to acknowledge the facts because of two factors identified by psychologists. One is called "belief perseverance," where folks refuse to give up a long-held belief despite new evidence. The second is called "confirmation bias," where people only accept evidence that supports their beliefs and ignore anything contrary.

****

A brief explanation about how science works might be helpful for readers to understand why some scientists can assert that a thing is true, while other scientists will look at the same data and say it is false, and nearly all scientists will hedge a bit on a conclusion. Scientists are human and like most humans they have both conscious and unconscious biases. The scientific method itself is constructed to overcome these biases by requiring rigorous proof of any findings and allowing other scientists to observe and investigate such evidence. This is necessary because human beings are remarkably good at convincing themselves of a great many things.

Questions and arguments are meant to be an integral part of the scientific process and are actually designed into it. The role of "devil's advocate" becomes important when validating observations and interpretations. Observational data can often have more than one explanation, so the concept of "multiple working hypotheses" is used. This basically means that everything is on the table until enough data are gathered to zero in on the most likely explanation. Depending on its nature, new data can alter the conclusions or even change the entire underlying hypothesis. Non-scientists often do not understand the details of this process and commonly dismiss valid disagreements among scientists as, "Those people don't really know anything."

Data sets can lead to a wide range of interpretations about what they mean, and debates among scientists are usually over the interpretation of the evidence. No one

---

[9] McIntyre, Lee, 2021, How to Talk to a Science Denier: Conversations with Flat Earthers, Climate Deniers, and Others Who Defy Reason: MIT Press, Cambridge, MA, ISBN: 9780262046107, 280 p.

questions, for example, that the western U.S. is subject to droughts, and these droughts have been increasing in frequency and intensity over the past several decades. Is this unprecedented and due to climate change, or is it part of a natural cycle that includes the dustbowls of the 1930s? Were the dustbowls themselves "natural" events or just an earlier manifestation of the climate change phenomenon? Two scientists can look at the same data and reach two different conclusions. The polite way to start a scientific debate is "I disagree!"

However, to the public, scientific findings may sound tentative, equivocal, or even wishy-washy. Scientists tend to be cautious about making absolutist statements because all results are subject to change in the light of new evidence. "Sticking to your guns" at all costs is not a very productive way to do science.

The history of science is littered with carefully constructed models that fell apart after someone came up with new observations, or sometimes just a better interpretation of existing data. For example, the ancient Greek philosopher Ptolemy viewed the Earth as the center of the universe surrounded by a nested series of crystal spheres that controlled the motion of the sun, moon, stars, and planets across the sky. During Greek and Roman times, and even through much of the Middle Ages, elegant models of cycles and epicycles were constructed to track and explain these celestial motions, including the complex Antikythera mechanism discovered in 1900 in the wreck of an ancient Roman cargo ship. Nicolaus Copernicus came along in 1510 and pointed out problems with Ptolemy's system of crystal spheres that simply could not be fixed mathematically. He put the sun in the center of the universe with the Earth and other planets revolving around it, and everything worked much better. Johannes Kepler later refined the shape of the orbits from circles into ellipses, and then Galileo used a telescope to actually observe celestial objects in orbit around another celestial object, in this case the moons of Jupiter.

In the early twentieth century, American astronomer Harlow Shapley demoted the sun from Copernicus' location at the center of the universe to just one star of the billions that make up the Milky Way galaxy. Shapley added insult to injury by showing that the sun is not even at the center of the Milky Way, but halfway out in one of the spiral arms. Edwin Hubble determined a few years later that the Milky Way was not the center of the universe either and in fact there was no center because billions of galaxies are retreating from one another at cosmic speeds as the universe expands. Hubble's findings prompted Belgian cosmologist Georges Lemaître in 1931 to rewind the expanding universe backward in time to a single point and propose the "big bang" theory that shapes our present-day view of the cosmos.

The point of this discussion is that science is not static but evolves over time as new evidence comes to light. It is not absolute and unarguable like religious doctrine or fixed and unalterable like legal documents. The only actual rule in science is "honor the data." Any valid new data must be honored and doing so often requires a shift in interpretation. Data on gravity, orbital mechanics, star lifecycles, galaxies, and cosmic redshifts have completely overwritten Ptolemy and his crystal spheres that dominated human knowledge of the universe for millennia.

I have found that one of the most difficult things to explain to the public is how scientists know what they claim to know. The common misunderstanding of scientific terms like hypothesis and theory makes this even more challenging. A theory is often thought of in popular terms as only a guess or speculation; thus some people will dismiss evolution as "just a theory." In a true scientific sense, by the time something becomes a theory, it has tons of observations, data, and evidence to back it up. A hypothesis is a conjecture based on data and observations, and a theory is an interpretation of the data, experimental results, and other evidence used to test the hypothesis. A scientific theory is an attempt to construct a coherent explanation from the evidence. In no sense is it a guess.

There is a trope favored in the media and popular culture about the lone scientist standing up against the establishment with his or her own contrary data and evidence that no one will take seriously. This is familiar to most people as the opening premise of almost every Hollywood disaster movie, where someone is warning about zombies, an oncoming asteroid, a volcano that is about to erupt, dinosaurs on the loose, or whatever, and being resoundingly ignored. Of course the rogue scientist turns out to be correct, and mayhem ensues. In real life, scenarios like this almost never happen.

Science is a painstaking, slow, and careful process. Discoveries are usually made gradually and incrementally, typically in collaboration with others. This is especially true in broad fields like environment, ecology, and climate science. Most scientific research these days is multidisciplinary, with a team of specialists in different fields working together on an objective and everyone contributing their area of expertise. There is just too much to know, and one person cannot know everything. Projects are too complex for a single person or even a small group of people to carry out. For example, I've been working on a combined carbon dioxide capture and storage project (described in Chap. 9) with a chemist, a mechanical engineer, an industrial engineer, a business economist, a microbiologist, two geneticists, a civil engineer, and two other geologists. There just aren't that many "rogue scientists" out there any more who are accomplishing anything.

It might surprise readers to learn that scientific ideas are never "proven" to be right. The only actual scientific proof is when they are proven wrong. If the researchers were careful with their measurements and math and gave due diligence to their interpretations and subsequent conclusions, the findings are assumed to be provisionally correct. However, they can be proven wrong at any time by other researchers running tests on the assumptions and trying to reproduce the results. If the results are not proven wrong after others have tested them, they are eventually considered to be correct. The more of these tests and cross-checks the scientific work passes, the more accepted it becomes. Even then, results are still subject to change if any new evidence comes to light. This is called peer review and it is a method for correcting scientific errors.

To save everyone from public embarrassment, scientific peer review is carried out before an article is published. The reviews are performed by recognized experts in the field (known as referees) who review the scientific work and return comments and corrections to the author and editor. Errors are addressed in a revision, or the article never sees daylight. The referee process is good but does not always catch every flaw, and if there is a problem with the article after publication, researchers can rest assured that someone somewhere will notice it.

I want readers of this book to know that it was peer reviewed prior to publication by two very competent referees, and their comments and suggested changes were incorporated into the revised text. I didn't object to their corrections, or feel slighted, or think they were trying to tell me how to rewrite the book. Just the opposite - I am grateful for the time and effort these two reviewers put in to read the entire text, and their suggestions improved the book significantly. Peer review should be taken with the spirit in which it is intended, and that is to improve the document, correct any flaws or typos, and ensure that the science is correct.

Many scientific articles issue "errata" statements after being published to correct small errors that have slipped through review. For major errors, the retraction of the entire scientific paper is sometimes required. A well-known example of a significant retraction a few years ago was an engineering process called "cold fusion" that was supposed to produce nuclear energy by tightly binding hydrogen atoms together chemically in a metal matrix until they fused. Scientists were skeptical when the article was published, because the only way humans knew how to make hydrogen fusion was with the intense heat and pressure of an atomic explosion. If cold fusion worked, it would be a major energy technology breakthrough. Despite multiple attempts to duplicate the cold fusion design described in the paper, no one else could ever repeat the results. The original work was found to contain some serious measurement flaws, and in the firestorm of recriminations that followed, the paper was withdrawn. The authors and their university were accused of rushing the article into publication to try cashing in on the technology.

An even more infamous paper was published describing a supposed link between human vaccines and autism. In this case, the "data" were found to have been straight-up faked, the study was thoroughly debunked, the paper was withdrawn, and the medical license of the author was revoked. Both of these papers created a great deal of interest when initially released, and the authors became world-famous, which seemed to be their motivation. However, once the facts caught up to them, they were forced to withdraw the papers from the literature. Each has suffered significant professional disgrace within the scientific community.

The point is that if anthropogenic climate change was indeed a hoax, a cold fusion-level of ruckus would have been raised by the scientific community over such false information. Instead, significant amounts of research have been published in refereed journals that show strong links between fossil energy use and climate change. Scientific evidence has been presented to support this connection and the critically important scientific papers contain solid data and findings. People who have repeated the measurements have been able to readily duplicate them. Still, because scientists generally avoid absolute statements in the conclusions, this hesitancy to say human influence on climate is 100% certain is interpreted by some climate skeptics and the news media as proof that the science is "unsettled."

Some commentators have suggested that scientists should just end the debate by saying that the climate crisis is absolutely certain, but that is not the way science works. The best I can offer is that the climate skeptics have not produced any compelling evidence as yet to refute the findings that humans are causing climate change. A search on Amazon shows that there are a number of books[10] by climate skeptics and "contrarians" attempting to debunk the whole idea of anthropogenic climate change, but as far as I can tell, none of these rise to the level of peer-reviewed scientific literature. Unless and until someone finally does prove this wrong and backs it up with solid science the findings stand.

I experienced something similar in recent years in the debate over hydraulic fracturing ("fracking"), a technique that uses pressurized water to crack open low permeability rocks like shale to obtain natural gas. Environmentalists were up in arms over this, claiming it was contaminating vast swaths of groundwater, exposing people to dangerous chemicals and making them sick. Multiple investigations were run by many different groups of scientists, including myself, to document these supposedly horrific environmental effects of fracking.

The evidence for this was essentially zilch. The frack fluids stay deep underground, never come anywhere close to drinking water aquifers, and while there are instances of surface spills and local environmental contamination, it is by no means systemic or widespread. Literally hundreds of papers were published that investigated the environmental effects of fracking, along with several large reports from the U.S. Department of Energy, U.S. Geological Survey, and Environmental Protection Agency. None of them reported any widespread contamination. More than a few good scientists approached this subject convinced that they were going to find all sorts of problems and were rather surprised when they didn't. But as honest brokers, they admitted there was no evidence to support the hysteria and that was that.

Not so for some environmentalists. A few were infuriated at the EPA for publishing a report that didn't condemn fracking. Others insisted that the practice be banned, evidence or no evidence, and it was in fact banned in a number of states, including New York. There is still a push to ban fracking nationally by people like Senator Bernie Sanders (I-VT) and Representative Alexandria Ocasio-Cortez (D-NY). I will concede that fracking is a climate concern because it has increased the production of natural gas and crude oil in the United States and maintained our dependence on fossil fuels for at least an additional decade. But a misguided fracking ban is likely to result in gas shortages and send electric utilities right back to coal, which is far worse than natural gas for the environment in general and for the climate in particular. There is no evidence that fracking as an oilwell completion practice is any more harmful to the environment than any other oilwell completion practice. Industry has been saying this all along, and it turns out that they were correct.[11]

My stance on fracking cost me a lot of environmentalist friends. Some people that I used to hike and camp with basically disowned me for not supporting their

[10] Wrightstone, G., 2017, Inconvenient Facts: The Science That Al Gore Doesn't Want You to Know: Silver Crown Productions, LLC, Itasca Books, Minneapolis, MN

[11] Soeder, Daniel J., 2021, Fracking and the Environment: Cham, Switzerland: Springer Nature Switzerland AG, 279 p. (https://link.springer.com/book/10.1007/978-3-030-59121-2)

side of the fracking debate. But what could I do? The evidence was against them. As Mahatma Gandhi said, "Never apologize for being correct." The facts always have a way of winning in the end. Fracking is discussed in more detail in Chap. 4.

It is important to understand that the evidence supporting human-influenced climate change is strong, and evidence against it is weak. The initial skeptics reviewed the data, repeated the observations, re-ran the models and reached similar conclusions. The strongest scientific theories are the ones that have been tested and re-tested and found to hold up. Fracking has been through this testing. So has climate change. Those who insist that they are still skeptical about either one have ulterior motives. It is not the science, because the science was done right.

Unfortunately, like the "ban fracking" enthusiasts, there have been a few cases where the proponents of anthropogenic climate change have pushed their conclusions beyond the bounds of the data. These were politically-motivated attempts to bias the results in favor of the climate agenda, and they were called on it by other scientists. Climate skeptics of course pounced on this as proof that all climate data are exaggerated and biased against the fossil energy industry. However, the literal handful of people championing a pro-climate agenda with biased data were a very small minority compared to the thousands of researchers performing good-faith climate-related investigations. The actions of a few unethical scientists by no means invalidate the bulk of the data showing that human induced climate change is real no matter how much the climate skeptics would like to discredit it.

A generally overlooked aspect to the climate debate is that climate skeptics could easily resolve the issue in their favor by using the scientific method. If they produced solid scientific evidence showing that the links between fossil energy, greenhouse gas, and the climate crisis do not exist, climate scientists would readily change their minds because the scientific method always honors the data. So far, however, no convincing data have been forthcoming. Those who claim to have evidence contrary to a scientific conclusion have only one obligation: Prove it.

The skeptics respond by saying that the arguments they do try to present are dismissed or covered up by mainstream science. The claim of a dismissal or cover-up is a standard tactic for conspiracy theories (and also for disaster movies, come to think of it). International scientific journals generally do not have a political agenda, and there are no shadowy groups trying to suppress information about the climate. To what end? Are the solar power companies supposedly in cahoots with the scientific journals to promote the climate crisis to sell more solar panels? This is as ridiculous as it sounds. Cover-ups are usually to hold onto money or power and climate had neither of these when the problem was first publicized in the 1980s. All the money and power was with the fossil fuel industry. So which side has an incentive to cover up something here?

The integrity of international scientific journals relies on them being trustworthy enough to publish true findings backed up by solid data no matter where those findings lead. Journals also have an editorial board or committee that makes decisions on what to publish, not just one Perry White-type editor spiking stories he or she doesn't like. Every publisher in the scientific publishing world that I know of would be eager to publish a paper with major, ironclad proof that anthropogenic climate change is false. This would be huge news and an enormous scoop much bigger than debunking cold fusion. The journal would gain prestige, advertisers, increased circulation, and more world class scientists submitting articles. Suppressing such findings for a supposed kickback from a solar panel manufacturer has far less value.

So where is the proof from the climate skeptics that this is all a hoax? Their supposedly contrary data have not been published in any peer-reviewed, refereed journals that would give it legitimacy, or if it has, it was not widely circulated. As far as I know, nothing significant has turned up in Nature, Science, EOS, AGU Atmospheres, Geological Society of America Bulletin, Environmental Science & Technology, International Journal of Greenhouse Gas Control, Climatic Change, or a host of other legitimate international scientific journals where this would certainly have made front page news.

Instead of scientific proof, we get websites from groups like the CO2 Coalition (https://co2coalition.org/) that requests contributors to "join us in our love for CO_2." The organization insists on its "Climate Facts" page that elevated levels of carbon dioxide in the atmosphere are perfectly normal, and in fact are beneficial for increasing the yields of corn and other crops. There are claims that greenhouse warming slows down and stops with increasing CO_2 concentrations (satellite data show it gets worse), the current warming trend is "good" for humans (anthropological evidence suggests that it is not), and over timescales of tens of millions of years, the current high CO_2 levels are not that unusual (although when compared to the last 400,000 years, they are indeed unusual). Some of these claims are backed up with modeling studies from relatively obscure journals while others are cited as "personal communications." I can't find much rigorous science in it anywhere. My contrary statements in the parentheses above are all backed up by peer-reviewed references in this book.[12]

This type of material used to appear mainly on blogs. A new tactic seems to be submitting it to "pay to publish" open access technical journals or producing it as self-published books to give it a veneer of acceptability. Many open access journals are technically rigorous and adhere to peer review standards. They charge a publishing fee to authors so the information can be freely disseminated, especially to lower income people in Third World nations who can't afford expensive journal subscriptions.

Unfortunately, some other journals have rushed to cash in on the open access model and appear willing to publish just about anything for money, especially those that promise a quick turnaround and need a large volume of articles. Self-published books these days have become clever at concealing the fact that they are self-published. It is hard for readers who are not engaged in the field to know which publications are legitimate and which are not.

An open access paper that I recently reviewed contained a claim that the absorption of infrared radiation by carbon dioxide actually cools the planet. This defies the

---

[12] See respectively, Kramer et al. (2021), Raymond et al. (2020) and Xu et al. (2020), and Climate. NASA.gov

laws of physics and demands some substantial data to back it up, which was not provided. Another paper stated that carbon dioxide affects the ozone layer and lets in more ultraviolet instead of infrared radiation. This completely misunderstands atmospheric chemistry and is just wrong. Although I'm tempted to ignore these articles, I have made a practice of reviewing them when asked. If a 500-word review letter laying out the inaccuracies and flaws in reasoning will get the article revised or rejected, it can help keep at least some of this nonsense out of the literature.

I remember a short story (by Asimov, I think) about a physics professor who discovers antigravity. He has real data that go against all the known laws of physics and no journal will publish his discovery. He can't get onto the speakers program at any conferences to talk about it, and he is shunned by colleagues. Finally, he attends a scientific conference as a member of the scientific society. He enters the lecture hall where an eminent researcher is talking about gravity, and he simply floats in silence above the audience in the back of the room. Once he is noticed, there is a great deal of commotion, with the elderly speaker demanding that he stop playing tricks and leave. He tells the gentleman that he is a card-carrying member of the society, paid his registration fee, there is no trickery, and he has every right to be there. When the society officials finally determine that he actually is floating in the air, they demand an explanation. He says they are observing antigravity, and challenges THEM to explain it.

That story has always stuck with me as an example of what it will take to overturn an overwhelming body of evidence. Climate skeptics are nowhere close.

Chapter 2
The Anomaly

Keywords Greenhouse gas · Keeling Curve · Energy trends

****

Let's examine some of the strong evidence that does support a connection between fossil fuel, greenhouse gas, and climate change. Since the Industrial Revolution, the use of fossil fuels to power civilization has increased, and the CO_2 content of the atmosphere has increased along with it. Actual, continuous measurements of atmospheric carbon dioxide began in 1957 on the flank of the Mauna Loa volcano in Hawaii at a trace gas observatory established by the U.S. National Oceanic and Atmospheric Administration (NOAA) and operated by the Scripps Institute of Oceanography. There are currently about 160 other stations worldwide making similar measurements of atmospheric chemistry, but the station at Mauna Loa has the longest and oldest data record.

No one paid much attention to the trace gas data for almost two decades, but in 1976 a group from Scripps and NOAA decided to analyze trends in the measurements to see if there were any interesting variations over time.[1] The iconic image shown in Fig. 2.1 is called the "Keeling Curve" after their 1976 publication of the analyses.

The Keeling Curve shows a steady increase in the concentration of CO_2 in the atmosphere of almost 100 parts per million (ppm) over the past 60 years. The sawtooth line represents the annual cycle of vegetation blooming in spring and taking up CO_2, then going dormant in the fall and returning the CO_2 back to the atmosphere. The solid, dark line up the center is the average annual concentration trend. It is important to note that the solid line is not straight. Laying a straightedge on it shows that the curve is steepening with time. Carbon dioxide concentrations in the atmosphere reached a record high of 421.37 ppm on May 11, 2022 (https://keeling-curve.ucsd.edu/). Sadly, this record is most likely only temporary.

---

[1] Keeling, C.D., Bacastow, R.B., Bainbridge, A.E., Ekdahl, C.A. Jr., Guenther, P.R., Waterman, L.S., and Chin, J.F.S., 1976, Atmospheric carbon dioxide variations at Mauna Loa Observatory, Hawaii: *Tellus*, v. 28, no. 6, p. 538–551

D. Soeder, *Energy Futures*, https://doi.org/10.1007/978-3-031-15381-5_2

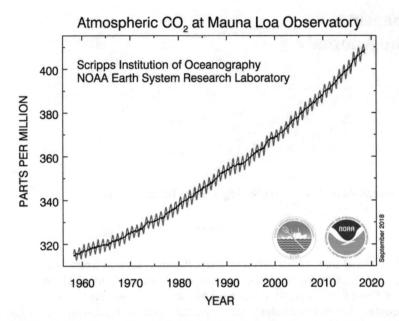

Fig. 2.1 The Keeling Curve of carbon dioxide levels in the atmosphere measured since 1957 at Mauna Loa in Hawaii. The sawtooth pattern represents seasonal changes in CO_2 as northern hemisphere plants bloom in spring and go dormant in fall. The solid line up the center is the annual average trend. (*Source: U.S. National Oceanic and Atmospheric Administration (NOAA) public domain*)

This is an anomaly. A few hundred parts per million may not sound like a big number, but when considered against the amount of gas making up the entire atmosphere, this translates into billions of tons of carbon dioxide. The trend is also alarming. The concentration of trace gases in the Earth's atmosphere does vary over time with events like volcanic eruptions, forest fires, gas releases from sediments, and other phenomena. However, these occur randomly at various times and in varying amounts. A steady, steepening trend in carbon dioxide concentration is unusual at best, and unnatural at worst. There is clearly something going on here that is both long-term and significant.

Where is all this CO_2 coming from and what is causing the anomaly? Some climate change skeptics suggest that natural sources such as volcanic eruptions might be responsible. Although volcanic eruptions may emit copious quantities of CO_2 into the atmosphere, a large number of large eruptions would be required to account for the Keeling Curve since 1957. There is no evidence that any sustained mega-eruptions have occurred within this time frame. Most volcanoes in fact erupt episodically rather than continuously, with significant dormancy periods between events. Large, individual eruptions would certainly have been noticed. The episodic nature of volcanic eruptions also would be expected to produce irregular, isolated upward spikes on the Keeling Curve, yet it shows a relatively smooth and steady increase in average CO_2 levels over time.

    The combustion of fossil fuel releases CO_2 that is not part of the atmosphere. This carbon was separated or sequestered from the atmosphere for tens of millions of years and stored deep underground as coal, petroleum and natural gas. Human mining and drilling activities brought the deep carbon to the surface and burning fossil fuel is adding it to the atmosphere, increasing the net CO_2 concentration. Most electricity in the world is generated by burning fossil fuels like coal and natural gas to make steam to turn a turbine and a generator. Most vehicles still run on liquid fossil fuels like gasoline or diesel fuel. Coal and natural gas are used directly in industrial processes ranging from steel-making to cement manufacturing. Both are also widely used for home heating and cooking.

    The trend of increased fossil fuel use since the Industrial Revolution (Fig. 2.2) matches the trend of increased CO_2 concentrations in the atmosphere shown in Fig. 2.1. These trends precisely explain the Keeling Curve anomaly. In my opinion, and in the opinion of 99% of other scientists, there is little doubt that human combustion of fossil fuel is the main cause of increased carbon dioxide concentrations in the atmosphere.

    Other proposed sources of GHG like forest fires, rotting vegetation, and even cattle flatulence would not produce the anomaly shown in the Keeling Curve. Vegetation is said to be "carbon-neutral" in that it is composed of carbon the plants had removed from the atmosphere while growing; releasing CO_2 as a combustion product simply puts this carbon back and does not change the net concentration in

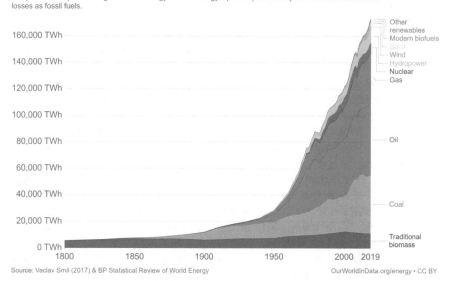

Fig. 2.2 Global energy consumption from 1800 to 2019 showing the dominance of fossil fuels. (*Source: Ritchie, H. "Energy" (open access) Published online at* OurWorldInData.org. *Retrieved from:* https://ourworldindata.org/energy)

the atmosphere. This is why biofuels like ethyl alcohol (ethanol) and biogas derived from plants are considered to be carbon-neutral.

How might the Keeling Curve CO_2 anomaly affect atmospheric temperatures? Anyone who has ever walked barefoot on the beach on a sunny summer afternoon knows that the sun heats the Earth's surface. The process is called insolation and the amount of solar heating at mid-latitudes can be greater than 300 watts per square meter during summer daylight hours. Short wavelengths of infrared radiation (IR) from the sun penetrate the atmosphere and heat the Earth. This heat energy from the warm Earth is then re-radiated back into space as longer IR wavelengths.

French physicist Joseph Fourier discovered back in 1824 that CO_2 was transparent to the shorter incoming wavelengths of IR, allowing these to reach the ground and heat it up, but then it absorbs the longer outgoing IR wavelengths radiated back from the ground. Thus, the atmosphere is warmed from below by heat radiated from the Earth and not from above by the sun. This is why air temperature decreases with altitude and high mountain peaks are perpetually snow covered. The atmospheric temperature gradient is also important for the condensation of clouds and rainfall.

Fourier referred to the atmospheric warming process as the "hothouse" effect, an old term for the glass-covered buildings we now call greenhouses. He determined that CO_2 acted to trap heat in a manner similar to the glass windows of a greenhouse, and today CO_2 and other heat-trapping gases are called greenhouse gases or GHG. Fourier also understood that some level of GHG in the atmosphere is necessary to act like a blanket and prevent temperatures on the Earth from plunging below freezing even in the tropics after sunset. Fourier's findings about the heat-trapping properties of carbon dioxide were verified experimentally by American physicist Eunice Foote in 1857 and confirmed for additional gases a few years later by John Tyndall in Ireland.

The physics of GHG have been understood for nearly two centuries. Some skeptics say that these old data points are unreliable, and we should re-measure all of this with modern instruments. First of all, the nineteenth century measurements were carefully done and were quite accurate. Secondly, more than a few people have re-measured this recently and obtained the same results. How many times does this need to be checked to convince the skeptics?

Climate skeptics claim there is no evidence that the atmosphere has warmed up as GHG levels have increased. However, the laws of physics state that as a material absorbs radiant energy, that energy is transferred to that material typically in the form of heat. This is easily proven with a variety of materials. Many times I have stood in front of a campfire and felt the metal coins and keys in my pocket grow quite warm from absorbing the radiated heat. As GHG concentrations increase in the atmosphere, the warming effect becomes greater. Any other supposition defies well-established physical laws.

Some skeptics also claim that the atmospheric concentrations of CO_2 are too low and question how something at the ppm level can possibly be responsible for the supposed warming. It is important to note that there are many substances where small quantities can have outsized effects. For example, the minimum lethal dose of arsenic is estimated to be less than 0.2 grams, about the size of a single grain of rice.

Physics data and climate models show that the increase in CO_2 concentration from the pre-industrial level of 280 ppm to today's 420 ppm is sufficient to produce the observed warming.

The IR absorption frequencies for CO_2 were measured with great precision at wavelengths of 4.3 micrometers (μm) and 15 μm at the University of Michigan back in 1932.[2] Water vapor and methane also have specific infrared absorption frequencies; for methane the main absorbing wavelengths are 3.5 μm and 8 μm. I had an instrument at the DOE National Energy Technology Laboratory (NETL) that used this principle to detect methane concentrations in the atmosphere (we were concerned about leaky gas wells) with an infrared laser tuned to the methane absorption frequencies. It accurately determined the concentration of methane from environmental background levels of 10 ppm up to the lower explosive limit of 5% by measuring the attenuation of the laser beam as it passed through the air sample. More methane blocked more of the laser light.

How do we know that increased CO_2 concentrations are actually heating up the Earth's atmosphere? Actual, direct measurements of longwave IR radiated from the ground into space have been made by satellites. An infrared interferometer spectrometer (IRIS) onboard the U.S. Nimbus-4 satellite launched in 1970 measured IR emissions from the Earth at wavelengths between 4 μm and 16 μm, bracketing the main CO_2 infrared absorption bands. The Japanese satellite Midori entered orbit in 1996 and used an instrument called the Interferometric Monitor for Greenhouse Gases (IMG) to obtain similar data. Comparison of these two data sets collected 27 years apart showed a measurable reduction in the intensity of IR radiation emitted from the Earth into space at wavelengths corresponding to CO_2 absorption frequencies. The decrease in IR radiation from the Earth corresponded to the increase of CO_2 in the atmosphere as measured on the Keeling Curve over the same period.[3] This was exactly like my laser sensor – more CO_2 blocked more IR radiation. The amount of incoming solar insolation is assumed to be constant at these time scales, so less heat radiated from the Earth back into space meant that more was being absorbed by GHG in the atmosphere. More recently, researchers independently confirmed that the radiation imbalance between the bottom and the top of the atmosphere had grown wider between 2003 and 2018, primarily due to rising concentrations of GHG.[4] Those who claim that higher CO_2 levels are not warming up the lower atmosphere must explain these data.

The CO_2 concentrations measured at Mauna Loa represent only about six decades worth of data. Despite the upward trend, climate change skeptics suggest that this is

---

[2] Martin, P. E. and Barker, E. F., 1932, The Infrared Absorption Spectrum of Carbon Dioxide: *Physical Review Journals*, v. 41, p. 291.

[3] Harries, J., Brindley, H., Sagoo, P., and Bantges, R. J., 2001, Increases in greenhouse forcing inferred from the outgoing longwave radiation spectra of the Earth in 1970 and 1997: *Nature*, v. 410, p. 355–357.

[4] Kramer, R.J., He, H., Soden, B.J., Oreopoulos, L., Myhre, G., Forster, P.M., and Smith, C.J., 2021, Observational evidence of increasing global radiative forcing: *Geophysical Research Letters*, v. 48, no. 7, e2020GL091585.

a natural increase, in line with variations of atmospheric chemistry over geologic time. The U.S. space agency NASA decided to investigate a longer timeline by analyzing air bubbles trapped in polar ice cores. The vertically-drilled ice cores provide miniature samples of ancient atmospheres captured by snowfall events hundreds of thousands of years in the past. The ancient snow layers piled one on top of another preserve a time record of atmospheric chemistry as bubbles in the ice – the deeper the ice core, the older the sample. The goal of the study was to obtain a long-term baseline of CO_2 concentrations along with data to see how these varied with ice ages and glacial cycles. The results show that carbon dioxide levels in the atmosphere have indeed changed over geologic time scales, but present-day levels of atmospheric CO_2 are off the chart and are a significant anomaly compared to the geologic past (Fig. 2.3).

Interactions between humans and Planet Earth are causing harm to the environment, and this in turn is causing harm to us. For those who doubt that we puny humans are capable of altering the environment on Earth, I suggest that evidence to the contrary is plentiful. Groundwater is being depleted by over-pumping aquifers, creating measurable changes in the rotation of the Earth. Coastal construction projects release plumes of sediment into the oceans, smothering coral reefs and altering shorelines. The pH of the oceans is changing as they acidify from excess atmospheric carbon dioxide dissolving into seawater as carbonic acid. Plastic pollution in the oceans is killing marine life. Microplastics as small as 1 μm in size have been

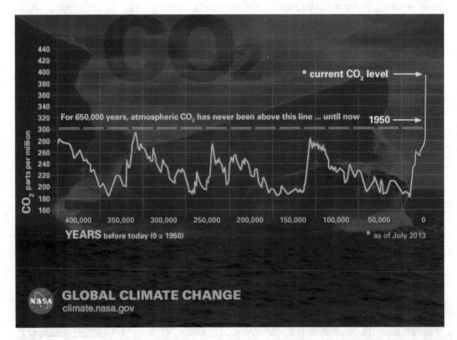

Fig. 2.3 Carbon dioxide levels in the atmosphere from 400,000 years ago to July 2013. (*Source: National Aeronautics and Space Administration (NASA) public domain*)

found from the summit of Mount Everest to the depths of the Marianas Trench. Radioactive fallout from atmospheric nuclear testing remains in sedimentary deposits from the 1950s.

Virtually every major river in the world has been dammed and impounded, affecting aquatic ecosystems and the movement of sediment. Groundwater and surface waters are contaminated with chlorinated solvents, hydrocarbon fuels, fire retardants, and a host of other chemicals. Runoff from excessive fertilizer use is creating algal blooms and anoxic, dead zones in the coastal oceans. Landfills are overflowing with non-recycled trash. Industrial aerosols above Houston, Texas emitted by more than 400 regional petroleum refineries doubled the amount of precipitation from a hurricane that stalled over the city in 2017 and dropped 1.25 meters (nearly 50 inches) of rain in 2 days, causing devastating floods.[5]

I do not believe that humanity is guaranteed a future on Earth. We can destroy ourselves through any number of scenarios, some of which are frighteningly easy. In addition to climate change, these include dangerous chemicals in the air and water, deforestation, the collapse of critical ecosystems, and that old standby, full-scale nuclear war. There are also multiple natural disasters waiting in the wings, including everything from drought, famine, floods, disease and pandemics, wildfires, asteroid impacts like the one that destroyed the dinosaurs, and supervolcano eruptions, one of which almost wiped out primitive humans some 75,000 years ago (the evidence is preserved in our DNA). People have got to do a better job with environmental stewardship on Earth, or we may well go extinct. I also think that we must become a multi-planet species so that all of our eggs are not in one fragile basket.

For too long, various religious and social philosophers have claimed that humans are special and therefore our success is pre-ordained. Such delusional optimism is dangerous and misleading, and it does nothing to address the problems. The universe is constructed with a firm set of rules and violating any of those rules has consequences. One can have all the faith in the world but driving a car off a bridge is going to end in a crash because gravity is one of those rules that structures the universe.

A corollary to the "humans are special" argument is that no matter what we do, God will not allow us to destroy ourselves. Everyone is entitled to their own faith and set of beliefs but in my opinion counting on a supreme being to miraculously save us from our own idiocy is not actually a workable plan. Many religions postulate that humans have a free will, and if we choose to destroy ourselves through deliberate foolishness, I think God may just stand back and watch free will in action.

There is an often-told story of a man trapped in a flood. The water rises and he is on his porch when the rescue boat shows up, yet he refuses to get in, saying that he trusts in God to save him. A little later he is on the roof surrounded by floodwaters and another boat arrives. Again he refuses to board it, saying God will save him.

---

[5] Pan, B., Wang, Y., Logan, T., Hsieh, J-S, Jiang, J.H., Li, Y., and Zhang, R., 2020, Determinant Role of Aerosols from Industrial Sources in Hurricane Harvey's Catastrophe: *Geophysical Research Letters*, v. 47, no. 23, p. e2020GL090014

Finally, the water rises so high that he is clinging to his chimney top when a helicopter flies over and drops him a rope. The man refuses to take it, proclaiming that God will save him. The man drowns and when he meets God, he says accusingly, "I trusted in you to save me, and yet you allowed me to drown." God replies, "Well, I did send two boats and a helicopter." When we know what needs to be done and yet choose to ignore the obvious solution, we deserve our fate.

The opposite of the delusionally optimistic are the overly pessimistic souls who think there is no point in doing anything because the end of the world is upon us. I think religious believers who are avoiding climate action because of a supposedly impending doomsday are foolish, and part of the problem. As a reminder, the end of the world has been predicted multiple times since the dawn of humanity, and every prediction thus far has been wrong. Given that batting average, if we create a complete and total mess of everything and once again the End Times don't arrive as scheduled, what do we do then? How do we explain our inaction to our grandchildren? On the other hand, if the End Times are indeed upon us, then I think the religious ought to be preparing for the last judgment by trying to clean up our wanton destruction of God's creations after the absolute failure of humanity to be responsible stewards of the environment. Because right now we have no excuses, and the Judge is not going to be happy.

So, what is to be done? First and foremost, it is important to avoid despair and remain optimistic that solutions are possible. To wit, every single environmental problem from plastics in the ocean to GHG-driven climate change is the result of human technology, and technology can be used to repair them all. A lot of very smart and clever people are working on technical solutions, and indeed such solutions already exist for virtually every environmental issue. They have not been widely implemented because of cost, economics, political inertia, resistance to change, and a million and one excuses about why it won't work. There will always be pushback to the changes required, but this can be overcome by citizen support, political will, the desire to shake up the status quo, and a willingness to forgo business as usual.

The overarching problem at present is climate change. We are sliding into a period of climate instability from our overheated atmosphere, and this must be addressed, or the other environmental problems won't really matter. The Earth is now committed to some amount of climate change with GHG concentrations at current levels, but we can and must prevent it from becoming even worse. Human civilization can adapt to modest levels of climate change, but if it goes to extremes, flooding coastal cities and rendering parts of the planet uninhabitable, it will be a much greater problem. The goal is to reach a new plateau of climate stability. This book will lay out some ideas and recommendations for ways to achieve that by eliminating our dependence on fossil fuel, fixing the damage that has already been done, and keeping the planet operating in a sustainable manner.

****

Whether we admit it or not, humans are an integral part of the terrestrial ecosystem, and any damage we do to it, we do to ourselves. The truth is that humans have never been the masters of Earth, but merely tenants. I don't think we've been very good tenants, and if we keep making a mess of things, there will be a reckoning soon.

Do humans have a right to impose an anomaly like climate change on nature? After all, the climate crisis doesn't just affect us. Plants and animals die in great numbers from the extreme droughts, heat waves, wildfires, ocean acidification, and powerful storms that come from climate change. Plants have no option to move if the ecosystem changes around them, and animals have no way of coping with excessive heat or drought except to tolerate it as best they can. Nearly half of the species that existed on Earth two centuries ago have become extinct, mostly because of the actions of humans.

Whether or not humans have the so-called "right" to dominate nature depends on your point of view as to what our role is supposed to be as a component of the Earth's ecosystem. Are we the undisputed masters of the Earth, as envisioned during Victorian times, which entitles us to do as we please with the plants and animals that surround us? Or do we take the more modern view that we are only occupants of Earth just like every other life form, here for a limited time and required to share resources with other living things? As the most intelligent life form on Earth, we understand that reckless, careless damage inflicted on the natural world can destroy fragile ecosystems, damage the environment, and wipe out entire species. No other animal on Earth seems to have achieved this level of awareness, and therefore the burden is on humans to be responsible stewards of the planet's resources because there is no one else.

A law professor at the University of Southern California named Christopher D. Stone published an article in 1972 titled "Should Trees Have Standing? Toward Legal Rights for Natural Objects."[6] Stone had a serious proposal that "forests, oceans, rivers, and other so-called 'natural objects' in the environment — indeed to the natural environment as a whole" be given legal rights. He stated that otherwise, "natural objects" were only valued in relation to their worth to humans and could be destroyed. The example often cited is that a forest has no intrinsic value to a human investor until it is cut down and sawn into lumber.

Some environmental groups have tried using this argument in lawsuits. It has not been successful so far in the United States, where the underlying philosophy was rejected by the Supreme Court. However, in a dissenting opinion, Justice William O. Douglas wrote that if "a ship has a legal personality, a fiction found useful for maritime purposes," and a corporation "is a 'person' for purposes of the adjudicatory processes … So it should be as respects valleys, alpine meadows, rivers, lakes, estuaries, beaches, ridges, groves of trees, swampland, or even air that feels the destructive pressures of modern technology and modern life."

---

[6] Stone, C.D., 1972, Should Trees Have Standing? - Toward Legal Rights for Natural Objects: *Southern California Law Review*, v. 45, p. 450–501

Other nations have recognized the Rights of Nature doctrine, with the first being Ecuador in 2008 when it added the Rights of Pachamama (Mother Earth) to the constitution, recognizing that nature has a right to be respected for its existence. It has the right to maintain itself and regenerate life cycles, structure, functions, and evolutionary processes. These rights are enforced by public authorities, but all persons, communities, peoples, and nations are expected to monitor and support the rights of nature.

Additional countries have begun introducing similar laws. India has new laws that include the legal rights of nature. Neighboring Bangladesh gave all rivers legal protection in 2019. In 2017, legal personhood was granted to the Whanganui River in New Zealand. Panama defines nature as "a unique, indivisible, and self-regulating community of living beings, elements, and ecosystems interrelated to each other that sustains, contains, and reproduces all beings." In the United States, despite the lack of a law at the federal level, more than 30 local governments and communities have introduced Rights of Nature laws.

These laws all essentially recognize that ecosystems have the right to exist and regenerate without any human interference. A human guardian acts on behalf of the ecosystem. These are usually an individual expert or a group that knows how to manage and care for the ecosystem. The ecosystem also has the right to defend itself in court.

Given both the science denial and the litigiousness of American society, this may be the only way to move environmental and climate issues forward. You have to love the lawyers.

****

Most of the problems with energy and the environment were caused by implementation of technologies in the past without any policy to either understand or address the associated impacts. Coal-fired electricity, for example, was developed late in the nineteenth century to utilize an abundant fossil energy resource to create electrical power for our civilization. This resulted in great things like electric lights and a variety of electrically-powered machines, but almost no one at the time considered possible future problems such as devastated landscapes from surface mining, acid mine drainage into streams, sulfur dioxide emissions and the resulting acid rain, disposal of toxic coal ash, and massive GHG emissions. As we switch energy to carbon neutral and climate friendly electrical generation, it is important to thoroughly investigate the possible "unintended consequences" of new energy sources to avoid different but potentially equally severe environmental impacts. For example, the manufacturing processes of wind turbine blades and solar panels both produce toxic chemical byproducts that must be carefully managed.

In my opinion, economics are as important as technology in the switch to climate friendly energy sources. If a non-GHG energy doesn't perform as well or better than existing fossil energy and doesn't cost the same or less to construct and operate, electrical utilities will not adopt it, no matter how "environmentally-friendly" they claim to be. The reality is that dividends are of greater interest to shareholders in investor-owned utility companies than the environment. Environmental advocates often ignore this, and then wonder why a new technology is not being implemented. I think an economic component should be included in any technology-policy remediation plan for climate change that will include transitioning away from fossil energy. Industrial policy incentives such as a steep carbon tax on emissions and generous tax credits for carbon capture and storage can significantly influence these economics in favor of carbon-neutral energy.

Economics are often used as an excuse for resisting change and keeping things the way they are. Calls to stop the war on coal, drill baby drill, and revive the old rustbelt manufacturing jobs are old thinking, stuck in the mud and trying to resurrect a past that is no more. Coal mining jobs peaked in 1923 as the coal industry adopted more machinery. Many manufacturing jobs have been replaced by cheaper imports or by automation; for example, robots can weld automobiles faster and more precisely than humans. Precision machining is performed by computer-operated lathes that make every item identical and perfect. Instead of a machinist at every lathe, one person can monitor an entire room full of these machines. Future products from electronic semiconductor chips to houses will be made autonomously by computers using additive manufacturing, also known as 3-D printing.

Old jobs have disappeared with new technology. There are no more elevator operators, blacksmiths, or railroad porters. Arguing that we should ignore the climate crisis for the sake of bringing back old jobs that no longer exist is the worst kind of backwards thinking imaginable.

Forward thinking brings new industries, which spawn new investments and new jobs. The application of technologies to combat climate change and develop sustainable, carbon neutral energy sources will create hundreds of new jobs in brand new professions. Consider this: a century ago, there was no such thing as a computer programmer, genetic engineer, solar panel installer, astronaut, or social media influencer. If you could travel back in time to the 1920s, I'd wager you would not be able to explain these jobs to the locals. New energy and resource technology will create many future jobs a century from now with names that we have never heard of and functions we do not understand.

The obstructionism of climate skeptics must be overcome to move forward. Their motives are to continue the profitability of the fossil energy industry indefinitely, no matter how much damage it causes to the environment. Their arguments essentially boil down to economics – that fixing the climate is too expensive. This is not true, especially when compared to the costs of not fixing the climate.

In my opinion, a successful switch from fossil fuel to sustainable alternatives must be gradual, over a period of perhaps a decade or two, and incorporate as much existing electrical and transportation infrastructure as possible to maintain a semblance of economic reality. Government policy will be required to force industry and utilities to adopt carbon neutral energy sources, even if it only involves a tax on carbon and tax credits for carbon free energy.

The first step of the transition is for the entire United States and eventually the world to be powered with 100% clean, non-fossil electricity that includes a sustainable mix of wind, solar, geothermal, nuclear, and biofuels. The next step is to power transport vehicles and everything else at the operational level with electricity, biofuels, and possibly hydrogen to displace the remaining fossil fuels used in cars and cooking/heating systems. A concurrent step while transitioning to carbon neutral energy sources is to reduce GHG concentrations in the atmosphere to pre-industrial levels by planting trees, growing algae, and building engineered devices to capture and sequester carbon. However, under no circumstances should the removal of GHG from the atmosphere be used as an excuse to continue burning fossil fuels. They have to go, period. These steps can prevent climate change from getting worse, eventually reverse the damage, and provide abundant, sustainable new sources of energy to power future human civilization. The details are described in later chapters.

Various computer models show that climate changes tend to add up and accumulate in the environment over time. A frightening aspect of the climate models is that they have consistently under-predicted climate effects. More frequent and intense droughts, massive wildfires, bigger storms, heat waves, and significant warming of the polar regions were not supposed to happen until mid-century. All of these are, in fact, occurring now and the actual changes being observed are worse than the model scenarios predicted.

A plan to mitigate climate change can't take forever. Most scientists estimate that we have a century at the outside to get things in order and more likely 30–50 years. Many governments that have pledged to go carbon neutral in the Paris Agreement say they are planning to do so in the 2030–2050 timeframe. China intends to be carbon neutral by 2060. India claims 2070. In the meantime, Australia is ramping up coal production for export to China, and the U.S. is approving more offshore oil drilling leases. There is a lot of talk, but not much action.

The new goal is to hold the global temperature anomaly to less than 2 °C, which will avoid the worst aspects of climate change. We have already passed 1.5 °C of warming, and the effects are becoming undeniable.

The time to act is now.

Chapter 3
The Energy Past

Keywords Coal · Drake well · Spindletop · Natural gas

****

How did we get so addicted to fossil fuels in the first place? To understand how our current energy system evolved, we need to look at the history of energy and technology development. A common saying in geology is that the present is the key to the past. In other words, you can understand how ancient deposits were formed by viewing modern-day processes and assuming the same activities took place in the distant past. In history, the past is the key to the present. Our present-day predicament with energy and climate is firmly rooted in the past practices of the Industrial Revolution and the forces that shaped our modern technological society.

Fossil energy is a relatively recent development in human history. Over many millennia, energy sources were limited to human and animal muscles for transportation, hunting, farming, and most small tasks. Fire from burning wood or charcoal was used for heating, lighting, cooking food, or smelting metals, and grain was processed using power from water wheels or windmills. Although ancient peoples were aware of the existence of coal and petroleum, these were not typically used as energy resources.

By the early part of the seventeenth century, England was facing a wood shortage. Population growth, expansion of cities, wars, and the construction of numerous ships for the British navy and merchant ships for voyages to the New World had decimated English forests. In response, King James I issued a royal proclamation in 1615 requiring that good English wood, "great and large in height and bulk" and with "toughness and heart" be reserved for ship building. As the monarch of an island nation, the king understood that preserving the best wood for ships was a national security issue. Other uses for wood were restricted, and the decree specifically outlawed the wasteful use of wood to "melt, make or causeth to be melted or made, any kind, form or fashion of Glass or Glasses whatsoever." The British glass-making industry, which up until then had been using wood fires to produce glass from silica sand and potash was stunned. Just as the demand for glassware and bottles was increasing, they were suddenly prohibited from manufacturing their products!

D. Soeder, *Energy Futures*, https://doi.org/10.1007/978-3-031-15381-5_3

In desperation, the English glass-makers reluctantly turned to coal as a substitute fuel. Wood was thought of as a clean-burning, noble fuel, while historically coal had been viewed as undesirable and dirty, giving off eye-watering, greenish-yellow smoke and foul smells when burned. Since Roman times, the mining of coal was considered by Europeans to be a form of vandalism or burglary from the Earth. This tradition had left English coal seams largely untouched for centuries and coal could be obtained cheaply and easily. Given the choice of either defying the king and going to prison or obeying the proclamation and going out of business, the glass companies decided to try substituting coal for wood.

After switching fuel sources for their furnaces, the glass makers found that the coal fires reached a higher temperature than wood, allowing them to create thicker, stronger, and more durable glass. Craftsmen developed new methods to take advantage of this discovery, and British glass soon became the best in Europe. The English glassmaking techniques were eagerly adopted by the French, who needed strong bottles for holding the recently invented champagne. Early champagne bottles were notorious for exploding under pressure, and wine inspectors routinely wore fencing masks to protect their faces from flying glass shards when working in French cellars. Sir Kenelm Digby of Buckinghamshire is credited with developing sturdy English glass wine bottles strong enough to contain champagne.

Once the word got out that coal was not only cheaper to burn than wood but also burned much hotter, other industries adopted it. Iron and steel makers needed high temperatures for casting cannon barrels and forging steel swords. Masons needed to calcine oyster shells at high temperatures to create mortar for their stonework. With the invention of steam power in the eighteenth century, coal became even more important as a fuel. The Industrial Revolution began in Great Britain during the mid-eighteenth century and transitioned the manufacturing of products from individual handcrafting to mass production using machines.[1] Steam from coal combustion was a major power source for the machinery in early factories. Coal was also used to power ships and a new invention, railroads. By the early twentieth century, most factory machines were run on electricity, which also commonly used coal as a power source.

Coal was formed over geologic time from the remains of woody plants that once grew in great coastal swamps and marshes. After the plants died and were buried, a lack of oxygen in the sediments preserved the organic matter. The sediments were subjected to increased heat and pressure by being more deeply buried as new sediment was deposited on top, and the organic matter underwent a process called thermal maturation where volatile compounds and water were driven off, leaving behind the non-volatile carbon.

Coal is graded by its degree of thermal maturity. The most thermally mature coal is called anthracite and consists of nearly pure carbon. Mid-grade coal is called bituminous and contains some volatiles; slightly less thermally mature coal is called

---

[1] Wrigley, E.A., 2018, Reconsidering the Industrial Revolution: England and Wales: *Journal of Interdisciplinary History*, v. 49, no. 01, p. 9–42.

sub-bituminous. The least thermally mature coal is known as lignite, and it may retain the original brown color of the plants instead of appearing black.

The energy in coal is derived from the living plants that used sunlight to photosynthesize. This process takes hydrogen and carbon derived from H_2O and CO_2 to create the carbohydrate structures of plants and returns free oxygen (O_2) to the air. The preserved remains of these ancient plants give this energy back when they are burned.

A significant amount of coal formed during the Mississippian and Pennsylvanian Periods in North America, the so-called "Era of Coal" between about 360 to 300 million years ago. (In geologic shorthand, the term for a million years is a mega-annum, abbreviated as Ma, a thousand years is a kilo-annum, or Ka, and a billion years is a giga-annum, or Ga. I'll use these abbreviations throughout the rest of the text.) Most eastern U.S. coals are Pennsylvanian in age. The Mississippian and Pennsylvanian Periods in Europe are combined into one, known as the Carboniferous, which contains most of the European coals. Western U.S. coals like those in Wyoming are much younger, having formed during the Paleocene Epoch of the Paleogene Period, between about 66 Ma and 56 Ma.

Wood was abundant as a fuel source in North America, and coal mining did not begin here until 1762 in the Mt. Washington neighborhood of Pittsburgh, an area known at the time as "Coal Hill" where the Pittsburgh coal seam was accessible.[2] Several mines began operations and coal production developed steadily in the Pittsburgh region through the nineteenth century, eventually peaking at 277 million tons in 1918. Coal was used for the iron, steel, chemical, and glass industries in the Pittsburgh area, and Pennsylvania's railroads and rivers transported coal to other industrial cities like Cleveland and Buffalo.

Coal production continued to increase at other locations in the Appalachians such as West Virginia and eastern Kentucky, as miners followed the Appalachian coal seams to the south. High grade anthracite coal was produced from the eastern ridges of the Appalachians, while lower grade bituminous coal was mined across the central and western parts of the region. Coal from the Illinois basin has been mined since the early twentieth century in southern Illinois and Indiana. At its pre-automation peak in 1923, the U.S. coal industry employed more than 883,000 miners. It has since learned how to produce more coal with less labor and more machinery.

Western coal in Wyoming was developed after the Second World War to meet increasing demands for domestic electricity. It consists of sub-bituminous coal and lignite with a lower heating value than eastern coals, but it also has a significantly lower sulfur content. The need to reduce sulfur dioxide emissions in the 1970s that were causing acid rain shifted electric power companies toward Wyoming coal, and it is currently the largest coal-producing state in the U.S. Peak coal production in the United States occurred in 2006 and has been declining since then, primarily due to

---

[2] Edmunds, W. E., 2002, Coal in Pennsylvania (2nd ed.): Pennsylvania Geological Survey, 4th ser., Educational Series 7, 28 p.

competition from abundant natural gas for electric power generation. Decreasing coal demand has caused coal prices to fall and many mining companies have declared bankruptcy.

Other countries that produce notable amounts of coal include China, which is by far the world's largest producer and user, primarily for power generation. China produces and uses more coal annually than the next nine largest coal producing nations combined. This has recently dropped back somewhat as China has suffered the health consequences of air pollution and the deaths of many coal miners in dangerous mines. China has begun substituting renewables for coal and has pledged to become carbon-neutral by 2060. India is the second largest coal producer, where increasing electrification and growing industrialization are driving demand. U.S. coal production is in third place behind India and ahead of Australia, which exports a significant amount of its coal to China. Indonesia, Russia, and South Africa are the next three notable coal producers, followed by Germany and Poland. Coal production in Europe has been falling for decades as Europeans switch to greener energy technologies.

****

Liquid fossil hydrocarbons were developed as fuels centuries after coal, although they were known to humans in antiquity from natural seeps of petroleum and bitumen in stream valleys, hillsides, and gullies (Fig. 3.1). Bitumen (also called asphalt) is a heavy, tar-like crude oil composed of a viscous mixture of long chain hydrocarbons. It is a common residual deposit in oil seeps after the lighter hydrocarbons have evaporated off. Crude oil and bitumen were used by the ancient Sumerians, Assyrians, and Babylonians for architecture, road construction, waterproofing ships, and medicines.

Like coal, petroleum is formed from the remains of plants. In this case the plants were not woody trees or shrubs, but microscopic algae and other soft plants. The algae used sunlight, water, and photosynthesis to take CO_2 from the atmosphere and grow. After they died, they sank to the bottom of the water column and accumulated in anoxic sediments, which preserved the organic material. As the remains of the algae were buried under more sediment, they underwent a thermal maturation process similar to coal. Algae typically contain an oily material called lipids, and heating this in the absence of oxygen produced petroleum. There was no "Era of Oil" in the geologic record like the Era of Coal in the Carboniferous. Petroleum occurs in rocks across nearly all geologic ages.

Natural gas is created during the thermal maturation process of both oil and coal, and it occurs with both. The primary component of natural gas is methane, the simplest hydrocarbon molecule composed of four hydrogen atoms attached to a carbon atom. The chemical formula is CH_4. In coal, methane is formed as free hydrogen reacts with carbon, and the gas will typically attach itself or "adsorb" onto the carbon molecules of coal.

Note that the process of trapping the gas in coal is called "adsorption" and not "absorption," which is different. In chemistry, absorption is a physical or chemical

Fig. 3.1 A natural oil seep in an otherwise dry gully, Salt Creek Oil Field, Wyoming. (Source: Photographed in 2019 by Dan Soeder)

process in which atoms, molecules, or ions enter another material and are taken up by it. A paper towel absorbs water by pulling it into the paper fibers of the towel. The phenomenon that takes place in coal (and also in organic-rich black shales) is adsorption, where a gas adheres electrostatically to a surface of organic molecules. The gas molecules are packed in tightly as a thin film on the organic surface, and the amount of adsorbed gas can be quite substantial. Dropping the pressure below a critical threshold releases the adsorbed gas, and some coal seams that were not gassy when first mined ended up producing copious quantities of methane once pressures dropped after water was drained from the coal seam. If the gas accumulates in an enclosed space like a mine, it can explode violently. Some terrible underground mine disasters have been caused by methane gas.

Natural gas from oil forms during both the early and late stages of thermal maturation. Early stage methane is a byproduct of microbial activity breaking down organic material in the sediment and is called "biogenic." A slightly higher but still relatively low thermal maturity is needed to form the complex hydrocarbon compounds that make up oil in what is known as the "oil window." As the thermal maturity continues to increase, the long chain hydrocarbons break down into simpler compounds like propane and butane, called "condensate." At very high thermal maturity, the only hydrocarbon left is "thermogenic" methane. Natural gas that occurs with condensate is known as "wet gas;" otherwise it is called "dry gas." If the gas occurs in the same geologic reservoir as oil, it is called "associated gas." Gas that occurs by itself is known as "non-associated gas." Don't worry about memorizing these terms – there won't be a quiz.

Oil was described by Roman scholars like Pliny, although the Romans had little practical use for it and regarded petroleum only as a curiosity. The medicinal uses of petroleum were defined in an encyclopedia of medicine written in the ninth century A.D. by a Persian physician named Ibn Sina. It included various preparations for eye diseases, reptile bites, respiratory problems, hysteria, epilepsy, uterine prolapse, and bringing on menstruation. Petroleum mixed with the ashes of cabbage stalks was said to be good for scabies. A petroleum concoction applied to the forehead was prescribed to warm the brain. The Chinese also used petroleum as a medicine and had an active distribution network as far back as 100 BCE.

The translation of Ibn Sina's medical encyclopedia into Latin spread the knowledge of petroleum to Europe, where it reached the scholar Constantinus Africanus in the tenth century.[3] He was the first person to name the liquid hydrocarbons "petroleum," which was derived from a Byzantine Greek word that literally means "rock oil." This is indeed an accurate description of the substance.

The Seneca tribe of the Iroquois Nation in North America had been collecting petroleum from natural seeps for hundreds of years, employing it as a salve, insect repellent, and cure-all tonic. Early European settlers called the black, gooey substance "Seneca Oil," and followed the example of the natives by using it as a medicine.[4]

By the mid-nineteenth century, numerous and often shady entrepreneurs were marketing "Seneca Oil" or its derivatives as a patent medicine for everything from ulcers to blindness. One of these petroleum-based patent medicines was a product sold by a man named Samuel M. Kier in Pennsylvania. Kier was originally in the business of supplying salt to Pennsylvania farmers for curing meat, pickling vegetables, and keeping livestock healthy. Pennsylvania is a long way from the ocean, the source of salt in coastal regions, so Kier operated a number of wells along the Allegheny River that produced saltwater from deep sedimentary rocks, which he then evaporated to precipitate out the rock salt. The wells also annoyingly produced a black oily substance that had to be tediously separated from the saltwater and discarded.

In 1848, Samuel Kier's wife developed tuberculosis. The doctor prescribed "American Medicinal Oil" as a cure, produced as a byproduct from a saltwater well in Kentucky. The woman's health improved (the role that "American Medicinal Oil" may have played in this is unclear) and Kier realized that this so-called medicinal oil was essentially the same black goo he had been discarding for years as a contaminant in his saltwater wells.

Kier decided to market the oil recovered from his wells as a stand-alone product, and in 1852 he launched "Kier's Genuine Petroleum, or Rock Oil" as a patent medicine. Like many other cure-alls of the time, it was advertised with wildly preposterous claims of "clearing the chest, wind-pipe and lungs," along with curing diarrhea,

[3] McDonald, G., 2011, Georgius Agricola and the invention of petroleum: *Bibliothèque d'Humanisme et Renaissance*, v. 73, no. 2, p. 351–364.

[4] Harper, J. A., 1995, Yo-ho-ho and a bottle of unrefined complex liquid hydrocarbons: *Pennsylvania Geology*, v. 26, no. 1, p. 9–12.

cholera, piles, rheumatism, gout, asthma, bronchitis, burns and scalds, neuralgia, ringworm, skin eruptions, deafness, and chronic sore eyes, among other things. Despite the overblown sales pitch, rock oil actually did have some value as a medicine, and it is still used today in the form of petroleum jelly, an effective ointment and salve. Petroleum also supplies many of the raw materials needed as chemical feedstocks by the modern pharmaceutical industry.

Kier never made much money on his medicinal petroleum, so he decided to look for other markets.[5] Rock oil could be burned, but in crude form it produced an unpleasant odor and heavy black smoke. Kier understood that if these problems could be resolved, petroleum could be sold as a lamp fuel. This was a potentially lucrative market.

The most widely used fuel for indoor lamp illumination in the mid-nineteenth century was whale oil, which burned cleanly and brightly. However, as sperm whales in the Atlantic Ocean were hunted nearly to extinction for their oil, it also became hideously expensive. In 1850, a barrel of whale oil commanded as much as $100 in the U.S. (about $3500 today; equivalent to $83 per gallon or $22 per liter). The economic incentives for finding a cheaper substitute were powerful.

A Canadian geologist in New Brunswick named Dr. Abraham Gesner developed a method of distilling or "refining" crude petroleum into a clean-burning liquid. Combining the Greek words for "wax" and "oil," Gesner called his resulting liquid "keroselain," but it soon became better known as kerosene. Samuel Kier realized that kerosene would make an excellent lamp oil and obtained drawings from a chemist in Philadelphia for an apparatus that would distill crude oil into kerosene. In the mid-1850s, Kier established a small refinery in the city of Pittsburgh to produce a few barrels of kerosene per day that he sold as a product called "Carbon Oil" for use as a lamp fuel.

The existing whale oil lamps at the time could not utilize Kier's Carbon Oil, so he developed and sold a lantern specifically designed to burn kerosene. Still in use today, Kier's familiar lamp design contained an adjustable cotton wick to transport kerosene to a combustion tube inside a tapered glass chimney where it would burn brightly at a high temperature without producing smoke. Kier never bothered to patent any of his inventions, content with a thriving business selling kerosene and lanterns, and other people began making the lamps. Instead of becoming a household name like Ford or Rockefeller, Samuel M. Kier is remembered today only as a footnote to the petroleum business.

Even more astounding, although it was not necessarily his intent when inventing the kerosene lamp, Samuel Kier was essentially responsible for saving sperm whales from extinction. Kerosene was substantially cheaper than whale oil, and as the popularity of kerosene-fired lamps skyrocketed, the demand for whale oil fell to zero. By the last decades of the nineteenth century, commercial whaling was no longer an income-producing occupation, and the whalers went extinct instead of the whales.

---

[5] Brice, W. R., 2008, Samuel M. Kier (1813–1874) – the oft-forgotten oil pioneer: *Oil-Industry History*, v. 9, no. 1, p. 73–96.

The subsequent replacement of the kerosene lantern by Thomas Edison's electric light several decades later caused little disruption in the petroleum industry because by then the refined rock oil had found many other roles. In fact, the most common use for Dr. Gesner's kerosene product these days is jet fuel.

Other companies also jumped into the medicinal petroleum business in the mid-nineteenth century. One of these was established by two Connecticut businessmen named George Bissell and Jonathan Eveleth who had purchased a northwest Pennsylvania farm with an oil seep along the aptly named Oil Creek. Bissell and Eveleth created the Pennsylvania Rock Oil Company in 1854 and transferred the farm property to the company.[6] A former railroad conductor named Edwin Drake had befriended Bissell in New York and invested some of his own money in the company. As a retired conductor, Drake could ride the rails for free and the two senior partners persuaded him to travel to western Pennsylvania to assess their oil prospect. Drake visited Titusville, Pennsylvania in December 1857 and reported that a substantial amount of oil appeared to be recoverable from the property.

Edwin Drake is credited with the idea of drilling a well to produce oil from the ground, although it is unclear exactly who came up with this. A competing narrative suggests that George Bissell sought shelter under an awning 1 day in New York City where he saw one of Samuel Kier's "Rock Oil" flyers that featured the derrick of a brine well and connected the two ideas. Kier was refining petroleum into kerosene at the time and creating new markets for oil, and many historians suspect that the Pennsylvania Rock Oil Company was aware of this. The demand for crude oil to supply Kier's kerosene lamp business certainly would have provided more justification for the cost of drilling an oil well than the small volumes of oil needed to make patent medicine. In any case, Drake was in charge of producing petroleum from the Titusville site, and he was ultimately responsible for drilling the well.

Drake hired William A. "Uncle Billy" Smith, an experienced saltwater well driller to drill the well. Uncle Billy constructed a steam engine house and a derrick on the flat floodplain of Oil Creek in May of 1859 (Fig. 3.2). The drill rig used a 6-horsepower (4.5 KW) steam engine and reached bedrock at a depth of 32 feet (10 m). Drake installed a cast iron pipe known as casing in the hole to stabilize it from groundwater infiltration and possible collapse.

Uncle Billy's rig was a vertical percussion rig known as a "cable tool." These work by essentially pounding a hole down into the ground, shattering the rock and bringing the fragments up to the surface. They are noted for being slow and noisy, but cable tool rigs are still in use today for shallow drilling in the Appalachian basin and elsewhere. The drilling continued at a rate of about three feet (1 m) per day through rock. (In contrast, a modern, hydraulic rotary drill rig can typically drill 2000 feet or 600 m of hole per day.) The drill reached its maximum depth of 69.5 feet (21.2 m) on August 27, 1859, and the borehole filled up with water. The next day Uncle Billy found oil on the water five inches (13 cm) below the top of the well.

---

[6] McKithan, C., 1978, Drake Oil Well, National Register of Historic Places Inventory -- Nomination Form 10–300, U.S. Department of the Interior, National Park Service, 10 p.

Fig. 3.2 Edwin Drake (at right with beard) poses with financial backer Peter Wilson in front of the derrick and engine house constructed at Oil Creek, Pennsylvania in 1859. (Source: U.S. Library of Congress public domain photo)

The Drake well produced 12–20 barrels (2–3 m³) of oil per day until 1861. (A barrel of oil is equivalent to 42 gallons, or 159 liters.) The derrick was exhibited at the 1876 Centennial Exposition in Philadelphia and replicas of the derrick and the engine house are located on the original wellsite, which is now designated as a U.S. National Historic Landmark. Edwin Drake is credited with drilling the first commercial oil well not only in the United States, but the world. Although oil had been recovered from saltwater wells for decades, the Drake well was the first one drilled specifically and deliberately for oil.[7]

To promote the Oil Creek drilling project, Bissell and Eveleth sent mail to Drake at his Titusville hotel addressed to one "Colonel Edwin Drake." Although he had never actually been in the military, "Colonel" Drake found that the title gained him more respect and attention from the locals than he would have received as plain old "Mr. Drake," so he allowed people to believe he was a retired soldier. As a side note, in 1959 on the centennial of his oil discovery, the Pennsylvania State Legislature posthumously made Edwin Drake a Colonel in the Pennsylvania National Guard, and his title became official.

Drake's business partners were stingy with capital, and he was constantly short of funds throughout the venture. After spending only $2500 on the effort (about $86,300 today, and laughable compared to the multi-million-dollar cost of the average modern oil well), Drake's financial backers ordered him to quit. However, before the letter from Bissell and Eveleth made it to the then-remote outpost of Titusville, Drake had taken out a bank loan to keep the operation going, co-signed by two local friends, R. D. Fletcher and Peter Wilson of Titusville. (Wilson appears with Drake in the Fig. 3.2 photograph.) By the time the cease-and-desist letter arrived, Drake had enough funding on hand to complete the well. He ignored Bissell and Eveleth and finished the drilling.

The project was called "Drake's Folly" by some locals because they thought he would never recover enough oil to offset expenses. When oil was found, it became a game-changer in the financial centers of New York and Boston, and Drake's Folly was no longer mentioned. People began to consider investing in petroleum, developing the industry and providing the capital needed to drill more wells.

The Titusville well touched off the first oil boom in the history of humanity. Titusville became a boomtown, and other towns such as Oil City sprang up out of nowhere. Thousands migrated into northwestern Pennsylvania for jobs and quick money on the oil rigs. Drilling started at other sites where oil seeps were present or where brine drillers had found oil in their salt wells. Natural resource booms from gold rushes to oil rushes seem to be a part of the human condition and are likely to continue as long as people discover new resources. Future booms on the moon, Mars, and in the asteroid belt likely will be high-tech versions of Titusville.

U.S. domestic crude oil production expanded to 10 million barrels (1.6 million m^3) per year by 1873, driven by ongoing industrial development in Europe. Manufacturing capacity in the U.S. was tiny and the main domestic market for petroleum was lamp oil. In contrast, Europe was embracing the Industrial Revolution, and British factories in particular were importing large quantities of cheap American oil for use as fuel and lubricants.

Along with expected materials like kerosene lamp fuel and lubricating oils, a number of unusual products were developed from petroleum, sometimes almost by accident. The so-called "Pennsylvania-grade" crude oil obtained from wells in the

---

[7] Harper, J.A., 1998, Why the Drake Well? *Pennsylvania Geology*, v. 29, no. 1, p. 2–4.

Appalachian basin tends to be paraffinic or waxy, which makes it a great lubricant. Deposits of "sucker rod wax" would build up on the oil well pump jacks and production had to be halted at regular intervals to scrape it off. A chemist from New York City named Robert Chesebrough obtained samples of the sucker-rod wax during a visit to the Titusville oil fields. Intrigued by its supposed healing properties, Chesebrough worked in his laboratory to purify the wax into a skin balm, filing patents for a material he called "petroleum jelly," and eventually developing it into a commercial skin care product he named "Vaseline." Petroleum jelly became popular among young ladies as a makeup base, and they soon discovered that mixing in a small amount of coal dust or lamp black made a primitive type of mascara for lengthening eyelashes.

In preparation for a date, a Miss Mabel Williams of Chicago was carrying out this particular operation to the fascination of her brother Thomas. He was inspired to develop a better-performing mascara using a Vaseline base that he sold via mail order. By 1917, Thomas Williams had established a cosmetics factory in Chicago, which he named Maybell Laboratories after the sister who had inspired him and branded his Vaseline-based mascara and other cosmetic products "Maybelline." A number of other unexpected byproducts have also come from the petroleum industry, including asphalt roads, fertilizers, and modern plastics.

The northwestern Pennsylvania oil rush resulted in wide swings in petroleum prices during the first decade of oil production. Within 2 years of the completion of Drake's well, booming production from the Titusville area caused the price of oil to drop from $10 to 10 cents a barrel.[8] Producers created the Oil Creek Association in 1861 to restrict output and maintain a price of at least $4 a barrel. They appear to have been the world's first oil cartel.

****

Most of the gas used for lighting and cooking in the nineteenth century and first half of the twentieth century was a manufactured fuel known as "town gas" made by heating coal and water in the absence of oxygen. The heat would break apart the water molecules into hydrogen and oxygen. The hydrogen would combine to create H_2 or hydrogen gas, and the oxygen would partially combust the carbon in the coal to create CO or carbon monoxide. Both of these gases will burn, and town gas was piped into residences and businesses. It is hard to believe these days that people actually had deadly carbon monoxide supplied to their homes. This was back when you could end it all by putting your head in the oven. Gas leaks would kill entire families quietly while they slept.

Natural gas is composed of non-toxic methane and is much safer than town gas. Natural gas was first discovered in Indiana in 1867, and subsequent drilling in the east-central part of the state and western parts of neighboring Ohio revealed that a large gas field was present to the north and east of Indianapolis, extending into

---

[8] Hildegarde, D., 1959, The Great Oildorado: The Gaudy and Turbulent Years of the First Oil Rush: Pennsylvania, 1859–1880: New York, Random House, 277 p.

Ohio. Named the Trenton Field, it triggered a gas boom and thousands of new wells were drilled. No national pipeline distribution system existed for natural gas at the time, so the resource had to be used locally. The Indiana gas soon attracted manufacturing industries to the Midwest, and towns with natural gas resources competed vigorously for new businesses. Unfortunately, much of the gas was wasted, because operators typically flared off a portion of the production at the wellhead in what was known as a "flambeau" to prove to investors that the gas was flowing. The numerous flambeaus reduced subsurface pressures in the Trenton Field, resulting in the immobilization of estimated 900 million barrels of oil associated with the gas. Almost all of the natural gas production from the Trenton Field ended by 1910, with the recovery of only about 10% of the petroleum.[9] High oil prices in the late twentieth century led to attempts at small scale oil production using advanced artificial lift technology.

Systems for capturing natural gas at a wellhead in the nineteenth century were primitive and unreliable, and the gas was sometimes present at high pressures that resulted in spectacular blowouts and flares. One of the more famous examples is the "Karg Well" drilled in 1886 in the town of Findlay, Ohio. The wellhead ruptured, releasing 12 million cubic feet (340,000 cubic meters) of gas per day and creating a plume of fire visible from 30 miles away (48 km) that towered a hundred feet (30 m) high for 4 months.

Oil production in Pennsylvania peaked in 1891, when the state produced 31 million barrels of oil. After three decades of dominating petroleum production in the United States, wells in the northern Appalachian basin began to decline at the turn of the twentieth century, and oil drillers started looking elsewhere for other prospects. The Anadarko and Arkoma basins in Arkansas, Oklahoma, and northern Texas became the next locations for prolific oil production in the U.S.[10] People also began looking at potential oil resources in the Gulf Coast area of Texas and Louisiana.

The Gulf Coast had been routinely dismissed as a potential petroleum prospect since the days of the Drake well. Despite the presence of sulfur springs and occasional seepages of flammable gas, there was little evidence of oil at the surface. The flat topography didn't appear to contain any of the rock types that drillers had come to associate with the presence of oil.

A self-taught geologist named Patillo Higgins had spent years trying to entice investors to drill for oil on salt domes, certain that the deformed sediments along the flanks of the domes contained trapped oil. Higgins was met with widespread skepticism until he ran across a mining engineer and saltwater driller named Anthony F. Lucas. Higgins convinced Lucas that a low hill called Spindletop three miles

[9] Gray, R.D., 1994, Indiana History: A Book of Readings: Bloomington, IN: Indiana University Press, 442 p. (ISBN 0-253-32629-X).

[10] Williamson, H.F., Daum, A., and Klose, G.C., 1981, The American Petroleum Industry: The age of energy, 1899–1959: Westport, Connecticut, Greenwood Press, 1981, 928 p. (earlier edition published by Northwestern University Press).

south of Beaumont, Texas might be the surface expression of a salt dome. Lucas decided to drill a well on Spindletop Hill to see if he could find oil.

Lucas made a lease agreement in 1899 with Higgins, built a derrick and began drilling in October 1900. He ran out of money early on and secured additional funds from John H. Galey and James M. Guffey, a couple of Pittsburgh oilmen. On January 10, 1901, the drill bit reached a depth of 1020 feet (311 m) and the drilling fluid began flowing backward out of the hole.

Workers fled as the flow increased, followed by a chuff of natural gas and then an eruption of oil in a "gusher" that reached a height of more than 150 feet (46 m). Today, such a loss of well control is known as a "blowout," and the well is said to be wild. The Spindletop gusher lasted for 9 days until the well was finally brought under control. Many newspaper photographers and even landscape painters captured iconic images of the event (Fig. 3.3).

Fig. 3.3 The oil gusher from the Lucas well in January 1901 at Spindletop Hill, Texas. (Source: Wikimedia Commons public domain; original photo by John Trost)

More important than the Lucas gusher itself was amount of oil produced. It was not lost on Lucas' financial backers Galey and Guffey that the initial oil production from the Spindletop well had been nearly 100,000 barrels per day, which was greater than the combined daily production from all of the other existing oil wells in America. The success of Lucas and Higgins resulted in other oil and gas drillers integrating geologic thinking into their strategy. If the Spindletop salt dome had oil, perhaps others did also. Salt dome oilfields were soon developed at Sour Lake in 1902, Batson in 1904, and Humble in 1905.

Salt domes like Spindletop form because sedimentary beds of salt, known as evaporites are less dense than the silica and carbonate sediments that overlie them. On occasion, the lighter salt will rise upward through these denser sediments in a columnar structure called a diapir. When it reaches shallow depths, the top of the salt spreads out like the cap of a mushroom into a dome shape. As it rises, the diapir drags the adjacent sediments along for a short distance, creating upward-tilting beds that terminate against the impermeable, vertical salt column. These tilted beds turn out to be very good traps for accumulating any petroleum in the sediment. Oil is lighter than water and tends to migrate upward into traps through pores in sedimentary rocks.

Petroleum geology became a science as geologists began to understand the origins of oil and gas, including the organic content and thermal maturity of source rocks, migration pathways for oil to accumulate in reservoir rocks, and traps and seals like folds and faults to contain the oil in place. The American Association of Petroleum Geologists (AAPG) was founded in 1917.

Oil well engineering also improved. New wellheads were designed to contain downhole pressures, and formulas for balanced drilling to avoid gushers were developed by adding minerals like barite to adjust the weight of drilling fluids. Spindletop and other gushers also led to the invention of the blow-out preventer (BOP), a hydraulic ram designed to close off a wild well. However, as the Deepwater Horizon incident demonstrated in 2010, even with these safeguards, if enough things go wrong at the same time, serious blowouts can still happen. (The failure of the BOP on the Deepwater Horizon well was one of the main causes of the blowout.) Specialized companies for dealing with wild well control, founded by legendary figures like Joe Bowden and Red Adair, can mobilize quickly in the event of a blowout and shut in a well.

Spindletop took the United States into the oil age.[11] The industry moved from producing small amounts of kerosene lantern fuel and lubricants to expanding into new markets. Petroleum in large quantities was economically feasible as a fuel for mass consumption and for displacing coal as the primary energy resource in the United States. Because it could be mechanically pumped as needed rather than having to be shoveled by hand like coal, oil soon displaced coal in transportation usage, and made significant inroads into electrical generation.

---

[11] Yergin, Daniel, 1991, The Prize: The Epic Quest for Oil, Money, and Power: Simon & Schuster, New York, NY, USA, 912p. (ISBN: 0671502484).

The Spindletop-Gladys City-Boomtown Museum on the campus of Lamar University in Beaumont, Texas has a replica of the Lucas well that re-enacts the gusher several times each week. It uses water instead of oil and only runs for 2 min versus 9 days, but it makes the point. The Gulf Coast remains an important hydrocarbon producer today and is a world-class petroleum province. Oil companies created to develop these resources include Gulf Oil, Texaco, Humble Petroleum, and Pure Oil. Many of them have disappeared through mergers, but their names remain monuments to the importance of oil and gas production on the Gulf Coast.

****

So-called "Big Oil" got started almost as early as the oil industry itself. In what has become common for American businesses from airlines to zip lines, visionary entrepreneurs create small companies to deliver a new and innovative product or service to niche markets within their industry. Soon after they are proven to be successful, one of the major players steps in and buys them out. The mergers and acquisitions typically narrow down the field to only a few major companies, with limited competition and high prices.

The grand master of Big Oil was and will probably forever remain John D. Rockefeller (1839–1937) and his Standard Oil empire. Rockefeller was born in New York and got started in 1863 at the age of 24 with a single refinery making kerosene in Cleveland, Ohio. He founded the Standard Oil Company in 1865 with partner Henry M. Flagler, who is probably better known for spending his Standard Oil fortune in the early twentieth century to build the "Overseas Railroad" from Miami to Key West through the Florida Keys.

Rockefeller grew his company during the largely unregulated nineteenth century by engaging in a practice known as "predatory pricing."[12] This worked by undercutting competitors' prices, sometimes at a loss, with the intent of driving their business into bankruptcy. Rockefeller would then buy up the failed business from receivership for pennies on the dollar and incorporate it into his empire, eliminating the competition. These ruthless tactics allowed Standard Oil to become the only game in town, at which point they could charge whatever prices the market would bear. Rockefeller always denied that he engaged in predatory pricing, and no one could prove it in court, but by 1880 the Standard Oil Company was a *de-facto* energy monopoly that had control over 90–95% of the oil refining capacity in the United States.

In 1882, Standard Oil attorney Samuel Dodd came up with a new idea that he called a "Trust." The Trust was an umbrella organization that controlled a large number of subsidiary companies. The Trustees appointed the directors and officers of all the subsidiary companies, effectively allowing the Trust to exert total monopolistic control over the component companies, self-deal, "compete" against itself,

---

[12]Tarbell, I.M. and Chalmers, D.M., 1966, The History of the Standard Oil Company: Dover Publications, Inc., 272 p. (ISBN: 978-0-486-13995-1).

and set prices across the board. All the profits went to the Trustees, and they then determined what dividends to pay shareholders.

Rockefeller and his partners thought Dodd had a grand idea and re-organized the Standard Oil Company into the Standard Oil Trust in 1882, with nine Trustees running the whole show. By careful design, the inner workings of the Trust were hidden from scrutiny behind corporate figureheads, legal maneuvers, and various paper constructs that made the umbrella organization essentially invisible to the public and impervious to regulators. Even its very existence was often questioned, an uncertainty that Rockefeller and his fellow Trustees no doubt encouraged. As one investigative reporter noted at the time, "You could argue its existence from its effects, but you could not prove it." The excessive concentration of economic power in the Standard Oil Trust was viewed by many Americans with alarm, even though Rockefeller claimed that he was only seeking the efficiencies of scale.

The nineteenth century business model of the Standard Oil Trust was largely based on the refining and sale of illumination oil for lamps, but with the advent of Thomas Edison's electric lighting, Standard Oil changed focus to supplying gasoline for automobiles. The gasoline-powered internal combustion engine had been invented in Germany by Siegfried Marcus in 1870 when he used a two-stroke engine to propel a pushcart (yes, the first ever gasoline-powered vehicle was basically a go-cart). A few years later, Nikolaus Otto, also in Germany, received a patent for the cleaner and more reliable four-stroke gasoline engine, and Karl Benz, a third German, built several identical copies of a gasoline-powered automobile in 1885, creating the first "production model" car. Early automobiles were hand-crafted and expensive, remaining little more than European curiosities for the wealthy known as "horseless carriages."

Henry Ford changed all that in 1908 by creating the Model T, built on a factory assembly line and made affordable to the average worker (Fig. 3.4). By the end of production in 1927 more than 15 million Ford Model T automobiles had been sold, and every single one of them required gasoline.

Sales of gasoline in the U.S. surpassed those of kerosene in 1919 and never dropped back. Gasoline demand resulted in Standard Oil adding exploration, production, and transport to its refining and distribution businesses. In today's terminology, the company became vertically integrated, controlling the supply from the oil well to the consumer's fuel tank.

Other businesses copied Samuel Dodd's trust model, and the growing problem of trusts caught the attention of Senate Finance Committee chair John Sherman of Ohio. The Sherman Anti-Trust Act used the constitutional authority of Congress to regulate interstate commerce as a way to dissolve trusts. It was signed into law by President Benjamin Harrison in July 1890.

Sherman's law was loosely worded, and armies of high-powered lawyers attacked and weakened it. A number of states then tried regulating trusts within their boundaries, with little success. For example, the dissolution of the Standard Oil Trust was ordered by the Ohio Supreme Court in 1892. In response, Rockefeller downgraded the host company for the Standard Oil Trust, the Standard Oil Company of Ohio, into a small subsidiary company called Sohio that only distributed finished

Fig. 3.4 The 1915 Model T Ford automobile, commonly known as a "flivver" or a "Tin Lizzie". (Source: Public exhibit photographed in 2021 by Dan Soeder)

petroleum products. The operations of the Trust were moved to other states to take them out of the jurisdiction of the Ohio court, and Rockefeller himself relocated to New York. In 1899, Rockefeller and his Trustees brazenly reconstituted the Trust by turning Standard Oil Company (New Jersey) into a holding company, and assigning the assets and interests formerly controlled by the Trust in Ohio to the New Jersey company.

President Theodore Roosevelt was able to use the Sherman Anti-Trust Act to successfully dissolve Northern Securities Company in Minnesota as part of his "trust busting" campaign in 1904. With this precedent in hand, President William Howard Taft invoked the Act in 1911 against the Standard Oil Company, ordering it to divest its major holdings. The Standard Oil Trust was broken up into 34 newly independent oil companies, and Rockefeller held significant amounts of stock in all of them. He was never hurting for money, and for a time he was the richest person in the world.

The sudden and unexpected death of wealthy financier J.P. Morgan in 1913 shocked both Rockefeller and his long-time nemesis, steel magnate Andrew Carnegie, and they decided to focus on charitable work for a more lasting legacy. However, even at this they remained rivals, competing to see who could give away the most money. Rockefeller eventually "won" the contest by out-living Carnegie, who died in 1919. Rockefeller continued to support charitable work until his death in 1937 at the age of 97.

John D. Rockefeller remains a controversial personality even today. His rapacious capitalism versus his generous philanthropy led one of his biographers to declare that "his good side was every bit as good as his bad side was bad."[13] His charitable organizations and foundations have continued, and several of his descendants went into politics. Another legacy of Rockefeller, the Sherman Anti-Trust Act, is still on the books. The federal government last invoked it in 2001 for a ruling against Microsoft Corporation.

One of the 34 companies to come out of the 1911 breakup of the Standard Oil Trust was Jersey Standard, derived from the original Standard Oil (New Jersey). Jersey Standard acquired a 50% interest in Humble Oil & Refining Company of Texas in 1919 to tap into oil supplies on the Gulf Coast. Jersey Standard introduced a new gasoline brand in 1926. They called it Esso, which most people didn't realize was a simple phonetic rendition of the initials 'S' and 'O' from Standard Oil (Fig. 3.5). Esso gasoline became recognized as the corporate brand for the company, and at a shareholders' meeting in 1972, the company name was changed to Exxon Corporation. Exxon acquired Mobil Oil in 1999 and became ExxonMobil Corporation.

Mobil had been founded in 1866 as the Vacuum Oil Company and was one of the businesses snapped up by the Standard Oil Trust. Vacuum Oil was re-established after the 1911 breakup and became Mobil Oil Corporation on the centennial of its founding in 1966. In a press release, the new ExxonMobil Corporation stated that a

Fig. 3.5 Depression-era Esso service station from 1934 preserved in the "New Deal" town of Arthurdale, West Virginia. (Source: Photographed in 2021 by Dan Soeder)

---

[13] Chernow, R., 1998, Titan: The Life of John D. Rockefeller, Sr.; New York, Random House, 750 p.

goal of the 1999 merger was improved efficiency. Somewhere, John D. Rockefeller was smiling.

ExxonMobil is one of the largest oil companies in the world. The other companies making up "Big Oil" (known in the industry as "the majors") tend to be multinational corporations headquartered outside of the United States. This, too, is part of the Standard Oil legacy. Rockefeller and his cohorts exerted so much control over petroleum markets in the U.S. that no one else could enter. The other large oil companies decided to produce, refine, and sell their petroleum products elsewhere.

In the late nineteenth century, the wealthy Rothschild banking family in France commissioned a pair of experienced British traders and shippers known as the Samuel brothers to transport kerosene from Russia to lucrative European markets. The Samuels obtained a tanker ship for the job named Murex after a type of seashell. The ship inspired them to found Shell Transport and Trading in 1897 and create a global transportation network for oil.

Royal Dutch Petroleum was started in the late 1800s after oil was discovered in the Dutch East Indies (now Indonesia). This was a fairly remote area at the time and getting the petroleum to European markets required a robust overseas transportation system. Royal Dutch Petroleum engaged Shell Transport and Trading for the task, and the two companies merged in 1907 to form the Royal Dutch Shell Group.

Royal Dutch Shell was incorporated in the United Kingdom and headquartered at The Hague in the Netherlands. The company had a dual-shareholder structure set up for investors to accommodate the different Dutch and British tax laws. After Brexit, the company decided in 2021 to consolidate into one country. As such, it officially changed its name to Shell plc in 2022 and moved the headquarters to London, scrapping the dual-share structure.

Shell plc will retain the familiar yellow scallop for their corporate logo that was originally adopted by Royal Dutch Shell. The scallop shell was used because the murex, the name of the Samuel brothers' original tanker ship, is a predatory marine snail known as the Venus Comb that possesses an elongated shell covered with fearsome spines (Fig. 3.6).

British financier William Knox D'Arcy formed the Anglo-Persian Oil Company after his prospector, George Reynolds, discovered oil in Persia (modern-day Iran) at Masjid-i-Suleiman in 1908. A pipeline and refinery were constructed, and D'Arcy supplied petroleum markets in Great Britain and Europe with his products. In 1910, the British First Sea Lord, Admiral John Arbuthnot Fisher, mandated that Royal Navy ships be modernized by converting them from coal to oil. Ships fueled by oil had a greater range than coal-powered ships, and oil could more efficiently be pumped aboard whereas coal had to be shoveled by hand.

Soon after this retrofit began, Sir Winston Churchill, the First Lord of the Admiralty at the time realized that a secure source of oil was necessary for future naval operations. Churchill worked out a deal in 1914 with the Anglo-Persian Oil Company to provide oil to the Royal Navy in return for a payment of £2 million and 51% British government ownership of the company. The ink was barely dry on the agreement when World War One broke out. The British government seized the assets of all German companies operating in Great Britain, including the British

Fig. 3.6 Not a great corporate logo: the spiny shell of the Venus Comb snail *Murex pecten*. (Source: Wikimedia Commons licensed under the Creative Commons Attribution 2.0 Generic license)

Petroleum brand that had been created originally by a German oil firm to market products in England. The Public Trustee sold the British Petroleum assets to Anglo-Persian Oil Company in 1917, and with the government holding majority owner-ship, the Anglo-Persian Oil Company became British Petroleum. These days, it is known simply as BP.

Along with ExxonMobil, Shell, and BP, the other majors include Chevron USA, a 1977 re-branding of Standard Oil of California, and Total S.A., a French multina-tional oil and gas company founded in 1924. ConocoPhillips, created in 2002 by the merger of Conoco Inc. and Phillips Petroleum Co., two midsize American oil com-panies known as "independents," and the Italian multinational company Eni S.p.A. are often included in the group for a total of seven multinational, major oil companies.[14] Most of the other large oil companies in the world are owned or con-trolled by national governments and focus mainly on supplying energy, income or both to their particular country.

Investor-owned petroleum companies have a long history of risking significant amounts of capital to develop and produce oil and gas. They occasionally have losses, drilling expensive wells that produce no oil, known as "dry holes." These are learning experiences and with a better understanding of geology and improved geo-physics, their success rate is better. Highly productive oil wells have provided lucra-tive profits since the days of Spindletop. There will always be a need for petroleum, even if only for petrochemical and pharmaceutical feedstocks. As long as the fossil fuel is not burned, it does not harm the climate. However, the non-fuel markets for petroleum are much smaller than the gasoline demand for cars, the diesel demand for trucks and trains, and the kerosene demand for jet aircraft. Non-fuel markets for coal and natural gas are even smaller. As such, adapting the existing business

---

[14]The Economist, 2019, Crude Awakening: ExxonMobil and the oil industry are making a bet that could end up wrecking the climate: *The Economist*, v.430, no. 9129, 9 Feb 2019, p. 9.

models of big oil, coal, and natural gas to the realities of climate change has been challenging. Fossil energy investors are primarily interested in dividends, not the environment.

****

Many oil companies are cognizant of reality and are attempting to restructure themselves for a carbon-constrained future as "carbon management companies" that include CCS as part of the cost of doing business. This is at least partly disingenuous because they don't count GHG emissions from customers using their products as part of their climate footprint. Some of the European majors are broadening the scope of their business plans by moving into the direct production of renewable energy. However, other oil companies defiantly state that their business will continue to be the production, transport, refining, and distribution of petroleum and natural gas, and promise to fight any regulations that address climate change.

This bravado may be inspiring to some, but these defiant oil companies risk becoming irrelevant by trying to sell an obsolete product to consumers who no longer want it. Witness the demise of Eastman Kodak when photography transitioned to digital, eliminating the need to buy film and pay for processing it into pictures when they are available to view and share instantly right on your phone. Blockbuster videotape rentals was another successful business that collapsed with the introduction of online video streaming. New technology and changing consumer tastes can kill a business almost overnight. Any reasonable response to climate change must include the elimination of fossil fuels. The handwriting is on the wall. If oil companies are to survive, they must become "energy companies" with new thinking and diversified products.

****

Electric power companies that use fossil fuel are investigating CCS to reduce or eliminate GHG emissions. CCS adds to the cost of electricity and requires a plan for storing or sequestering the carbon away from the atmosphere, usually by underground injection. The U.S. Department of Energy (DOE) has been promoting carbon capture, utilization, and storage (CCUS), with the idea that finding a profitable use for the captured CO_2 will offset the costs of carbon capture. Both terms appear in the literature. Unfortunately, the market for captured carbon dioxide is very limited, and there little commercial incentive for CCUS. Most of the current CCS activity is being funded by the government.

If a pipeline can be arranged, oil companies may take the captured powerplant CO_2 and inject it into old oil fields in a process called enhanced oil recovery (EOR) to re-pressurize the reservoir and recover additional oil. Although this does provide a commercial use for the captured CO_2, it is not actually beneficial for the climate if the captured emissions are employed to simply produce even more fossil fuel.

Coal producers are also seeking to stay relevant by potentially supplying rare earth elements (REE) from processed coal. These elements are important for electronics, computer screens, magnets in electric motors, and have a number of critical defense uses. REEs are associated with volcanic ash layers, known as "ton-steins" that often occur in coal. They are mined with the coal and then separated out during the "clean coal" preparation process that removes inorganic mineral matter from coal before it is burned to reduce the amount of ash. The discarded material is slightly enriched in REE that can be extracted by chemical leaching or other methods, although the economics are marginal. There is only one operating REE mine in the U.S. at Mountain Pass, California, and over three quarters of the world's production of REE comes from China. The United States considers it a national security issue to develop domestic REE supplies, and DOE is supporting research on the recovery of REEs from coal.

Mitigating and responding to climate change cannot be left up to the industry. Climate alarms were first raised in the 1980s, and since then neither the "free market" nor "economics" have made the slightest difference. The fossil fuel industry has expanded and profited as GHG levels soared. In a capitalist economy that relies on private investment to fund business, the main goal of such businesses is maximizing the dividends returned to investors. This means expanding, growing, and producing and more of their products to sell. Requiring the fossil fuel industry to reduce sales to mitigate climate change goes completely against their business models and explains why they have done nothing about climate change on their own.

Free market advocates claim that if industry needs a "correction," new businesses will step up and offer those products and the old industries will fade away. So yes, telephone lines have been replaced by mobile phones, and telephone companies like AT&T have fully embraced cell phone services as their new business model. I don't see the same kind of forward thinking and adaptability in the fossil energy industry. There have been a few feeble efforts to deal with climate issues, but overall the stubbornness of the industry to adapt suggests that it will not realign itself to a climate-constrained world on its own. In my opinion, the only way the industry will respond to climate change is when government policies are enacted that reward compliance and punish non-compliance. For example, a stiff carbon tax added to the sticker price of new gasoline-powered vehicles will encourage people to buy electric. If enough people go electric, the industry is going to be stuck with a lot of oil, just like Kodak got stuck with a lot of film.

The nineteenth century antics of the Standard Oil Trust provide a clear example of what can happen when industry is left to the "free market." Fossil energy companies detest government oversight, and the industry has whined since Rockefeller's day about being subjected to "burdensome regulations." However, if they continue to refuse to respond to the climate crisis on their own, a regulatory framework may simply be necessary.

Chapter 4
The Energy Present

Keywords Oil embargo · Shale gas · Fracking

The U.S. dependence on fossil fuels became painfully apparent during the 1973–74 oil embargo by the Organization of Petroleum Exporting Countries (OPEC). Even though the actual reduction in oil supply was only about 10%, the resulting social and economic disruption influenced U.S. foreign policy for the next 50 years. The development of shale gas and tight oil two decades later using horizontal drilling and hydraulic fracturing was the direct result of the embargo. By 2019, the U.S. was producing more oil than Saudi Arabia, and more gas than Russia. The abundant natural gas from shale led many electric utilities to convert power plants to natural gas from coal. Coal use suffered as a result, with the coal companies claiming it was due to an EPA "war on coal." The truth is that gas is cheaper, much cleaner, and 50 percent more efficient at generating electricity.

****

Over the past two centuries, fossil fuels have proven to be of great benefit to humankind. First coal, and then oil and gas made possible the Industrial Revolution, the Age of Steam, and the invention of the railroad, steamship, automobile, airplane, electric light, electric motor, and a thousand other devices that led to our modern, technological society. Nearly every technology today uses fossil fuel either directly, such as automobiles and jet aircraft, or indirectly from electricity created by burning coal or natural gas.

Fossil fuel technology has increased productivity, significantly reduced the time needed for people and goods to move around the world, and improved human health, comfort, and longevity. Steam power from coal replaced the drudgery of human or animal muscles and allowed the development of new manufacturing techniques for a variety of goods. Coal, fuel oil and natural gas substituted for wood in heating and cooking, sparing many forests. Cheap kerosene replaced increasingly rare and expensive whale oil as a lamp fuel for illumination in the late nineteenth century, eventually driving the whaling industry out of business and saving sperm whales from extinction. By performing work far more efficiently than human or

D. Soeder, *Energy Futures*, https://doi.org/10.1007/978-3-031-15381-5_4

animal muscles, fossil fuel-powered machines have allowed a much greater variety of people to carry out tasks that formerly required brute strength. For example, a small woman operating a diesel-powered front-end loader can move more material than a hundred burly men with shovels. Fossil fuel and the machines it powers have become a great equalizer.

Energy from fossil fuels is pervasive in our society. Despite thousands of wind turbines and the 60-year history of nuclear power, the United States, with the largest economy in the world, still obtains about 70% of its electricity from fossil fuels. The second largest national economy, that of China, obtains 86% of its electricity from fossil fuels.[1] Replacing these with energy resources that emit zero GHG will be an enormous challenge.

But replace them we must. Despite all their advantages and contributions to human civilization, powering our technology with fossil fuel has enormous downsides, of which climate change from GHG is only one, albeit a big one. Coal mining damages the landscape and oil spills damage the oceans. The bottom line is that the supply of fossil fuels is finite. We are going to run out sooner or later and the earlier we replace them, the better.

Other significant environmental problems abound. For example, the excess CO_2 in the atmosphere from fossil fuel combustion is dissolving into the oceans as carbonic acid (H_2CO_3) and lowering the pH of seawater through a process called ocean acidification. Shelled creatures are adapted to ocean water where the pH is neutral to slightly alkaline and use this chemistry to build their shells by extracting calcium carbonate ($CaCO_3$) from seawater. Calcium carbonate forms the mineral calcite, which dissolves in acid (one of the tests geologists use for limestone is a drop of acid – if it fizzes, it is made of calcite). Ocean acidification is killing shelled marine animals like snails, clams, and corals by literally dissolving them away.

Coal mining, especially surface or "strip" mining damages landscapes and watersheds, while underground mining is dirty and dangerous. Exposure to coal dust in the mine often gives long-time miners a chronic, emphysema-like condition called "black lung." Many of the retired miners I know in West Virginia suffer from this. Sulfur in coal creates sulfur dioxide (SO_2) when burned. This combines with rainwater (H_2O) to form sulfuric acid (H_2SO_4) in the air and falls as acid rain. Before sulfur dioxide controls were instituted, the rainfall downwind of coal burning power plants could be as potent as weak battery acid. Many limestone, marble, and concrete buildings in eastern U.S. cities will bear the scars of acid rain for many years to come. Coal ash also contains toxic metals like mercury and selenium that can seep into groundwater and enter the environment.

A brutal surface mining process called "mountain top removal" or MTR is often used to extract coal from the Appalachian highlands. The coal-rich regions of the Appalachian plateau consist of a series of flat, isolated tables of rock surrounded by deep, water-cut ravines. Horizontal coal seams are present in the upper parts of

---

[1] USEIA (U.S. Energy Information Agency), 2020, Country Analysis Executive Summary: China. 18 p.

these tables under a few dozen feet (meters) of sedimentary rocks called overburden. The MTR method uses explosives and heavy equipment to strip the overburden off the coal seam across the entire table. This waste material is unceremoniously dumped into the surrounding stream valleys and the coal is then excavated from the exposed seam. Once the coal is removed, the area is abandoned, and the site is left to weather and erode. Sulfur and iron minerals in the dumped rock leach out and damage the surrounding streams. There is usually no remediation of the highly disturbed landscape by the coal company. I've been on these lands to collect drill core samples, and the MTR sites that have been mined out often resemble the lifeless surface of the moon.

The coal industry is rarely held responsible for subsequent problems to groundwater and surface water quality, damage to terrestrial and aquatic ecosystems, or risks to the health of nearby human populations. The required cleanup is typically performed by the state and federal governments. West Virginia has a substantial remediation program for streams affected by "acid mine drainage" or AMD. Iron sulfide minerals associated with the coal are mobilized by water draining from abandoned surface and underground mines. The sulfides oxidize to sulfates, which react with water to create sulfuric acid. The iron is left behind in the stream bed as iron hydroxide, and AMD streams can often be recognized by their orange "rust" color (Fig. 4.1).

The federal government, typically the U.S. Geological Survey or the Army Corps of Engineers monitor flow in the streams to assess the potential for flash floods in these disturbed and modified watersheds. Water quality improvements in the stream

Fig. 4.1 Rocks and sediment in Left Fork Little Sandy Creek stained by orange iron hydroxide from the upstream Kingwood coal mine (now closed), Preston County, West Virginia, USA Source: Photographed in 2021 by Dan Soeder

are performed by the state environmental agency, usually with a combination of state and federal funds. The acidic stream is often treated by adding powdered limestone at various locations (Fig. 4.2). The limestone neutralizes the acid, acting as a "buffer" to return the stream to a more normal pH.

State public health agencies are charged with addressing human health problems, and the responsibility for restoring fish populations, forests, and a stable landscape falls on other government agencies. Neither the mining company that caused the damage nor the power plants and steel mills that used the coal typically end up paying the cost to repair the environment. The remediation for society's use of coal are externalized costs paid by state and federal taxpayers.

Years ago, coal mine operators were required to restore surface mines (called "strip mines" by the public but not by industry) after the completion of mining operations to leave behind real estate that was useful for other purposes. Restoration of old mines has been done in a number of places to produce large expanses of flat land near hilly Appalachian towns where flat land is valuable. Many of these old surface mine sites near cities like Morgantown and Clarksburg in West Virginia and in the Pittsburgh area have been repurposed into shopping centers, big-box retail, commercial buildings, office parks, and residences. In many cases, the high wall of the mine is still visible if one knows where to look.

The original requirement to restore the "topographic contours" of the land was resisted by industry, especially for mines in rural areas where there was no inherent real estate value in restored lands. The industry threatened to eliminate jobs, close

Fig. 4.2 State-owned equipment treating acid mine drainage in Birds Creek by adding powdered limestone to buffer the pH, Preston County, West Virginia, USA
Source: Photographed in 2021 by Dan Soeder

down mines, and move operations elsewhere if forced to meet these standards. Politicians in coal-dependent states like West Virginia and Kentucky gave in, fearing that a loss of coal mining jobs if a coal company moved out of state would result in the loss of their own job in the next election. These were largely empty threats because there are only a limited number of places to obtain coal and moving a coal mining operation to New Jersey, for example, is not going to work out well. Still, most elected officials refuse to call their bluff.

Many coal country political campaigns contained a platform for saving "good coal jobs" without ever mentioning the actual costs of those jobs to society and the taxpayers. In 2019, there were about 14,000 working coal miners in West Virginia. In contrast, both the medical industry and the retail industry employ many more people in the state, with some 12,500 people in WV working for Walmart alone. Nobody digs coal with a pick and shovel anymore. The coal industry has readily adopted automated machinery and heavy equipment that allow greater amounts of coal to be produced with less labor, limiting the total number of coal jobs, especially in surface mines. Coal mining reached a high level of efficiency and profitability with the implementation of MTR techniques, especially when there are no associated cleanup or restoration costs. Some coal company operators became billionaires while large swaths of Appalachia were devastated.

In the past decade, many coal jobs disappeared and many mines were closed as the industry complained that the Obama EPA had declared a "war on coal." What actually happened is that the shale gas boom from 2008 to 2018 resulted in abundant natural gas supplies that became available for generating electricity. Electrical generation with gas is cheaper, cleaner, less GHG-intensive, and significantly more efficient than coal-fired power generation. Thus, when shale provided massive, long-term natural gas supplies, the electric utilities switched from coal to gas and the demand for coal dropped sharply. This was good for the environment and was also a wake-up call for coal-dependent states like West Virginia to diversify their economy.

****

Petroleum production in the United States continued at a rapid pace in the 1920s and 30s, and especially during World War II, when it became critical to the war effort. After the war, the pent-up consumer demand for automobiles, appliances, and many other goods substantially increased fossil energy consumption. Early models of refrigerators, televisions, air conditioners, and other electrical appliances were flying off shelves. None of these were noted for being particularly energy efficient. Likewise, motor vehicles from the late 1940s to the mid-70s were built for comfort, performance, and durability, not high gasoline mileage.

The construction of better roads and the new interstate highway system gave people more incentives to drive. New institutions that catered to the automobile, such as drive-in restaurants, drive-in theaters, drive-up bank tellers, and others sprouted during the 1950s and encouraged people to use vehicles more often and burn more fuel. The post-war migration from the cities into the suburbs left behind most of the public transit options like buses, streetcars, and subways (and often even sidewalks). Living in the suburbs essentially required ownership of at least one motor vehicle to survive. Whereas in the past only the wealthy had indulged in owning more than one vehicle, many suburban families with driving-age children found it necessary to possess multiple cars.

Large-capacity passenger trains and ocean liners were gradually replaced with faster but much smaller jet aircraft, increasing the amount of energy used per passenger per mile. The speed of jets also encouraged more people to travel, known as the "jet set" in 1960s vernacular, crowding once-remote vacation destinations with hordes of tourists. Additional demands on the petroleum supply arose in the latter half of the twentieth century from the use of petrochemicals to manufacture newly invented plastics, fertilizer, synthetic rubber, and other materials.

The upward spike in post-war petroleum demand occurred while the U.S. oil and gas wells drilled in the 1920s and 30s were facing declining production. By the 1950s, new domestic oilfields were becoming increasingly difficult to find, and the big oil companies began exploring for oil in more challenging regions like northern Alaska and the offshore Gulf of Mexico. Some of the more wasteful practices of the previous century, such as oil well gushers or the burning of natural gas in a flambeau were coming back to haunt the industry. Operators began investigating and applying new techniques like infill drilling, waterflooding, reservoir re-pressurization, and new types of artificial lift on old fields to get more oil out of the ground.

Driven in no small part by Standard Oil's lockdown of the American market, the major oil companies went to all parts of the globe to obtain crude oil. Petroleum became a truly international commodity; produced, refined and sold at many different places around the world. The Samuel Brothers had shown with their Shell Trading Company tanker ships that local production of crude oil was unnecessary, because it could be profitably transported long distances over the oceans.

Post-WW2 discoveries of enormous crude oil reserves in the Middle East, North Africa, Lake Maracaibo in Venezuela, and elsewhere took advantage of the most technologically-advanced drilling and completion techniques. These included "gas drive" production that maintains a natural gas pressure head on the petroleum and pushes oil to the surface without the need for artificial lift devices like pump jacks. Gas drive oil wells are common in Saudi Arabia, where one needs merely to open a valve and the oil flows. Production from these new wells was significantly cheaper and more efficient than wells drilled in the United States using old pre-war technology. Crude oil produced from some of the most remote and inhospitable regions of the world was shipped to refineries with the final products ending up in global markets across Europe, Asia, and North America.

The multinational oil companies got involved early on with the drilling and development of these big, new petroleum reserves. However, once the locals developed enough expertise to drill and produce the oil themselves, they realized that the so-called Western "experts" were no longer needed. They also became aware of the lopsided profit sharing plans forced on them by colonial governments and determined that they could keep a lot more money in the country if they produced the oil themselves. Countries like Algeria, Libya, Saudi Arabia, Venezuela, Brazil, Mexico, and others set up nationalized oil companies owned by the government. Some of these nations were newly independent, awash in post-colonial sentiments, and only too happy to show foreign companies the way to the exit. The majors largely stepped back and adopted the role of middlemen, loading foreign crude oil onto tankers and transporting it to refineries. The finished products were then sold through the established company distribution systems. By the late 1960s, the benzene, toluene, ethylbenzene, and xylene molecules (BTEX) making up the gasoline in the tank of an American car could have come from anywhere in the world.

After the Second World War, the United States became the largest oil-consuming nation on Earth. By 1960, U.S. domestic oil production was unable to meet demand and for the first time in history, American oil companies began importing petroleum from overseas to make up the difference. As U.S. oil production continued to decline and demand continued to rise, the volume of imported petroleum increased. By 1970, about a third of the petroleum supply in the United States consisted of imported oil. At the time, almost no one thought that this mattered, until it very much did.

∗∗∗∗∗∗∗∗∗∗∗∗

On October 6, 1973, the Jewish holy day of Yom Kippur, Egyptian and Syrian armies invaded Israel, followed by armies from Iraq and Jordan. The Israeli Defense Forces (IDF) responded with a counterattack, there was a dust-up lasting several weeks, and hostilities ended on October 25, 1973, with a ceasefire brokered by the United Nations.[2]

Although the Yom Kippur War was an Arab-Israeli conflict, there were a number of underlying complications. The United States and the Soviet Union were rather intense Cold War adversaries at the time and transformed the participants into proxies. Egypt was resupplied and armed by the Soviets, while the Americans supported Israel. U.S. President Richard Nixon asked Congress to provide $2.2 billion ($13.5 billion in 2021 dollars) in emergency military aid to Israel. Along with military equipment and money, the U.S. also provided the IDF with a substantial amount of clandestine aid, including both strategic and actionable intelligence from U-2 spy plane overflights of Egypt and Syria to observe troop strength and movements. As the Israeli counterattacks grew increasingly more effective, a frustrated USSR threatened direct military intervention in the war, prompting the U.S. to move to an elevated state of nuclear readiness and placing nearly everyone in the world on edge.

A few weeks into the war, Arab nations learned the extent to which the United States was actively supporting Israel against them and concluded that selling crude oil to the U.S. was clearly aiding an ally of their enemy. The logic of war and the dictates of military strategy demanded that oil exports to the U.S. be stopped. At a meeting of the Organization of Petroleum Exporting Countries (OPEC) in Kuwait on October 20, 1973, Libya introduced a resolution for members of the cartel to

[2] Rabinovich, A., 2004, The Yom Kippur War: The Epic Encounter that Transformed the Middle East: New York, Schocken Books, 543 p. (ISBN:0805241760).

embargo oil exports to the United States. The embargo encompassed other countries as well that had supported Israel, including the Netherlands, Portugal, and South Africa. Interestingly, of the four Arab nations that invaded Israel, only Iraq was an actual member of OPEC. Nevertheless, most of the Arab members of OPEC joined the embargo and halted crude oil shipments to the United States.

To avoid creating a global oil surplus and a price collapse caused by the pile-up of supplies they were no longer sending to their biggest market, OPEC instituted a series of production cuts that resulted in shortages and nearly quadrupled the global price of oil. Numerous internal debates and disagreements ensued within OPEC about the effectiveness of the embargo and whether or not the United States of America had "learned its lesson" for supporting Israel. The oil embargo was offi-cially lifted in March 1974. Although some of those who lived through it remember it as lasting much longer, the embargo was actually in effect for barely 5 months.

The OPEC cartel was created in Baghdad, Iraq by five oil-exporting nations in September 1960: Iran, Iraq, Kuwait, Saudi Arabia, and Venezuela. Its stated goals are to coordinate petroleum policies among members to secure "fair and stable" prices and provide an "efficient, economic and regular" supply of petroleum, not unlike the goals of the Standard Oil Trust half a century earlier. Other OPEC mem-bers that have joined the five founding nations (some have come and gone) include Qatar, Indonesia, Libya, United Arab Emirates, Algeria, Nigeria, Ecuador, Angola, Gabon, Equatorial Guinea, and Congo. The OPEC global headquarters is in Vienna, Austria.

In 1973, U.S. oil imports were averaging about 5–6 million barrels per day out of a total daily petroleum consumption of about 17 million barrels per day. Of the 30–35% of the total U.S. oil supply made up of imports, only about half originated in OPEC countries, while the remainder came from non-OPEC sources like Mexico or the North Sea. Even if all OPEC members had agreed to embargo oil exports to the U.S. (and not all of them did), oil supplies would only have been cut by about 15%. The actual supply reduction was closer to 10%.

Although the U.S. still had 90% of its oil supply, the OPEC oil embargo precipi-tated one of the greatest crises in American history. It is difficult to overstate just how much trauma, drama, concern, panic, and angst the OPEC embargo inflicted on the social fabric of the U.S., except maybe to point out that it influenced American foreign policy for at least the next 40 years.

The immediate impact of the embargo was a steep increase in gasoline prices, severe gasoline shortages, and panic among car-dependent citizens. Lines of vehi-cles many blocks long would form at service stations that had fuel, and purchases were typically limited to ten gallons (38 liters) or less to prevent hoarding (Fig. 4.3). These were the days before the internet and smart phones, so the only ways to dis-cover which service stations had gasoline available were announcements on radio stations or word of mouth. People would often spot a gasoline tanker truck in transit and follow it to a service station. Most citizens were stoic and polite in this era of greater social graces, but if someone tried to cut in line out of turn, all hell could break loose.

Fig. 4.3 Vehicles lined up waiting for gasoline during the 1973–74 energy crisis
Source: D. Falconer, U.S. National Museum of American History; public domain

Although gasoline had been rationed during the Second World War, few people complained then because most felt that it was their patriotic duty to save fuel for the war effort. A number of spot shortages occurred in places where fuel demand outpaced supply during the post-war economic expansion, but these were localized and short-lived. The OPEC embargo was different. It created a national crisis that not only included shortages and price hikes for gasoline, but also heating oil, jet fuel, and diesel fuel, impacting homeowners and straining the finances of the airline and trucking industries. Before the embargo, electrical power plants often used fuel oil, but with tight supplies of petroleum products focused on transportation fuels to keep America moving, many generating facilities were converted to coal or natural gas, leading to shortages of those commodities.

The OPEC embargo was not just about oil but was actually a full-blown "energy crisis" that led to long-term repercussions in American politics and policies. I was in college when all of this happened, and I remember it well. The need for secure energy supplies was one of the factors that influenced me to study geology. Few people at the time were aware of the inter-dependency of energy resources in the United States. Shortages and supply uncertainties of one kind of energy would often result in shortages of others as substitutions were made.

In 1975, the U.S. Congress banned oil exports from the United States to retain as much domestic oil within the country as possible. The 21st Century success of shale oil and gas development through the use of hydraulic fracturing made the U.S. the world's largest gas producer in 2009, and the world's largest petroleum producer in

2013. U.S. domestic producers, awash in a surplus of crude oil produced by fracking in North Dakota and Texas and with prices falling as a result lobbied Congress heavily to rescind the ban. The oil export ban was finally rescinded in 2015, a solid 40 years after it was first introduced.

At this writing, domestic energy supplies in the United States have been robust and stable for more than a decade. Nevertheless, fears about the use of oil as a weapon, the protection of foreign oil fields and tanker transport routes, and the real or imagined potential for crippling energy shortages continue to influence U.S. government policy.

Although the OPEC oil embargo had been imposed for political reasons to demonstrate Arab displeasure at U.S. support for Israel, that support never wavered. The embargo backfired because rather than forcing U.S. policy changes in the Middle East, the withholding of petroleum exports angered many American citizens. There was a certain air of entitlement at the time that other nations should supply the United States with unlimited energy. Americans were outraged that Arab nations would dare to hold back "our" oil. Commentators said the U.S. economy was being "held hostage" by OPEC. Arab grievances over unlimited and unquestioned American support for Israel almost never even entered domestic discussions in the U.S.

The United States largely viewed the energy crisis as a technology issue that could be solved with more technology. America was still on a high-tech high from the recent success of the Apollo moon missions, and a common saying was, "If we can land a man on the moon, we ought to be able to gas up our cars." Some of this sentiment is still with us today among those who think we can continue burning coal as long as we capture the carbon dioxide or fix climate change by blocking sunlight with high altitude aerosols. There are technological solutions to climate, but these must be implemented with an eye on the "law of unintended consequences." Keep in mind that our technological solutions to the energy crisis were domestic shale gas and tight oil resources produced by fracking, which worked but got us decades more dependence on fossil fuels. So, instead of trying to find a technological fix for coal combustion, maybe we should do away with it altogether and develop something else as an energy source.

The U.S. Department of Energy was established in 1977 to increase the domestic energy production of the United States in partnership with industry through the development and application of technological solutions. They also have other duties like running the national labs and looking after the nation's nuclear arsenal, but domestic energy is their core mission. With hard lessons learned about putting all of our energy eggs into one imported oil basket, DOE began developing an "all of the above" energy strategy that is still with us today. Every possible energy resource under the sun, including the sun itself, has been and continues to be investigated as a power source. These include nuclear, solar, wind, geothermal, oil and natural gas, coal, and biofuels, among others.

The current success of natural gas and oil production from shale is a direct result of DOE shale projects from the late 1970s and 1980s that investigated the use of hydraulic fracturing to open up the rocks.[3] The high volume hydraulic fracturing currently being used to produce oil and gas from shale was developed in the 1990s by George Mitchell of Mitchell Energy (now part of Devon Energy). Mitchell had been interested in shale for a long time. In fact, the first shale core that I collected on my first job out of school as a contractor to DOE on the Eastern Gas Shales Project back in 1979 came from a gas well in southeastern Ohio drilled by Mitchell Energy. George Mitchell's persistence after DOE lost interest in shale in the late 1980s made it an eventual success.

Drilling and fracking technology in the 1980s was unable to achieve anything more than hit or miss gas production with shale. Traditional, vertical wells simply do not get into contact with enough rock volume, and a single frack only allows for two vertical cracks to extend into the formation in opposite directions from the wellbore along the axis of greatest principal stress.

Directional drilling, developed by the majors for deep water offshore oil development became widely available in the 1990s, and Mitchell Energy adopted it for drilling long, horizontal wells known as laterals into shale. Mitchell's laterals were able to remain within the target formation for kilometers, contacting huge volumes of rock. Mitchell was able to stage a series of hydraulic fractures along the length of the laterals to create multiple, high permeability flowpaths into literally cubic kilometers of gas-bearing shale. Although gas migrates very slowly out of each square meter of shale exposed on the fracture face, the staged hydraulic fractures created many, many square meters of shale fracture faces.

The economics of the technique were driven by the large volumes of gas that could be recovered. A looming shortage of U.S. domestic natural gas in the 1990s that resulted in steep price increases also helped the technology develop. Mitchell's shale gas was competing against the high cost of developing infrastructure to import liquefied natural gas (LNG) from North Africa into the U.S. East Coast. Importing LNG is always an expensive option, and shale gas stepped up to fill the supply gap before more than a few of the import terminals could be built. These are now used for exporting abundant U.S. natural gas from shale as LNG.

Shale gas proved to be amazingly successful, and by 2009, the United States had surpassed Russia to become the largest natural gas producer in the world. However, as domestic natural gas supplies turned from a shortage into a glut, prices fell sharply. Gas well drillers became more efficient at controlling costs and options like the LNG exports mentioned above helped to ease some of the oversupply. Gas prices still remained stubbornly low, and the industry focus shifted to fracking shale formations that contained light, tight oil, like the Bakken in North Dakota and the Eagle Ford in Texas. Liquid hydrocarbons like petroleum are more profitable to produce and easier to export than natural gas. Tight oil from shale quickly pushed North Dakota into the number two spot for petroleum production in the U.S., behind

---

[3] Soeder, Daniel J. and Borglum, Scyller J., 2019, The Fossil Fuel Revolution: Shale Gas and Tight Oil: Cambridge, MA: Elsevier Publishing, 336 p.

only Texas. The United States surpassed Saudi Arabia in 2013 to become the largest crude oil producer in the world.

The success of fracking in the late 1990s provided a massive surplus of both petroleum and natural gas in the U.S. by the first decade of the twenty-first century, thus ending the energy crisis once and for all. The U.S. began exporting oil and gas overseas and some of the old-timers at DOE celebrated the achievement of a long-sought goal. However, it is important to keep in mind that shale gas and tight oil resources are large but not infinite. The U.S. position as the world's leading producer of petroleum and natural gas is only temporary.

DOE has evolved as an agency and is now responding to the climate crisis by focusing on carbon dioxide capture and sequestration from the atmosphere. The DOE Office of Fossil Energy recently changed their name to Fossil Energy and Carbon Management (FECM). There has also been a stronger emphasis on carbon-free energy resources, including traditional wind and solar, but also engineered geo-thermal systems, new nuclear technology, energy storage (batteries), and the potential role of hydrogen. These issues are discussed in their own chapters later in this book.

****

Along with precarious economics and the emission of climate-damaging green-house gas as a combustion product, oil and gas also present a number of additional environmental hazards. Natural gas is composed almost totally of methane (CH_4), which is an infrared-absorbing, greenhouse gas like CO_2. Methane is in fact much more efficient at absorbing longwave IR emitted from the ground than CO_2, but it also has a much shorter residence time in the atmosphere. The Environmental Protection Agency (EPA) considers CH_4 to be about 28 times stronger than CO_2 at trapping heat over a 100 year time period.

Leaks in natural gas production, transmission, and distribution systems are a major source of methane emissions to the atmosphere. Gas that leaks directly into the atmosphere is known as a "fugitive emission." (The subsurface migration of gas in groundwater aquifers is called "stray gas." People often use these terms incor-rectly.) Fugitive emissions can come from uncapped wells, poorly-sealed compres-sors, leaks in interstate transmission lines, deteriorated old gas distribution systems under city streets, or virtually anywhere else in the natural gas system. Stray gas can also turn into a fugitive emission if it bubbles out of groundwater in a well, spring, or seep and enters the atmosphere.

Gas utility companies are often lax about locating and sealing leaks, mainly because the low price for natural gas doesn't justify the cost of sending out repair crews. It is cheaper to lose 1–2% of their throughput from fugitive emissions than to pay the costs for repair crews to ensure that the system is leak-tight. This must be improved if natural gas is to serve a role in the energy transition process.

As long as it stays inside the pipelines, natural gas has significant environmental advantages over other fossil fuels. Because the methane molecule carries four hydrogen atoms to each carbon, combustion of natural gas emits half the CO_2 per

Btu (British thermal unit[4]) as coal, and about 2/3 that of petroleum. GHG emissions from the U.S. have fallen significantly over the past decade as utilities replaced coal with natural gas.

Gas leaks are also an explosion hazard if gas accumulates in the proximity of a possible ignition source such as pilot lights in confined spaces like basements. Natural gas concentrations in air between 5% and 15% are combustible. There are explosions and loss of life almost every year from gas leaks.

Petroleum, on the other hand, has its own set of environmental hazards as a liquid that can spill. Oil spills often devastate ecosystems for years, as demonstrated by the Exxon Valdez spill of 257,000 barrels of oil (41 million liters) in Alaska's Prince William Sound in 1989. Wave action and microbial degradation typically break down spilled oil over time, but the cool climate and rocky beaches of Prince William Sound have preserved the oil below the beach surfaces and extensive patches still remain to this day.

The much larger Deepwater Horizon blowout in 2010 spewed nearly 5 million barrels of oil from a wild well into the Gulf of Mexico over a period of 87 days. The oil covered an area in the Gulf roughly the size of Oklahoma and fouled more than a 1000 miles of shoreline. Although much of the surface oil broke down over a period of about 3 years, a number of studies[5] have suggested that nearly half of the spilled oil still remains in the Gulf, either on the sea bottom, or floating as a plume at some intermediate depth in the water column. Oil is often thought to be lighter than water and always floats, but crude is usually a mixture of short and long chain hydrocarbons. These can separate out by density when dispersed in a spill situation and the heavier, tar-like molecules will sink if they are denser than water.

As long as petroleum continues to be produced from offshore wells and transported over the oceans, the risk of spills in the marine environment is present. Oil spills are common enough that at least one brand of American grease-cutting dishwashing detergent advertises its usefulness for gently cleaning off waterfowl oiled by a spill. There have been a number of other well-known oil tanker disasters besides the Exxon Valdez.

The Amoco Cadiz was a supertanker that ran aground on rocks in March 1978 off the coast of Brittany, France. The rocks punctured the ship's hull and her cargo of Saudi Arabian and Iranian crude oil leaked into the ocean. Rough weather prevented attempts to offload the cargo of oil onto other ships, and the fierce waves broke the ship in two. Her entire cargo of crude oil was lost into the sea.

In 1979, a tanker named the Atlantic Empress collided with another tanker in the Atlantic Ocean about 10 miles (16 km) from the islands of Trinidad and Tobago during a tropical storm. Both ships caught on fire and the Atlantic Empress sank, losing its entire cargo of oil. The other ship was towed to Trinidad. An oil tanker

---

[4] A British thermal unit (Btu) is the quantity of heat required to raise the temperature of 1 pound of liquid water by 1 °F. The metric unit is a calorie, or the energy required to raise the temperature of 1 gram of water by 1 °C. One Btu equals about 250 calories.

[5] Ramseur, J.L., 2015, Deepwater Horizon Oil Spill: Recent Activities and Ongoing Developments: Congressional Research Service Report 7–5700, 21 p.

called the Castillo de Bellver caught fire in 1983 off Cape Town, South Africa, eventually breaking in half and sinking. Some of her crude oil cargo burned in the fire, but the remainder was lost into the South Atlantic Ocean. Despite efforts to reduce risk by building tankers with double hulls and improving fire control and navigation systems, moving crude oil around the world in these gigantic ships will never be perfectly safe. If the ship breaks in half on the rocks, whether or not it has a double hull makes little difference.

Oil spills on land have been equally disastrous. The 1990 collapse of the Soviet Union left a lot of petroleum and natural gas infrastructure at risk due to a lack of maintenance. A large oil spill in Uzbekistan occurred in 1992 from a well blowout that sent millions of barrels of oil into a small valley. A corroded oil pipeline in the Russian Arctic leaked oil into a diked containment for 8 months in 1994. The dike eventually gave way, spilling tens of millions of gallons into the Kolva River and tainting about 72 square miles (186 km^2) of fragile tundra and wetlands.

One of the most serious energy-related disasters on land in North American history occurred when a trainload of Bakken petroleum derailed in the Canadian town of Lac-Mégantic in 2013, catching fire and incinerating part of the town.[6] This is a light, tight crude oil produced from the Bakken Shale in the Williston basin of North Dakota, Montana, Saskatchewan and Manitoba. Bakken crude oil has the consistency of diesel fuel and is just as flammable. The petroleum is too light for refineries in the Rocky Mountains or California to process and there were no significant pipelines out of this rather remote part of central North America, so the oil was typically transported by truck or train to refineries on either the Gulf Coast or Atlantic Seaboard that could handle it.

The Lac-Mégantic disaster happened after an oil train stopped for the night at a designated crew change point about 7 miles (11 km) west of the Quebec town. The train was composed of 72 tanker cars, each loaded with 30,000 gallons (113,000 liters) of Bakken crude oil, and because of a provision in Canadian railroad laws and the desire of the company to save money, there was only one crew member aboard – the engineer. At the end of the shift, the procedure was to park the train and leave it until the next shift arrived to continue the route. A nearby siding was filled with boxcars, so the engineer left the train on the main track, which is unusual but not illegal, set the brakes and left. The track here has a descending grade of 1.2% and the town of Lac-Mégantic (population 6000) is downhill.

Shortly before 1 AM, the brakes on the untended train failed, and it began rolling toward Lac-Mégantic, gradually picking up speed. When it reached the town 15 minutes later, the train was moving at an estimated 65 miles per hour (105 km/h). It derailed on a sharp curve near the center of Lac-Mégantic that has a designated speed limit of 10 mph (16 km/h), sending 63 of the 72 oil tank cars off the track. Most of these ruptured, and the spilled oil immediately caught on fire. Witnesses

[6] Campbell, B., 2018, The Lac-Mégantic Rail Disaster: Public Betrayal, Justice Denied. Toronto, Ontario, Canada: James Lorimer & Company, 200 p.

reported a fireball three times higher than the downtown buildings. The disaster killed 47 people and destroyed 36 buildings.

The controversial Dakota Access Pipeline (DAPL), subjected to years of protests and hung up in numerous legal battles was built in response to the dangers of oil trains. The intent was to provide a much safer pipeline route for moving Bakken petroleum out of the Willison basin. The pipeline began operations in 2017 and carries 750,000 barrels of Bakken crude oil per day a distance of 1172 miles (1875 km) from North Dakota to Illinois.

In addition to being a fire hazard and an ecological threat, petroleum also emits volatile organic compounds (VOCs) that can cause human health problems if inhaled or absorbed through the skin. Many VOCs will react with sunlight to form photochemical smog.

About half of the oil and three quarters of the natural gas currently produced in the United States use hydraulic fracturing[7] to create permeable pathways into the rocks that allow hydrocarbons to flow more easily into a production well. The process works by using elevated static water pressure to overcome rock strength and crack open the rock. The water then carries sand into the growing fracture to prop the crack open after the pressure is released. A variety of chemicals are used in hydraulic fracturing operations to reduce friction, inhibit corrosion and scale buildup, control downhole microbial growth, and clean up the wellbore. These chemical additives are a small percentage of the total amount of fluids injected downhole, which consist mostly of water and suspended sand, but because of the large injection volumes involved in a massive hydraulic frack, substantial quantities of chemicals are often supplied to well locations. Many of these are acids or water-soluble organic compounds that can be significant groundwater contaminants if the chemicals spill or leak. The biocides used to control downhole microbes are especially hazardous.

As mentioned earlier, regional contamination of groundwater from below by upward-migrating frack chemicals was an alarming possibility raised by anti-frack activists during the early days of the fracking boom. This has not occurred and is in fact difficult to impossible to achieve for a number of physical reasons, primarily the absence of a mechanism to drive the fluids upward for long distances against gravity. However, local water contamination from chemical leaks or spills at the surface is not unusual. Groundwater near fracked wells has been found to contain diesel fuel, friction reducers, chemical inhibitors, biocides, and other compounds that have been spilled or leaked at well sites and infiltrated downward through the soil and into the water table.

Fracking has allowed gas and liquid hydrocarbons to be produced from resources that were previously not economical, achieving a goal that the U.S. DOE has had since the 1973–74 OPEC oil embargo, namely, to develop domestic oil and gas supplies and free the United States from dependency on imported oil. However, the

---

[7] Soeder, Daniel J., 2018, The successful development of gas and oil resources from shales in North America: *Journal of Petroleum Science and Engineering*, v. 163, p. 399–420.

favorable economics of shale gas has also made fossil energy a preferred option. Electric utilities moved away from coal and into abundant and cheap natural gas for power generation instead of pricier renewables. For all of its domestic desirability in the late 1970s after the oil embargo, the successful fracking of shales to produce substantial amounts of natural gas and petroleum has added significantly more fossil fuel to the global supply and pushed the world ever closer to a climate catastrophe.

Here is my small guilt trip for contributing to climate change: Back in the 1980s, I was a research scientist at the Institute of Gas Technology (IGT) on the campus of the Illinois Institute of Technology in Chicago. (It has since changed names to the Gas Technology Institute (GTI) and moved out to Des Plaines, Illinois near O'Hare Airport.) This was less than a decade after the OPEC oil embargo, and the Department of Energy was funding us and a lot of other folks to study "unconventional" natural gas resources in shales, tight sands, and coal seams. In 1988, I published a paper[8] in a Society of Petroleum Engineers (SPE) journal based on an investigation that showed the gas potential of the Marcellus Shale to be much greater than the values accepted at the time. The IGT laboratory studies indicated that the shale formation had a strong adsorbed gas component and could contain 26 cubic meters of gas per cubic meter of rock at a typical range of subsurface pressures, more than 50 times the official DOE assessment.

This is a lot of gas. I had "discovered" the Marcellus and I thought it was quite exciting at the time, but I was pretty much the only one. This was years before George Mitchell worked out a technique using lateral drilling and staged hydraulic fracturing to economically recover large quantities of gas from the Barnett Shale in Texas. My paper sat unnoticed on the shelf for nearly two decades.

In 2006, geologist Bill Zagorski at Range Resources was thinking about applying some of George Mitchell's drilling and fracking techniques from the Barnett on potential gas shales in the Appalachian basin. He ran across and read my 1988 paper and decided to take an otherwise unsuccessful Silurian dolomite well in southwestern Pennsylvania and re-complete it in the overlying Devonian Marcellus Shale. It was successful, and Range Resources and other companies began developing Marcellus Shale gas throughout Pennsylvania and West Virginia. The Marcellus soon became the largest gas producing formation in North America.

I was working on groundwater issues for the U.S. Geological Survey (USGS) at the time when someone at Shell tracked me down and called to discuss my 1988 paper. When I asked this person why they were interested in a decades-old paper, they told me that Shell was considering getting into the new gas production from the Marcellus Shale. I seriously thought at first that this was an elaborate practical joke, and no one was more surprised than I was to learn that shale gas had become successful. My formerly obscure SPE paper has now been cited well over 500 times in numerous shale gas research articles. These days, when advising students I tell them

---

[8] Soeder, D.J., 1988, Porosity and permeability of eastern Devonian gas shale: *SPE Formation Evaluation*, v. 3, no. 2, p. 116–124.

to always do their best at everything, because you never know what might turn out to be important someday.

The success of the Marcellus and other gas shales produced a glut of natural gas in a steady supply that reduced prices. The switchover by power plants from coal to natural gas improved efficiency, using less fuel to generate the same amount of electricity, and as mentioned previously, substituting gas for coal has reduced U.S. GHG emissions. On the other hand, the success of shale gas locked the U.S. into several additional decades of fossil fuel use, and also increased fugitive emissions of methane gas into the atmosphere.

The first time I met Bill Zagorski at a conference, he said "I read your paper!" If he hadn't found my paper, would he still have targeted the Marcellus? Most likely, yes. In my conversations with him, he told me that my paper was just one piece of evidence, and he had other indicators that the Marcellus might be productive. In any case, oil and gas development is always a crap shoot and Zagorski was willing to take a chance. Despite all the fancy geophysics, lab analyses, and computer models, a common saying among petroleum geologists is that "you don't really know what's down there until you get down there." Before they drilled the well, Zagorski's previous target in the Silurian dolomite was supposed to be a sure thing. After it failed and with nothing to lose, he pursued the shallower Marcellus as a long shot. Funny how things work out sometimes.

So like it or not, I had a hand in the climate crisis. Probably minor but still there was some impact from my work on the production of shale gas. The success of horizontal drilling and staged hydraulic fracturing in shale and other tight rocks pioneered by George Mitchell has provided an abundance of both natural gas and petroleum in the United States. We can declare unequivocally that both the energy crisis and our dependence on imported oil are over. Solved and done. Yay!

Unfortunately, it was mostly solved with fossil fuels, leaving us with a climate crisis instead. Solving this one will require us to be just as clever and resourceful as Mitchell. Except where he spent years puzzling out a solution, we already know how to fix it. The "clever and resourceful" part is less technical and more policy oriented and will be required for getting governments and people to act.

* * * * * * * * * * * *

Humanity needs to break our addiction to fossil fuels. In addition to the environmental arguments against their use, fossil fuels are an unsustainable and finite resource. They will eventually run out, and we are going to have to wean ourselves off of them sooner or later. Once it's gone, it's gone. There is no reason to burn every last drop of oil or every small lump of coal before moving on to other energy resources. By eliminating fossil fuels as soon as possible, we can move into a new energy future while maintaining a technological civilization and mitigating some of the impacts of climate change.

Conventional oil and gas wells have a limited lifespan measured in terms of years to a few decades. The maximum flow of hydrocarbons from an oil or gas well occurs right after completion of the well and represents the highest rate of production the well will ever see. This is known as the Initial Production or IP. As hydrocarbons are removed from the ground, downhole pressures drop and other fluids like saltwater migrate into the reservoir rock, causing a reduction in the amount of oil and gas flowing into the well. This fall-off in production over the life of the well is known as the "decline curve." Studies of decline curves show that all oil and gas wells decline, some gradually and others more abruptly. However, no oil or gas well will produce forever.

Back in the 1950s, a Shell geophysicist named M. King Hubbert developed the concept of "peak oil."[9] His investigations were initially undertaken to learn how the new marvel of atomic energy might affect the sustainability of the oil and gas business. During the course of his studies, Hubbert found that the amount of petroleum produced from any given oilfield over time followed a bell-shaped curve, climbing as the field was developed, peaking when the field reached full development, and then declining as pressures dropped and the residual oil became an immobile phase and stopped flowing. This meant that new oil resources would constantly need to be discovered to keep up with demand and new wells constantly drilled to maintain production. If not, wells would decline, and the world would run out of oil fairly quickly. Hubbert predicted that peak oil production for the world would occur in the 1970s. His timing was a bit off, and it appears that global peak production for conventional oil wells took place around the turn of the millennium. Mitchell Energy's fracking technology has supplied additional tight oil and gas reserves from shales, moving the production peak out a few years, but only postponing the inevitable.

Coal has also gone through a decline in production. Peak coal production in the U.S. was in 2006, and it has been falling ever since. The timing places the development of shale gas as the cause for the coal production peak. Electric utilities began a wholesale switchover from coal-fired power plants to natural gas power plants in the 2007–2008 timeframe as vast amounts of Marcellus Shale gas came online, and it is reflected in the declining production of coal.

Mining costs have risen as the highest grade, highest quality seams were mined out, especially those with easy access to economical transportation options like river barges. Coal from the more remote mines must be transported by rail, or even by truck. High-grade metallurgical coals, like the Pittsburgh seam in the Appalachian basin are largely gone, with a few remnants such as pillars that once supported the roof of an underground mine occasionally visible in road cuts (Fig. 4.4). Most of the other easy to reach coal seams are also gone. Although coal proponents like to say that the U.S. has "hundreds of years" worth of coal, the truth is that most of what is left can only be mined with difficulty at higher costs, if at all.

Because of sustainability, human health, and climate problems, fossil fuels will have to be phased out sooner or later. In my opinion, continuing to promote the use of fossil energy for the sake of short term profits is increasingly irresponsible as the externalized costs become ever more apparent. Fossil energy must be replaced with

---

[9] Hubbert, M. K., 1956, Nuclear Energy and the Fossil Fuels: API Drilling and Production Practice, v. 95, p. 1–57.

Fig. 4.4 A remnant of the Pittsburgh coal seam at the center of a road cut exposure near Morgantown, WV
Source: Photographed in 2021 by Dan Soeder

sustainable, renewable, carbon neutral energy as soon as possible, because every year we delay is a year deeper into the climate crisis. Delays also put us at risk for another potential energy crisis, where if we do suddenly run out of an energy resource, governments will be scrambling to solve the problem. It is apparent from the response to the 1973–74 energy crisis that a full-blown panic is not the best environment for rational decision-making. Our current energy mix in the U.S. is stable for now. We should use this opportunity to develop new energy resources in a deliberate and careful manner.

The bottom line is that the energy we are using today was largely defined by decisions made in the past to develop gasoline-powered automobiles and coal-fired electricity. The future of energy requires a more deliberate approach because we cannot continue down the path of making endless fossil fuel profits at the expense of the environment for much longer. I think the choice for the future is to either have abundant, sustainable, clean energy to run a complex technological civilization, or there will be no technological civilization.

Chapter 5
Greenhouse Gas and Climate Change

Keywords PETM · Arrhenius · Arctic acceleration

****

Our dependence on fossil fuels as a modern energy source is substantial. As mentioned previously, most of the electricity generated in the United States (about 70%) uses the combustion of fossil fuels like coal and natural gas to make steam to turn a turbine and run a generator. Other nations also create significant amounts of electricity using fossil fuel, including China, Japan, India, and Russia. Even the European Union, which receives about 28% of its energy from renewables and nuclear still relies on fossil energy for the remaining 72% of its electricity. Most vehicles in the United States (98%) still run on fossil fuel hydrocarbons like gasoline or diesel fuel. On a global scale, only one vehicle out of 250 (0.4%) is electric. The Earth has one atmosphere and every bit of GHG produced from fossil fuel adds to the total.

The Earth has some natural feedback mechanisms that maintain the climate at a more or less steady equilibrium. For example, if a large volcano like Yellowstone erupts, it can add significant amounts of GHG to the atmosphere. Initially the ash, dust, and aerosols thrown high into the stratosphere will block sunlight and cool things down. After these particulates settle out, the climate will warm from the CO_2 and SO_2 emitted by the eruption. Over time, these will combine with rainwater and fall as weak carbonic and sulfuric acids. The acids react with the minerals making up basalt, a very common rock type, and release calcium and magnesium ions into seawater. These combine chemically and biologically with dissolved CO_2 to form carbonate minerals like calcite or dolomite that lock down the carbon dioxide as a solid. Calcium also reacts with SO_2 to form calcium sulfate, $CaSO_4$, another solid mineral better known as gypsum, the main ingredient in plaster and drywall. The elevated CO_2 concentration will also be reduced by enhanced plant growth. These processes gradually draw down the GHG levels and return the atmosphere to climate equilibrium over geologic time scales. Unfortunately, humans are way too fast for this.

The most extreme climate change event recorded in the relatively recent geological past was the Paleocene-Eocene Thermal Maximum or PETM, which occurred

D. Soeder, *Energy Futures*, https://doi.org/10.1007/978-3-031-15381-5_5

about 56 Ma.[1] The PETM appears to have been triggered by a series of large volcanic eruptions combined with methane released from sea floor sediments that led to an abrupt increase in atmospheric GHG over a time period of about 6 Ka. The resulting global temperature rise was up to 8 °C, abnormally warming the planet for the next 150–200 Ka. The polar ice caps melted completely away and did not get re-established until the Oligocene Epoch.

Although the PETM warming event was "abrupt" by geologic standards, taking place over 6000 years, it was painstakingly slow from a human perspective. All of recorded human history barely extends back that far. The age of Stonehenge is about 5 Ka, and the Egyptian pyramids are about 4 Ka. The current rise in atmospheric CO_2 from the human combustion of fossil fuel has taken place in just over 250 years, a very fast blink to a geologist (refer back to Fig. 2.3).

The steam engine was invented at the beginning of the eighteenth century and put into widespread use over the next 100 years. Most climate scientists date the rise in anthropogenic GHG emissions to the First Industrial Revolution, which occurred in Great Britain during the mid-eighteenth century. Humanity's use of fossil fuel as an energy resource began to grow steadily upward at this point in history. The increase in fossil fuel and all energy use grew much steeper after the Second World War (refer back to Fig. 2.2). By the time the continuous trace gas monitoring program was started at Mauna Loa in 1957, we were well on our way toward creating a major GHG anomaly.

For several tens of thousands of years into the future, the effects of human-induced climate change may become as intense as those of the PETM if we do nothing. The Earth will eventually correct the climate naturally once GHG inputs stop increasing and the carbon cycle returns to some sort of equilibrium. Based on the geologic history of the PETM, if we insist on continuing with business-as-usual, the Earth will bring the climate back into balance some 200 Ka or so after the collapse of human civilization. This balance will not be achieved naturally on a meaningful human timescale, which is why we must do it ourselves.

****

The heat-trapping properties of carbon dioxide gas, as defined and confirmed by Fourier, Foote, and Tyndall were of interest to a nineteenth century Swedish chemist named Svante Arrhenius. Professor Arrhenius was well-known in scientific circles for his discovery that passing an electrical current through water will separate positively and negatively charged dissolved ions. His description of this process earned him the Nobel Prize in 1903.

In the mid-1890s, Arrhenius and other contemporary scientists were intrigued with recent geologic discoveries that the Earth had been through a series of major Ice Ages and the climate was anything but stable over geologic time periods. Professor Arrhenius hypothesized that changes in the concentration of

---

[1] Jardine, P., 2011, The Paleocene-Eocene Thermal Maximum: *Palaeontology Online*, v. 1, Article 5, p. 1-7.

heat-absorbing atmospheric gases like carbon dioxide might have been responsible for the climate changes that caused the Ice Ages. We now know that the Ice Ages were brought about by many factors, including the mechanics of the Earth's orbital cycles, changes in solar heat output, and variations in ocean currents, along with changing carbon dioxide levels in the atmosphere. The climate we are in at present is an "interglacial" period within a larger ice age that began approximately three million years ago.

Arrhenius didn't have any way to directly study the potential effects of different carbon dioxide concentrations on the climate, so he constructed and published what many consider to be the first mathematically-based climate model.[2] It consists of a fairly concise and constrained mathematical treatment showing how variations in atmospheric CO_2 concentrations allow the Earth to warm or cool. Arrhenius got it mostly right, at least on a broad scale. Referring back to the ice core CO_2 data in Fig. 2.3, the glaciers in the last Ice Age peaked about 20 Ka during a low concentration of atmospheric CO_2, and the previous warm interglacial period occurred around 120 Ka during the last CO_2 high concentration - which we have now greatly exceeded.

Carbon dioxide in the atmosphere is not the sole cause of climate shifts or Ice Ages, but the concentration of the gas does have an effect on the heat budget of the Earth. A significant source of CO_2 known back in Arrhenius' day was fossil fuel combustion. He calculated that burning the total world production of coal from 1896 (around 500 million tons) and releasing all of the combustion products into the air would increase CO_2 levels in the atmosphere by about one part per thousand. If the actual annual rate at which CO_2 increased in the Earth's atmosphere were this high, our planet would resemble Venus by now.

Obviously Arrhenius overestimated the Earth's atmospheric CO_2 concentrations because we are definitely not Venus with its massive CO_2 atmosphere exerting 90 times the surface pressure of Earth's and keeping the planet's temperature at a scorching 850 °F (455 °C). Venus in fact contains about the same total mass of CO_2 as the Earth, but it is all in the atmosphere. In contrast, ours is largely in the oceans, or locked down as carbonate rocks, organic carbon in plants, soils, sediments, and fossil carbon. The interchange of carbon on Earth among the atmosphere, oceans, biosphere, and soil is called the carbon cycle, and it is normally in equilibrium. The addition of fossil carbon to the atmosphere has unbalanced it.

In any case, the attempt by Professor Arrhenius to quantify the fate of carbon dioxide released from fossil fuel combustion is noteworthy. In his day, the terrestrial carbon cycle was poorly understood, and he could not have been expected to account for CO_2 absorption by the oceans or incorporation into vegetation. The modern rate of CO_2 buildup in the Earth's atmosphere as shown in the Mauna Loa data back in Fig. 2.1 is about 2 ppm per year. This is surprisingly low, considering that current coal consumption is more than 17 times higher than in Arrhenius' day at some 8.7 billion tons per year (7893 million metric tons). The lesson here is that the

---

[2]Arrhenius, S., 1896, On the influence of carbonic acid in the air upon the temperature of the ground: *Philosophical Magazine and Journal of Science* (London, Edinburgh, and Dublin), Series 5, v. 41, no. 251, p. 237-276.

interaction of GHG with climate is complicated. We are blasting CO_2 into the carbon cycle of Earth, messing with the atmosphere, oceans, and climate without really understanding all of the variables or outcomes, some of which could be quite bad.

It is important to note that although Svante Ahrrenius got many of the details wrong, his work along with that of Fourier, Foote, and Tyndall shows that many of the links between fossil fuel, greenhouse gas and climate change were understood or at least suspected in the nineteenth century. The point is that this is not new science or a recent discovery by any means. Ahrrenius did predict in his 1896 paper that the uncontrolled release of CO_2 into the atmosphere from the combustion of coal would lead to temperature increases and climate disruptions. At least in a general sense, he seems to have called that one correctly.

Carbon dioxide is not the only greenhouse gas that can affect climate. Methane (CH_4) is the primary component of natural gas and can also be produced by anaerobic microbial activity. As mentioned previously, methane absorbs about 86 times more longwave IR radiation than an equivalent amount of CO_2; concentrations in the atmosphere are presently around 1800 parts per billion (ppb), more than 200 times lower than CO_2. Methane is also lighter than air and rises, trapping less heat near the ground like CO_2 and it oxidizes easily into H_2O and CO_2 after a relatively short residence time in the atmosphere. Thus, scientists assign CH_4 about 86 times the global warming potential of CO_2 over a period of 20 years, but due to attrition from oxidation this drops to about 28 times the warming potential of CO_2 over a time interval of a century. Some atmospheric methane has natural sources like swamp gas and cattle flatulence, but a significant amount originates as fugitive emissions from leaking natural gas infrastructure. Another GHG often released with biogenic methane is a metabolic product called isoprene that is rarely included in climate models.

Other very potent GHGs are the so-called ozone-depleting substances that include compounds such as chlorofluorocarbons (CFCs), widely used at one time as refrigerants, and halon used in fire extinguisher systems. These substances were banned worldwide in 1987 and levels in the atmosphere have been slowly declining. However, with atmospheric lifetimes of 52 years for CFC-11 and 102 years for CFC-12, it will take some time before they are gone.

The U.S. House Committee on Energy and Commerce has called for the U.S. to achieve net zero carbon emissions by 2050. This follows the recommendation of the IPCC and aligns with similar net zero-GHG emission goals in France, Germany and Japan. Despite the political characterization of climate change as a "Chinese hoax" by some in the U.S., China expects to reach net zero emissions a decade later, in 2060, and India plans to reach net zero in 2070. The Chinese have been highly dependent on coal for electrical power generation but are shifting toward much cleaner natural gas to not only reduce GHG emissions but also to improve the dismal air quality in their expanding cities. Chinese conventional natural gas resources are largely "sour gas" that contains corrosive and toxic hydrogen sulfide, and the country has been importing LNG to displace coal. China is also seeking to develop its potentially large shale gas resources, along with an active and robust CCS program larger than the one in the United States.

The U.S. Supreme Court ruled in favor of Citizens United in 2010, finding that the limits on union and corporate political expenditures mandated by the 2002

Bipartisan Campaign Reform Act violated the First Amendment's protection of free speech. This ruling allowed political action committees (PACs) to receive essentially unlimited corporate funding, and Washington was soon flooded with tons of money from fossil energy interests and conservative PACs. Senator Sheldon Whitehouse (D) of Rhode Island, who has become one of my personal climate heroes, has said that Congressional efforts to address climate change were slowly but steadily plodding along up until then, but "fell apart" soon afterward. Climate action since the Citizens United ruling has been tied in knots wrapped up in zip ties and bound with wire.

A dysfunctional and combative Congress over the past decade has made little progress on climate proposals like the Green New Deal. Sponsored by Representative Alexandria Ocasio-Cortez (D-NY) and Senator Ed Markey (D-MA), the Green New Deal contains what I think are both good and not so good ideas for combating climate change. In my opinion, these should be debated and hashed out to find compromises for developing a meaningful revision that will pass Congress, but instead they are not even talking about it. In the meantime, things continue to get worse. The droughts, storms, heat waves and other extreme weather events from climate change are having negative health impacts that take an increased toll on human populations.[3] It is past time to stop squabbling and start taking action.

****

In the world of corporate environmental stewardship, GHG emissions are tallied in three categories. Scope 1 emissions are those that come directly from company operations, such as the diesel fuel burned by a company bulldozer while leveling a worksite or a cook firing up the gas burners on a franchise pizza oven. Scope 2 emissions are those that come from outside the company but are tied to company operations. An example of scope 2 would be the GHG released by the utility supplying the electricity that a factory uses to operate punch presses on an assembly line. Scope 1 and scope 2 emissions related to the manufacturing of a product are known as "embodied" emissions that resulted from the creation of that product.

Scope 3 are value chain emissions from the company products themselves. These can be "upstream" emissions from the suppliers that manufacture and deliver the materials needed to make the product, and "downstream" emissions from customers as they use the company's products, for example, the gasoline burned in new car. The total of these emissions is called the "carbon footprint" of the company, and many businesses have stated that they would like to reduce emissions with the goal of reaching a net zero carbon footprint. This is easier said than done.

The process of tallying up GHG emissions is called a Life Cycle Analysis, or LCA. As the name implies, the LCA counts scope 1 and 2 emissions beginning with the construction and establishment of the facility or business, emissions produced as

---

[3] Romanello, M., et al., 2021, The 2021 report of the Lancet Countdown on health and climate change: code red for a healthy future: *The Lancet*; DOI: https://doi.org/10.1016/S0140-6736(21)01787-6

the business is operated and products are manufactured, scope 3 emissions from those products that are sold and utilized by customers, and finally emissions involved with eventually shutting down and decommissioning the facility. LCAs are typically produced by outside consultants who have a more objective view of company operations than insiders who are too deeply involved and may have a stake in the outcome. Also, the LCA is a rather specialized field of expertise, and most companies do not have people on board who can do this. Many investors these days want to see an LCA for any companies claiming to be "green" and environmentally sensitive. Trying to fake it until you make it (i.e., calling yourself green when you are actually not) is called "greenwashing."

Hundreds of companies have made net-zero pledges but very few have publicly revealed any definitive, long-term plans to decarbonize. Fossil energy companies in fact have been proudly announcing cuts in scope 1 and 2 emissions on their way to achieving net zero carbon without mentioning scope 3. This appears to be fairly common; less than a third of corporate net-zero targets are addressing scope 3 emissions. It is also deceitful, especially from a fossil energy company where the major contributors of GHG emissions to the company carbon footprint are scope 3 emissions from customers who burn their products in cars, boats, and aircraft, and release combustion products into the atmosphere.

It is blatant greenwashing for the fossil fuel industry to ignore scope 3 emissions from their products. Scope 1 emissions basically come from the fuel used for drilling or mining operations, refining, and transportation to market. Scope 2 emissions consist of little more than the GHG released by a power company to light and air condition their corporate offices. Scope 3 emissions dwarf these and are possibly being ignored because the only way for a fossil energy company to reduce scope 3 is to capture and sequester the GHG produced from the combustion of its products, or to stop producing coal, oil, and gas altogether. Given the efforts being expended to keep climate legislation tied up in knots, it is apparent to me that the fossil energy industry is not quite ready to become titans of renewable, clean energy.

****

Climate models show that climate change will manifest itself in a number of ways. A model prediction that is readily observable today is called "Arctic acceleration." This states that polar regions will be more strongly affected than other parts of the Earth and will see the effects earlier. One climate-related effect is the melting of polar sea ice. Year-round Arctic sea ice has steadily declined since 1979 at a rate of about 3% per decade.

Sailors since the early 1600s have searched for the so-called Northwest Passage, a sailing route that connects the Atlantic and Pacific Oceans around the northern coast of North America through the Arctic Ocean along the shores of Canada and Alaska. Clogged by year-round ice in the past, the first sea crossing of this route was achieved by Norwegian explorer Roald Amundsen between 1906 and 1909. It was a landmark voyage for Amundsen, but of little practical value to commercial shipping because the journey took 3 years and navigated through routes that were too shallow for most ships. Other attempts to sail the route were made throughout the twentieth century, with the first commercial tanker ship traversing it in 1969 to determine if Alaskan North Slope oil could be transported out of Prudhoe Bay. This proved impractical, and in 1977 the Alaska Pipeline was completed to carry North Slope oil to the ice-free, southern Alaska port of Valdez.

In recent decades, climate change has thinned Arctic sea ice and created long stretches of open water from late August into September. This has allowed many ships to routinely navigate the northern route, avoiding the Panama Canal and shortening the voyage from western Europe to eastern Asia. However, Canada has been disputing the international waters designation for the Northwest Passage, claiming that the route traverses a substantial stretch of Canadian territorial waters. Other debates about who owns what in the Arctic are beginning to heat up among the nations that border the Arctic Ocean: Canada, the United States (Alaska), Russia, Norway, and Denmark (Greenland).

In fact, Russia has been promoting an alternative to the Northwest Passage they call the Northern Sea Route along the Siberian coast. The selling point is that shipments from Asia to Europe can avoid sailing through the Red Sea and the congested Suez Canal. The Russians have been refurbishing old Soviet bases and adding new ones along their Arctic coastline, including the ultra-modern *Arktichesky Trilistnik* (Arctic Trefoil) base in the Arkhangelsk Region. This alarmed the United States and NATO allies, who have responded with their own military build-ups. The Russian Kola Peninsula alone holds three military bases including the headquarters for Russia's Northern Fleet and its associated nuclear weapons depots. The potential for a nuclear confrontation or accident in the Arctic is worrisome.[4] As polar sea ice continues to melt and the Arctic Ocean becomes more accessible, discussions will probably intensify over various claims for mineral rights, fisheries, and freedom of navigation.

Melting sea ice has little to no effect on sea levels, because it is made up of water already in the ocean. Ice sheets and glaciers on land, however, are a different story. At the peak of the most recent ice age around 20 Ka, so much water was tied up in massive continental ice sheets that the sea level had dropped by as much as 120 m (393 feet) below modern levels. Large areas of the present-day continental shelves were dry land. Long Island, for example, represents the terminal moraine of the farthest advance of the Connecticut Valley Lobe of the Wisconsin Glacier. Fossils of wooly mammoths (an elephant-like land creature) have been found in sea bottom sediments on the former glacial outwash plain off the coast of New Jersey.

Oceanographers have identified deep canyons in the east coast continental shelf carved by the Hudson River, the Delaware River and the Susquehanna River during this low stand of sea level. As the glaciers melted over the next 13,000 years, these rivers drained meltwater from the retreating ice sheets and were raging torrents compared to their flow today. The Ohio, Missouri, and Mississippi Rivers also

---

[4] Boulègue, Mathieu, 2022, The militarization of Russian polar politics: Addressing the growing threat of tension and confrontation in the Arctic and Antarctica: Research Paper, 50 p., Chatham House, Royal Institute of International Affairs, London; DOI: 10.55317/9781784135256

carried off large volumes of glacial meltwater. Sea levels stabilized about 6000 years ago, but if the remaining terrestrial ice sheets and large glaciers in polar regions melt under climate change, this would result in additional sea level rise.

Sea level has been unstable throughout most of the Earth's history and coastlines are especially vulnerable to climate change. During the last interglacial period about 120 Ka, sea levels were 4-6 meters (13–20 feet) above present-day levels suggesting that even without GHG and climate change, we could expect some additional sea level rise. Climate models predict that the oceans will rise as much as two meters above present sea level by the year 2100. This would flood parts of many large coastal cities like New York, London, Tokyo, Hong Kong, Shanghai, Mumbai, Sydney, and others. If the entire ice volume in Greenland and Antarctica were to melt, the oceans could potentially rise by as much as 80 meters (260 feet) above present-day sea level.[5] Melting of the vast East Antarctic ice sheet alone would raise sea levels by 65 m (213 feet), while the smaller West Antarctic ice sheet adds 8 m (26 feet), and Greenland ice adds 6.5 m (21 feet). Large volumes of fresh water pouring into the oceans from melting ice caps also will likely disrupt ocean currents and influence the paths and intensity of many storms. Along with physical flooding at the surface, saltwater would enter underground infrastructure and cause catastrophic damage, shorting out electrical circuits and communications, and flooding subways, sewer lines, and other utilities. Coastal freshwater aquifers would face saltwater intrusion from rising seas that could increase the salinity in large volumes of groundwater to levels that are unsuitable for drinking.

Under even the warmest future climate scenarios, the complete melting of vast polar ice sheets in Greenland and Antarctica is expected to take centuries to millennia. Coastal regions should have plenty of time to build seawalls, dikes, raise land surfaces, or relocate populations before an area is inundated by rising seas. However, there is a small but not insignificant risk that sea levels could rise much more rapidly if a large volume of land ice suddenly entered the ocean. Ice does not necessarily need to melt to displace water and raise sea levels. Think about what happens when you drop an ice cube into a glass of water filled to the brim. Exactly.

The ice sheet in Greenland sits in a large, bowl-like depression formed in the island bedrock by the weight of the ice. It is expected to stay there like ice cream in a dish as it melts over centuries, probably forming a ring-like freshwater lake around the perimeter of the ice sheet. The land will slowly rebound as the ice disappears, but if the last Ice Age glaciers are a guide, this will take thousands of years (parts of North America are still rebounding today from the 20 Ka Ice Age glaciers).

Antarctica, especially West Antarctica, is a much different configuration. The Antarctic ice sheets slope upward inland from the coast to central highland areas like the Transantarctic Mountains that cut across the continent. From nearly anywhere on the Antarctic ice sheet, it is literally downhill to the sea. The ice has depressed the underlying bedrock like the Greenland ice sheet, and this has put coastal areas at the base of the ice sheets below sea level.

Scientists are concerned that some of the outflow glaciers on the Antarctic ice sheet have become unstable due to an influx of warmer water into the Southern

[5]Poore, R.Z., Williams Jr., R.S, and Tracey, C., 2000, Sea level and climate: U.S. Geological Survey Fact Sheet 002–00, Reston, VA, 2 p.

Ocean that is melting them out from underneath. Many Antarctic outflow glaciers are grounded below sea level. Melting them from below destabilizes the anchoring points and could allow a large volume of ice to float free and slide off into the water.

Parts of the ice sheet along the coast may only be held back by ice shelves. Some of these, like the Ross Ice Shelf, have been deteriorating in recent years due to the warmer ocean. Icebergs the size of small U.S. states have been breaking off the ice shelves and migrating into the open seas. There has been little effect on sea levels so far because the ice shelves are already floating on the ocean. The bigger worry is that they may allow land ice to enter the sea if they disintegrate.

The massive Thwaites Glacier in West Antarctica is roughly the size of Florida, and if the entire ice mass became ungrounded and slid into the ocean, it would quickly raise global sea levels by about 2 feet (60 cm). The grounding line of Thwaites has retreated nearly 10 miles (16 km) inland in the last 30 years, and the glacier is anchored by one last ridge. Recent studies have shown that cracks are growing in the ice shelf that holds back the glacier, and some investigators think it could collapse within a decade.[6]

Of perhaps greater concern is that Thwaites Glacier in turn holds back a number of neighboring glaciers and ice streams. If it disappears, the fear is that these other parts of the ice could follow it quickly into the sea, destabilizing the entire West Antarctic ice sheet and potentially pitching the whole thing off the continent and into the ocean rather quickly. This volume of ice would rapidly raise sea level by eight meters (26 feet), taller than most two-story houses. Such an event would be an unprecedented disaster for people inhabiting coastal regions.

Climate skeptics will often point to a 2015 NASA report[7] that indicates the Antarctic ice sheet is actually gaining ice, rather than losing it, and claim this proves global warming is a hoax. This is a classic example of cherry picking the data, or only using the data that support your argument and ignoring those that don't. The ice sheets in Antarctica and Greenland are always both gaining and losing ice. The NASA report states that altimetry data show that the ice gain exceeds losses in East Antarctica and the interior of West Antarctica, but loss exceeds gain elsewhere on the continent. The study agrees with other data that show an increase in ice discharge at the Thwaites and Pine Island regions of West Antarctica. The stability of the ice is a more important factor than absolute gains or losses.

Ironically, a warmer atmosphere and a warmer ocean can be expected to increase precipitation in the central regions of Antarctica, because warmer air carries more moisture. "Warmer" in Antarctica is still well below freezing, and the increased snowfall adds mass to the ice sheets. Satellite data indicate that the Antarctic ice sheet experienced a net gain of 112 billion tons of ice per year from 1992 to 2001, but that slowed to 82 billion tons of ice per year between 2003 and 2008. Precipitation on the Greenland ice sheet has increased as well, although some of this is now falling as rain, which has never before been recorded in parts of Greenland.

---

[6] https://www.science.org/content/article/ice-shelf-holding-back-keystone-antarctic-glacier-within-years-failure?

[7] https://www.nasa.gov/feature/goddard/nasa-study-mass-gains-of-antarctic-ice-sheet-greater-than-losses

A better indicator of retreating ice is mountain glaciers, many of which have receded precipitously over the past century. The U.S. Geological Survey undertook a photographic reconnaissance in Glacier National Park a decade ago to duplicate photographs of specific glaciers that had been taken in the early twentieth century (Fig. 5.1).

The vantage point, camera angle, sun angle, time of year, and landscape backgrounds were matched as closely as possible to the old photos and provide a dramatic documentation of how far these mountain glaciers have melted.[8] This is happening worldwide, and in another 100 years mountain glaciers may be gone altogether. Climate skeptics have not explained how these glaciers are shrinking without climate change.

****

Translating climate change into daily weather patterns is a complex task. The precise effects of anthropogenic CO_2 on climate are difficult to predict because atmospheric warming by GHG is complicated by many other factors. For example, a warmer atmosphere will also hold more water vapor, which is another GHG. Water vapor condenses out as clouds, trapping heat radiated from the Earth below, but also exerting a cooling effect by blocking sunlight from above. The actual impacts of water vapor on climate are vigorously debated and the subject of intense research.

The severity and frequency of droughts are increasing, especially droughts in the northern hemisphere associated with heatwaves. The strength and frequency of major storms like typhoons and hurricanes are also increasing. Even tornadic thunderstorms appear to be on the rise, occurring in areas with little previous history of such events. Changing climates are steering powerful storms into locations that are ill-equipped for dealing with such weather and increasing the vulnerability of human populations. Climate scientists say these drier droughts and wetter storms result from the "intensification" of the hydrologic cycle. A phenomenon known as "weather whiplash" is being seen more often in the last decade, where an intense and prolonged drought or heat wave is followed by massive floods. This has happened in 2022 alone in places like Texas, Europe, China, and especially Pakistan, where a historic heat wave was followed by historic flooding that killed thousands. Climate scientists like Daniel Swain at UCLA say these events are likely to continue even if we end emissions immediately because the conditions are now "baked in" to the climate. Of particular concern to Swain is California, where a megastorm from the Pacific could cause unprecedented flooding and turn the Central Valley into an inland sea. He says it's a question of "when" not "if." As climate scientist Andreas Prein explained to the Washington Post, "The infrastructure we have is really built for a climate we are not living in anymore."

An additional complicating factor is that many areas of the Earth now have a lower albedo or reflectivity after losing highly reflective sea ice and glaciers. Most

---

[8] https://www.usgs.gov/centers/norock/science/repeat-photography-project

Fig. 5.1 Repeat photographs of Grinnell Glacier in Glacier National Park taken 106 years apart. (Source: U.S. Geological Survey (USGS) public domain)

people know that dark colors absorb more heat than light colors. This is easily tested by resting your hand on the fender of a black car versus a white car on a warm, sunny day. Along with atmospheric warming from GHG, the wind-blown deposition of black carbon soot from coal and diesel combustion onto snow and ice darkens it and contributes to ice melting. Wind-borne microplastics also play a role. Once the ice melts, the underlying land or water surface is now darker and absorbs more sunlight, contributing to warming in a feedback loop that melts even more ice. Coastal cloud patterns have changed due to warming oceans, further decreasing the overall albedo of the Earth. This darkening of the Earth over the last two decades has been observed and documented in the decreasing brightness of "Earthshine" on the moon.[9]

The legendary British statistician George E. P. Box (1919-2013) famously said that all models are wrong, but some models are useful. Mathematical models can help explain the physical world, and while I don't necessarily agree that they are "wrong" (which seems a bit harsh) they are an imperfect description of complex processes. No computer model can fully reproduce the granularity of nature, so there will always be rounding errors on values, artificial boundary conditions, and averaging across cells. As long as the model structure is sound, cause and effect relationships between cells are accurate, and the values used for inputs are reasonable, models can indeed be quite useful. However, it is important to understand their limitations and not try to take them beyond those limits.

The outputs from computer models are called "realizations" and the conditions predicted by a model for future climate is known as a "scenario." Climate models indicate that global temperatures are steadily increasing, and that the rate of this increase is due to human-generated GHG. When the models are run with GHG levels held to pre-twentieth century values, scenarios of global temperatures actually show a slight cooling trend (Fig. 5.2). So-called "tipping points" for a warming climate include the collapse of the Greenland and West Antarctica ice sheets, the loss of coral reefs, and the thawing of Arctic permafrost. These are significant changes that will require a long time to reverse.

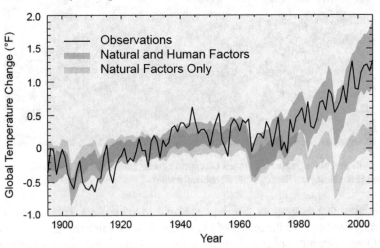

Fig. 5.2 Realizations from climate model scenarios showing global temperature trends with and without the influence of anthropogenic greenhouse gas. The solid black line represents observations. (Source: U.S. Environmental Protection Agency (USEPA) public domain)

---

[9] Goode, P.R., Pallé, E., Shoumko, A., Shoumko, S., Montañes-Rodriguez, P., and Koonin, S.E., 2021, Earth's Albedo 1998–2017 as Measured From Earthshine: *Geophysical Research Letters*, V. 48, No. 17, 8 September 2021; e2021GL094888

Figure 5.2 is a useful model for comparing the temperature trends with and without human factors and showing that the observed data follow one of the trends and not the other. The two temperature trends appear to start diverging around the year 1960. This was also about the time that fossil energy use began to increase steeply, as shown back in Fig. 2.2.

Some reasons for this may include the fact that the human population has more than doubled from around 3 billion to nearly 7.5 billion between 1960 and 2020. Total anthropogenic emissions of CO_2 during this period have increased by almost a factor of six from about 6 gigatons per year to more than 34 gigatons per year.[10] On a per capita basis, the amount of CO_2 emitted per person has increased from about 2 tons per year in 1960 to 4.5 tons per year in 2020. Not surprisingly, humanity produces more CO_2 annually than any other manufactured chemical compound. As a result, atmospheric CO_2 concentrations have risen by more than 100 ppm over this 60-year time interval, or about 30%, disrupting the climate and causing global temperatures to diverge from the natural trend. The divergence is increasing over time, indicating that the longer we delay implementing climate policies, the harder this will be to correct.

A weather variable known as convective available potential energy (CAPE) increases with temperature and leads to the development of severe weather events. Models have found that weather proxies like CAPE show an overall increase in intensity by as much as 20% per degree C of warming.[11] This means that both the frequency and intensity of severe weather events, like strong thunderstorms, tornadoes, hurricanes, blizzards and so on can be expected to increase as higher levels of heat energy become trapped in our atmosphere.

Climate models predict that a long-term drought in the western United States will decrease available water and cause more frequent wildfires. This scenario has been a reality for the past few years as the Colorado River and Lakes Mead and Powell reach record lows. Cities like Las Vegas are ripping up lawns and prohibiting sprinklers. Los Angeles has banned car washing. Denver has water restrictions. Salt Lake City is watching the Great Salt Lake slowly dry up and expose arsenic-bearing sediments that may produce deadly dust storms. Wildfire seasons begin earlier and last longer. In my opinion, some of the model scenarios are less of a prediction than a news headline, and they do suggest that things can get even worse.

Models predict the U.S. Midwest will experience wider extremes between drier droughts and wetter floods, potentially threatening this important agricultural region. More powerful and wetter hurricanes will hit the southern U.S., and coastal flooding due to sea level rise will be more likely from Texas to Maine. Tornado outbreaks will be more severe and occur in places where they have been rare in the past. Heat waves will become more intense and more frequent. When these events

---

[10] https://www.statista.com/statistics/264699/worldwide-co2-emissions/

[11] Lepore, C., Abernathey, R., Henderson, N., Allen, J. T., and Tippett, M. K., 2021, Future global convective environments in CMIP6 models: *Earth's Future*, v. 9, No. 12, e2021EF002277.

are multiplied across the world, especially on populations with far less resources than Americans, the seriousness of the climate crisis becomes apparent.

Climate and weather changes over the past half century are not just some video game fantasy ginned-up by computer models. The World Meteorological Organization (WMO) has reported that the number of recorded, actual annual weather disasters increased by a factor of five over the past 50 years.[12] Disasters that have taken many lives include droughts, severe storms such as typhoons and hurricanes, floods, and extreme temperatures. The number of deaths could have been far higher but improved early warning systems and the implementation of rapid response teams have resulted in only one third the number of deaths per disaster compared to 50 years ago. So at least that's good.

****

The Arctic polar vortex is a strong band of winter winds that blow around the North Pole at high altitudes in the stratosphere. A polar jet stream in the troposphere below the vortex blows from west to east and normally keeps the very cold air of the vortex trapped at the pole. Warm air pushing north can disturb the jet stream, creating waves that cause the polar vortex to become disrupted, splitting it in two and pushing masses of very cold air southward off the pole. This has always been the cause of cold weather outbreaks in mid-latitude regions during northern hemisphere winters, but Arctic acceleration has both intensified the outbreaks and made them more frequent. Climate skeptics like to point out that abnormally cold winter weather is a sign that "global warming" is not real; what is actually happening is that climate change is driving warm air into and disrupting the polar vortex, resulting in some of the worst outbreaks of winter weather on record.

Abnormally warm air masses moving into polar regions are a manifestation of Arctic acceleration. Record high temperatures of 38 °C (101 °F) and associated wildfires were recorded in northern Siberia in June 2020. The following winter, North America, Europe, and western Russia experienced a number of severe cold waves. As depicted in Fig. 5.3, a wedge of unusually warm air pushed into Greenland in February 2021 and split the polar vortex into two lobes. One lobe descended over eastern Europe and Russia, resulting in significant levels of snowfall for Moscow. The other mass of frigid Arctic air headed south across the center of North America, bringing record cold temperatures and accumulating snowfall as far south as the Texas-Mexico border.

Southern Texas is as far south as central Florida, and people, buildings, electric power grids, and water systems were in no way prepared for cold of this magnitude. Power failed, exposed water pipes froze and burst, many homes were without heat for days, and 210 people died, most from exposure to the cold or from carbon monoxide poisoning as they tried to heat their homes with propane grills or other appliances not meant for indoor use.

---

[12] https://public.wmo.int/en/media/press-release/weather-related-disasters-increase-over-past-50-years-causing-more-damage-fewer

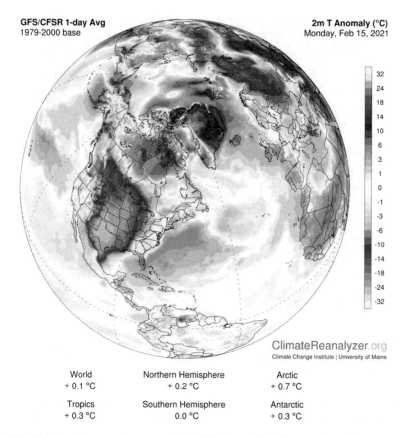

World	Northern Hemisphere	Arctic
+ 0.1 °C	+ 0.2 °C	+ 0.7 °C
Tropics	Southern Hemisphere	Antarctic
+ 0.3 °C	0.0 °C	+ 0.3 °C

Fig. 5.3 Global surface temperature map from 15 February 2021 shows a dome of warm air over Greenland disrupting the polar vortex and sending very cold air masses into North America and Russia. (*Source: Image from Climate Reanalyzer (*https://ClimateReanalyzer.org*), Climate Change Institute, University of Maine, USA, open access; data from U.S. National Weather Service*)

The following summer, the U.S. Pacific Northwest and British Columbia were subjected to a massive heat wave from late June through mid-July 2021. This event produced some of the highest temperatures ever recorded in this normally cool region, where very few homes or buildings have air conditioning. It set a new record on June 29, 2021, for the highest temperature ever measured in Canada of 49.6 °C (121.3 °F) at Lytton, British Columbia. The intense heat further dried out the already dry vegetation in the area, and a wildfire near Lytton burned the entire town to the ground a few days later. Large wildfires fueled by a combination of heat and drought also occurred during 2021 in Washington, Oregon, Idaho, and California.

The heat wave damaged road and rail infrastructure, forced business closures, triggered extensive melting of mountain snowcaps, and resulted in wide-scale damage to crops. Marine life also suffered, including millions of mussels attached to shoreline rocks that were literally cooked inside their shells when exposed to the hot air temperatures during low tide. The combined U.S. and Canadian death toll from

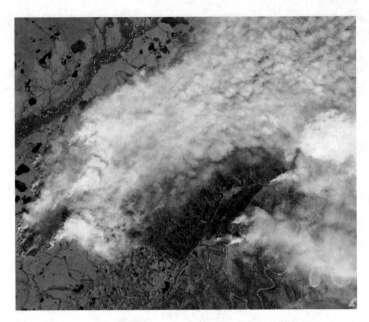

Fig. 5.4 Satellite image acquired July 4, 2021, from the Operational Land Imager (OLI) on Landsat 8 shows smoke plumes and burn scars from large fires burning near Penzhina Bay in the Russian Far East at the northern end of the Kamchatka Peninsula (width of image is about 60 km or 37 miles). (*Source: NASA Earth Observatory, public domain*)

the 2021 Pacific Northwest heat wave is estimated at around 1200 people. Climate scientists determined that the heat wave was a 1000-year weather event, meaning it had a probability of only 0.1% of occurring in any given year. They also concluded it was made 150 times more likely by climate change.[13]

Russia and eastern Europe also suffered through a heat wave in May and June 2021, with temperatures higher than 30 °C (86 °F) measured above the Arctic Circle and record high temperatures for June set in Moscow and St. Petersburg. For several days, temperatures on the shore of the Barents Sea reached 25–30 °C (77–86 °F), equaling those on the beaches of Italy and southern France. The heat and dry conditions resulted in large wildfires in Siberia (Fig. 5.4). Significant wildfires also broke out in unusual places such as Greece, Algeria, and Türkiye. There were large fires in the western United States, as drought conditions continued to intensify, and many western states imposed water restrictions.

The year 2021 set a number of records, none of them good. Data from NASA and NOAA ranked it as the sixth-warmest year on record. Even the presence of a La Niña event in the tropical Pacific during 2021, where sea surface temperatures tend to be cooler than average, didn't move it out of the top ten hottest years. At this writing in the spring of 2022, both India and Pakistan were under an unusual

---

[13] https://www.worldweatherattribution.org/western-north-american-extreme-heat-virtually-impossible-without-human-caused-climate-change/

pre-monsoon heat wave that should have occurred only once every 300 years. It has been 3 years since their last one.

Heat waves have the potential to cause severe disruption. These kill based on duration, daytime temperature extremes, and excessively warm nighttime temperatures. The first sign of distress is heat exhaustion, marked by heavy sweating, dehydration, muscle fatigue and cramps, and foggy thinking as the body struggles to self-cool. A person suffering heat exhaustion should be cooled down immediately and given fluids with electrolytes. If untreated, heat exhaustion can be followed by heat stroke, where the body can no longer control its temperature. This is a serious medical condition with an 80% fatality rate if not dealt with immediately. It is marked by a lack of sweating, an increase in skin temperature, rapid breathing, a racing heart, headache, confusion, and the possible loss of consciousness. Heat stroke is defined medically as a core body temperature of 104 °F (40 °C). Treatment requires cooling the person down to reduce the body temperature into more normal ranges.

Extended heat waves have a cumulative effect, wearing down a body's heat defenses as a person overheats day after day. If a warm, sultry night prevents someone from recovering sufficiently, the next hot day takes an even greater toll. It is important for people without air conditioning to utilize community "cooling centers" commonly set up during heat waves to reduce their body temperature at night and recover for the next day.

Normal human body temperatures are maintained when an outward-flowing heat gradient is able to transport metabolic heat from the body core. The maximum skin temperature where this can occur is around 35 °C (95 °F) where there is still enough of a thermal gradient from the body's core temperature of 98.6 °F to the skin to effectively transport heat. The evaporation of perspiration keeps the skin below this critical value at elevated temperatures, but it requires low humidity to work. This is why hot temperatures in a desert are more comfortable than hot temperatures in a rain forest.

Relative humidity is measured by the temperature difference between a dry thermometer (dry bulb) and a thermometer with the bulb encased in a water-saturated wick (wet bulb). Water takes up heat as it evaporates, and this so-called "evaporative cooling" on the wet wick in dry air drops the wet bulb temperature below that of the dry bulb. As humidity increases, evaporation becomes less efficient, and the difference between wet bulb and dry bulb temperature decreases until the two temperatures become identical at high humidity levels (95% to 100%). In other words, there is no evaporative cooling at high humidity because there is almost no evaporation.

Thus, at wet bulb temperatures above 35 °C or 95 °F, the evaporative cooling mechanism of perspiration is no longer effective on human skin. This is a physiological and physical limit, and no amount of rest, fluids, or shade will help the body cool under these conditions.[14] These levels of heat and humidity are simply beyond

---

[14] Raymond, C., Matthews, T., and Horton, R.M., 2020, The emergence of heat and humidity too severe for human tolerance: *Science Advances*, v. 6, no. 19, 8 May 2020; DOI: 10.1126/sciadv.aaw1838

what the human body can tolerate, and they will kill. Locations that repeatedly experience temperature and humidity in this range will be rendered uninhabitable by humans, including some areas of the tropics and subtropics before the end of the twenty-first century. Populations will have to relocate, and the rest of the world will have to be prepared to take in the resulting refugees.

Because the critical wet bulb temperature of 35 °C is an upper limit, temperatures below this can still put people in distress and even kill them. If the combination of heat and humidity hits an area of the globe that does not normally experience hot temperatures, the local populace may be very ill-equipped to deal with them. One of the worst heat waves on record struck western Russia and eastern Europe in 2010, reportedly killing an estimated 55,000 people. Worldwide deaths from extreme temperatures – both hot and cold – average around five million annually, but heat-related deaths are rising.

The ratio of extreme heat events to extreme cold events has been changing in favor of heat. In 2009, there were about twice as many record hot days as record cold days. Climate model simulations showed that in a "business as usual" scenario for GHG emissions in the twenty-first century, the ratio of record highs to record lows is expected to increase, with about 20 record hot days to each record cold day by mid-century, and a ratio of roughly 50 to 1 by the end of the century.[15] There will be many more intolerably hot days in our future.

There is almost always an environmental justice issue tied to climate change, and heat waves are no exception. Extreme heat hits low-income neighborhoods and communities of color the hardest, where people are more likely to be homeless or living in poorly ventilated buildings. Many do not have air conditioning, and even if they do, they are often reluctant to turn it on because they can't afford the cost of electricity. The elderly and those with chronic health conditions are the most at risk and are often unable to access medical treatment when suffering heat stress. Some 37% of global heat-related deaths have been attributed to heat waves associated with climate change, and deaths among poorer people without access to some kind of mechanical cooling like air conditioning were in the majority.

Heat waves are worse in crowded cities because of what is called the "urban heat island" effect. This is caused by concrete, brick, stone, and steel absorbing more heat during the day and then radiating it at night. Adding to the effect are the lack of green vegetation to cool the air through evapotranspiration, and the heat contributed by motor vehicles and ironically, air conditioners. AC units cool a room by transferring the heat outside and in a city with closely packed buildings, all this heat discharge can raise temperatures by several degrees. A study published in 2021 found that air temperatures were measurably higher in low-income neighborhoods and communities of color compared to rich neighborhoods. The poorer areas of a city

---

[15] Meehl, G.A., Tebaldi, C., Walton, G., Easterling, D., and McDaniel, L., 2009, Relative increase of record high maximum temperatures compared to record low minimum temperatures in the U.S.: *Geophysical Research Letters: Climate*, v. 36, No. 23, December 2009

can have temperatures as much as 7 °F (3.8 °C) higher than wealthier locations during summer months.[16]

Weather events that include both extended droughts and heatwaves have increased in frequency, duration, and severity worldwide in recent years. The northern hemisphere appears to be more strongly affected than the southern, possibly because there are more land masses north of the equator and more temperature-moderating ocean to the south. The high temperatures of intense heat waves are amplified by high humidity and pose serious threats to human health.

According to NOAA, when ranked against the twentieth century average, the world has not experienced a "cooler than average" year since 1976. When NASA climate scientist James Hansen famously warned Congress in 1988 that human-caused climate change was underway, 1988 was the warmest year on record. Some climate skeptics still hold up 1988 as the year for record warmth to make the argument that things are cooler now. However, the record for 1988 has long been exceeded. These days it ranks a paltry number 28 in NOAA's data set.

The 9 years from 2013 through 2021 are among the top ten warmest years on record, according to NOAA. The world in 2021 was 1.2 °C (2.2 °F) warmer than the pre-industrial average. We are perilously close to the Paris Agreement's goal of limiting warming to 1.5 °C above pre-industrial levels, and we are highly likely to exceed it in the next few years. As a result, we can expect more sea level rise, severe storms, floods, polar vortex disruptions, droughts, wildfires, and heat waves. Welcome to the future.

A study by the World Bank found that climate change could force 216 million people in the world to move within their countries by 2050.[17] Such internal climate migration might include moving inland to higher ground from flooded coastal areas, moving out of the path of intense storms, moving away from drought-stricken areas to wetter locations, or moving farther north or south from tropical or subtropical regions to escape deadly heat waves. Climate change will drive migration because it impacts people's livelihoods and results in a loss of livability at certain locations.

Weather events are likely to get worse. A recent study found that a relatively small amount of surface warming has a disproportionate effect on humidity, accelerating evaporation and placing significantly more water into the atmosphere. This could produce a substantial increase in extreme precipitation, especially in the energy powering tropical thunderstorms.[18]

---

[16] Benz, S.A. and Burney, J.A., 2021, Widespread Race and Class Disparities in Surface Urban Heat Extremes Across the United States: *Earth's Future*, v. 9, no. 7, July 2021

[17] Clement, V., Rigaud, K.K., de Sherbinin, A., Jones, B., Adamo, S., Schewe, J., Sadiq, N., Shabahat, E., 2021, Groundswell Part 2: Acting on Internal Climate Migration: World Bank, Washington, DC. © World Bank. https://openknowledge.worldbank.org/handle/10986/36248 License: CC BY 3.0 IGO.

[18] Song, F., Zhang, G.J., Ramanathan, V., and Leung, L.R., 2022, Trends in surface equivalent potential temperature: A more comprehensive metric for global warming and weather extremes: *Proceedings of the National Academy of Sciences*, v. 119, no. 6, p. e2117832119; DOI: 10.1073/pnas.2117832119

Extreme weather events from the climate crisis must be predicted accurately. Getting people wound up about the "storm of the century" only to have it turn out to be a gentle rain shower makes them skeptical of the credibility of weather forecasts and leads to public complacency. The danger is that when a real storm hits, people may ignore the warnings as yet another false alarm and get blindsided. A study published in 2009 found that tornadoes occurring in areas with high false-alarm ratios actually kill and injure more people, everything else being constant.[19] Crying "wolf" once too often leads people to ignore the tornado sirens and make little effort to seek shelter. When a real tornado hits the area, such inaction can quickly become deadly. Better weather forecast tools will reduce both false alarms and unpleasant surprises.

Over the past 5000 years of recorded human history, and probably since agriculture was established some 10,000 years ago, people have tended to concentrate in climates with mean annual temperatures of approximately 13 °C (55 °F). Climate change is projected to shift the geographical position of this favorable temperature zone significantly poleward over the next 50 years. Up to a third of the global population (one to three billion people) living in tropical and subtropical regions may be experiencing mean annual temperatures greater than 29 °C (84 °F) by the late twenty-first century.[20] Such temperatures at present are limited primarily to the Sahara Desert. The most strongly affected locations are likely to be among the poorest regions in the world, where the ability to adapt is limited. High temperatures may displace many people and result in mass migrations to cooler locations. The U.N. High Commission on Refugees (UNHCR) confers refugee status on people under rules established by the 1951 Refugee Convention. This provides protection for those fleeing their homes because of race, religion, nationality, political opinion, or membership in a particular social group. At present, the UNHCR has no legal refugee status for the estimated 22 million people forced to migrate annually because of natural disasters.

The United Nations predicts that sub-Saharan Africa could see as many as 86 million climate refugees by 2050, with another 49 million in East Asia and the Pacific, 40 million in South Asia, 19 million in North Africa, 17 million in Latin America, and five million in Eastern Europe and Central Asia. Immediate actions to reduce global GHG emissions and support climate-resilient development in these areas could reduce migration by as much as 80%.

****

[19] Simmons, K.M. and Sutter, D., 2009, False Alarms, Tornado Warnings, and Tornado Casualties: *Weather, Climate, and Society*; Collections: Tornado Warning, Preparedness, and Impacts; p. 38–53; DOI: https://doi.org/10.1175/2009WCAS1005.1

[20] Xu, C., Kohler, T.A., Lenton, T.M., Svenning, J-C, and Scheffer, M., 2020, Future of the human climate niche: *Proceedings of the National Academy of Sciences* (PNAS). v. 117, no. 21, p. 11350-11355

The discussion of climate refugees leads to an ugly aspect of the climate crisis called "ecofascism." This is essentially an odious white supremacist belief that climate refugees must be prevented from entering traditionally "white" countries. The refugees are assumed to be people of color from the poorer regions of the world who are said will dilute the "racial purity" of the countries in question, and may even eventually become the majority population, relegating whites to the minority. Although this tripe has been spread widely across the internet as a new idea, ecofascism is actually based on old German Nazi and Italian Fascist doctrines about protecting the racial purity of the fatherland. Ecofascism uses the climate crisis only as a mechanism to provide a plausible source of refugees who are deemed unworthy of asylum.

I find ecofascism deplorable and I wonder why some people presently in the majority are terrified of the prospect of becoming a minority. Is it because of how they themselves treat minorities? Are they worried about receiving the same kind of treatment if they were to become a minority? I suppose that could be frightening, but it also says a lot about how we behave toward others.

It is important to recognize that wealthy populations living in temperate latitudes are not immune to climate change and could easily end up as climate refugees themselves. Global climate change is just that: global. A recent study indicates that tropical storms, hurricanes, typhoons, and other tropical cyclones can be expected to form farther north and south of the tropics, and to travel into the mid-latitudes more frequently than in the past.[21] Large, temperate coastal cities like New York, Tokyo, Shanghai, and London could be facing direct hits from powerful storms. None of these cities are prepared for this kind of onslaught. The resulting damage to infrastructure and the inability to get aid and supplies into congested urban areas could result in hordes of people fleeing these cities and becoming climate refugees.

It is not just big cities. Large wildfires in the western U.S. and elsewhere over the past few years fueled by drought have destroyed entire small towns, turning the inhabitants of these communities into instant climate refugees. An example given earlier was Lytton, British Columbia, where the entire town burned to the ground in 2021 a few days after setting a record high temperature record. If the drought in western North America continues to deepen, wildfires will be larger and more frequent, and more towns and cities will be lost. Likewise, in a really severe drought, agriculture may become impossible in some areas. Many farmers may be forced to relocate from excessively dry regions like California to other places in the country, sort of a reverse "Grapes of Wrath" when people migrated to California from the Oklahoma dust bowl of the 1930s. The climate crisis can strike anyone anywhere, including ecofascists. Climate refugees from the Third World could easily find themselves in the company of rich strangers.

We need to address climate change for the health and well-being of the planet. Putting a racist spin on this that the only reason to tackle the climate crisis is to stop

---

[21] Studholme, J., Fedorov, A.V., Gulev, S.K., Emanuel, K. and Kevin Hodges, K., 2021, Poleward expansion of tropical cyclone latitudes in warming climates: *Nature Geoscience*, https://doi.org/10.1038/s41561-021-00859-1

the migration of non-white refugees is offensive and unproductive. Two important things to remember are that nobody chooses to become a refugee, and anyone can become a refugee in the blink of an eye. People have either lost their homes for some reason or they are fleeing from dangerous or deadly circumstances. Sometimes both. It's not a vacation trip. Treating refugees with compassion and helping them as much as possible is the right thing to do and the proper path forward. There is also an issue of karma here – if everyone on Earth is at risk of becoming a climate refugee, we all could conceivably require the same kind of help someday. Is it not better to do unto others as you would have them do unto you?

****

Greenhouse gas emissions from China are the highest in the world, mostly from coal-fired electricity. The power needs of the manufacturing-based Chinese economy are enormous and growing. In a distant second place for GHG emissions is the U.S.

No solution to the climate crisis will be viable without a major reduction in GHG from China. The Chinese are acutely aware of the problem. I was invited recently to publish an extremely condensed version of the concepts in this book in a Chinese geoscience journal as an attempt to delineate the issues and solutions.[22]

China does not have the natural gas resources to replace coal like the U.S. although they are trying to develop shale gas resources in the southern part of the country. The local tectonics and stress fields have made it challenging to frack these rocks successfully. Much of the Chinese climate strategy involves installing CCS on coal fired power plants to reduce emissions, while also developing carbon-free sources of energy to replace fossil fuels. The huge hydroelectric facility at Three Gorges Dam on the Yangtze River was completed in 2006 to help alleviate some of the need for coal-fired electricity. It produces about 22.5 gigawatts (GW) of power equivalent to about ten large coal-fired plants.

The United States and China set up a joint committee on emissions prior to the COP26 Climate Summit in Scotland in 2021. The goal of the joint committee is for the two nations to work together and learn from one another with the objective of reducing GHG emissions sufficiently to keep the Paris Agreement temperature limit within reach. China and the U.S. will seek policies and technologies to decarbonize industry and electric power through CCUS, energy storage, better grid reliability, and green hydrogen. In addition, both nations pledge to increase renewable energy, encourage green and climate-resilient agriculture, construct energy efficient buildings, and develop low-carbon, sustainable transportation. The committee also intends to cooperate on reducing emissions of methane and other non-CO_2 greenhouse gases, address emissions from international civil aviation and maritime activities, and implement near-term policies to reduce emissions from coal, oil, and natural gas.

---

[22] Soeder, D.J., 2021, Greenhouse gas sources and mitigation strategies from a geosciences perspective: *Advances in Geo-Energy Research*, v. 5, no. 3, p. 274-285, doi: 10.46690/ager.2021.03.04

GHG emissions from the United States have been well-documented by the EPA as part of the U.S. commitments under the United Nations Framework Convention on Climate Change.[23] The highest U.S. emissions come from the transportation segment of the economy, which contributes nearly 30%. Close behind it is the electric power segment, followed by industrial emissions and the residential/commercial use of fossil fuels. Total net GHG emissions from the United States were 5.8 gigatons in 2019, about one sixth of the world total. CO_2 is the primary greenhouse gas contributing to U.S. emissions, accounting for about 80% of the total in 2019. Other greenhouse gases include methane (10%), nitrous oxide (7%) and CFCs and other fluorinated gases (3%). Nitrous oxide or NOx is a greenhouse gas formed as a compound of nitrogen and oxygen in high temperature combustion. Total CO_2 emissions from the United States have declined 13% between 2005 and 2019. The EPA attributes much of this to the shift from coal to natural gas for electric power generation, along with growth in the use of renewable energy. However, this is still not totally good news - emissions in 2019 were still 4% higher than in 1990.

In addition to fossil energy combustion, other sources of GHG emissions include non-combustion emissions from industrial processes (primarily CO_2 off-gassed during the production of cement and CO_2 released after scavenging oxygen from iron ore during steelmaking), along with land use changes such as the conversion of forest land to agricultural or urban use. Land use management can also provide carbon sinks for the removal and long term storage of atmospheric CO_2, such as the conversion of land to forests. Such carbon sinks captured about 12% of total gross U.S. greenhouse gas emissions in 2019.

****

The COP26 climate summit in Scotland in 2021 was well-covered by the news media and brought a lot of climate-related issues before the public. The media focus on climate was also an opportunity for climate skeptics and fossil fuel advocates to spout misleading statements and false claims on social media. The purpose of these claims, as always, was to sow confusion and create artificial uncertainty to delay any action on climate. Delays allow the fossil fuel industry to continue selling their products and keep revenues flowing in. The BBC, Associated Press, CNN and other legitimate news outlets ran fact checks on many of these statements and debunked them. Some of the more prominent claims include the following:

1. "A grand solar minimum will offset global warming." The sun is a slightly variable star and does go through maximum and minimum phases as part of a natural 11-year cycle. Sunspot numbers, solar flares, and magnetic activity increase during a solar maximum and then fade at the following minimum. The activity at solar maximum can have a substantial effect on space weather and the intensity

---

[23] U.S. Environmental Protection Agency, 2021, Inventory of U.S. Greenhouse Gas Emissions and Sinks, 1990-2019: Report EPA 430-R-21-005, Washington, DC, 791 p. (https://www.epa.gov/ghgemissions/inventory-us-greenhouse-gas-emissions-and-sinks-1990-2019)

of the aurora borealis on Earth, but it does not directly affect the climate. Likewise, a solar minimum is not nearly enough to offset the effects of GHG. The most recent solar minimum at this writing was reached in 2019, and by 2021 the sun had begun climbing toward the next solar maximum. The 11-year solar cycle may have some effect on the long-term global temperature average, but it is definitely not enough to offset the overall upward trend (refer back to Fig. 5.2).

It is true that long-term variability in solar output and the orbital mechanics of the Earth, in particular the Milankovitch Cycles that naturally change climate over thousands of years were the major drivers of the Ice Age glacial pulses. Climatologists do include these factors in their models, but there is no compelling evidence showing that an increase in solar insolation or a change in the Earth's orbit over the past century can account for the observed rise in global temperatures. The IPCC has noted that between 1750 and 2005, the steady warming of the sun has increased insolation on the Earth by about 0.12 watt/m^2. This is less than 10% of the 1.6 watt/m^2 net forcing attributed to anthropogenic GHG and miniscule compared to the summer mid-latitude insolation of 300 watt/m^2.

2. "Global warming is good and will make parts of the Earth more habitable, and besides, more people die from frostbite than heat stroke." More people die from frostbite than heat stroke at the moment, but as the climate warms we would simply be trading a drop in cold-related deaths for an increase in heat-related deaths. There have already been large losses of life from killer heat waves, including Russia and eastern Europe in 2010 and western Canada and the U.S. Pacific Northwest in 2021. These are likely to get worse. It is also important to note that cold is easier to deal with than heat. Frostbite and hypothermia can be avoided by bundling up properly or building a fire. But as mentioned earlier, if temperatures rise above a certain level, no amount of shade, fluids, or lightweight clothing will make them survivable. Heat can kill quickly. People hiking in the hot Nevada desert have died within a mile (1.6 km) of their vehicle.

The notion that global warming is good and carbon dioxide is our friend is another attempt by the so-called "climate realists" to get us to accept our fate, understand that there is no substitute for fossil fuels, and make the best of things. Their ultimate goal is to continue producing and selling oil, gas, and coal at a profit and of course that requires no restrictions on GHG. This effort hopes to convince everyone that higher levels of CO_2 will be beneficial, with greater crop yields and balmy temperatures in places that are now cold. It's more likely to disrupt agriculture before it does anything for crop yields; a recent study indicates that changes in average temperatures and rainfall patterns may require the "corn belt" of the Midwest to migrate northward into Minnesota and Canada by 2100.[24] Warming of the planet may indeed make some parts of the Earth more habitable,

---

[24] Burchfield E.K., 2022, Shifting cultivation geographies in the Central and Eastern US: *Environmental Research Letters*, v. 17, no. 5, 14 p. 054049

but it will also make other places uninhabitable. The end result will be large-scale disruptions of society and mass migrations of climate refugees.

3. "Restricting fossil fuels will stunt economic growth and make people poorer." This is hogwash. For starters, when the externalized costs of carbon are included, fossil fuels are anything but cheap. They are also not sustainable and are going to get more expensive as they become scarce. Sooner or later we will need substitutes. Working on developing those substitutes now while we still have an intact and solid economy is the smart option. Waiting until the climate crisis has damaged infrastructure, disrupted lives, closed businesses, and created food shortages before trying to develop alternative energy would be akin to the oil-starved Germans making synthetic liquid fuels from coal late in the war. Their process was expensive, inefficient, and polluting, but of course they were desperate. We probably don't want to wait until we get that desperate before finding replacements for fossil fuels.

Non-fossil energy such as wind and solar is available anywhere. Powering the developing world with these resources instead of Western fossil fuel technology will give them resilient, indigenous sources of energy that don't require expensive, giant power plants and miles of vulnerable transmission lines. We don't need everyone else to repeat our mistakes.

4. "Renewable energy is unreliable, as shown by the massive electricity grid failure in the 2021 cold snap that left millions of Texans without power." The Texas blackouts were caused not by renewables but by the Texas electrical grid, which prides itself on independence from the U.S. national grid and is barely connected to it. Thus when the Texas grid has a problem, it is unable to obtain electricity from other states. The high demands imposed on it by the cold snap brought it to a crisis. Texan fossil fuel power plants turned out to be much less reliable than renewables during the cold snap. The valves on natural gas lines froze up and couldn't be opened and iced-over coal piles couldn't be moved. Wind turbines kept operating fine in the Texas chill, as they have operated successfully in climates far colder, including Minnesota and even Antarctica.

Texas' neighbors, New Mexico, Oklahoma, Arkansas, and Louisiana all experienced the same cold weather (refer back to Fig. 5.3). None of these surrounding states suffered a loss of power or a loss of life anywhere proportional to what happened in Texas. Despite pledges to improve reliability after the 2021 cold snap, the Texas electrical grid is still largely disconnected from the national grid and continues to have problems.

5. "Carbon dioxide is only a trace gas in the atmosphere. Water vapor is also a greenhouse gas and far more abundant than CO_2, thus the increase in atmospheric CO_2 from 284 ppm to 420 ppm is irrelevant." For starters, small concentrations of things can have big effects. As pointed out earlier, the minimum lethal dose of arsenic is less than 0.2 grams, about the size of a single grain of rice. Carbon dioxide plays a significant role in climate dynamics by absorbing infrared radiation and warming the atmosphere. The effects have been well understood by people like Fourier, Foote, Tyndall, and Arrhenius since the nineteenth century, and nothing we have learned since has proven them wrong. Human

combustion of fossil fuels is by far the largest contributor of atmospheric CO_2; by some calculations the annual input of anthropogenic CO_2 to the atmosphere exceeds volcanic sources by more than 130 times.

Climate skeptics often argue that water vapor is a far more abundant and powerful greenhouse gas than CO_2 and claim that climate scientists routinely leave it out of their models. Actually, the first digital climate models, produced in the 1960s, were primarily focused on water vapor, but found that CO_2 had to be included to get the results to match observations.[25] Carbon dioxide independently adds heat to the atmosphere by absorbing different wavelengths of infrared than water. This causes more water vapor to enter the atmosphere and amplifies the greenhouse effect. However, water vapor can also condense into clouds and reflect sunlight away from the Earth, cooling it instead of warming it. Water vapor complicates climate models, but it is included along with other GHG to obtain a realistic output scenario.

6. "According to records at the U.K. Met Office Hadley Center, 1998 was the world's warmest year and the Earth has been cooling since then." Climate skeptics have latched onto this as "proof" that global warming is over. As mentioned previously, when NOAA declared 1988 as the hottest year on record, the skeptics latched onto that as well. Both "record" years have since been left in the dust by much hotter years in the new millennium.

There was actually a lull in the steady increase in global temperature trends right around the turn of the millennium, but this did not change the overall trend. The cause of the lull is still vigorously debated at academic conferences with the two top candidates being aerosols in the upper atmosphere or changes in ocean circulation patterns. The whole issue became moot once 1998 was no longer the hottest year on record. That title now belongs to 2016, with 2019 in second place, and in fact all ten of the current hottest years on record have occurred within the twenty-first century. Ironically, the U.K. Met Office Hadley Center has collected most of these data.

7. "The supposed 99% consensus among scientists on anthropogenic climate change is irrelevant because science isn't settled by popularity contests, and the real truth is hidden because climatologists conspire to lock away their data." These types of ideas tend to come from people who have no idea how science works. Scientists are remarkably disagreeable and often over-share. We are skeptics who do everything we can to prove each other wrong.

When I worked for the USGS out at Yucca Mountain in Nevada (more on that in Chap. 7), I questioned a formation boundary that geologists from Los Alamos National Laboratory had placed on the volcanic ignimbrites making up the bedrock. It appeared to me that there wasn't a clear change in the rocks, which would be expected if they were laid down by separate volcanic eruptions. Most of these deposits develop soils called paleosols on the exposed surface that can

---

[25] Manabe, S., and Wetherald, R.T., 1967, Thermal equilibrium of the atmosphere with a given distribution of relative humidity: *Journal of the Atmospheric Sciences*, v. 24, no. 3, p. 241-259.

be used as a marker bed when another layer of ash gets deposited on top. There was no paleosol at their declared boundary and it all looked like one unit to me. The two Los Alamos geologists made a trip out to the Nevada Test Site, and we spent an interesting afternoon tramping around on the outcrop as they explained their choice for selecting the boundary. In the end they convinced me that some subtle differences in texture showed that it really was two separate ash beds and I agreed with their boundary. These types of debates between scientists are very common and there are no hard feelings when someone asks questions. Questions help the researchers understand where some of their reasoning and interpretation of the data might not be clear and they can focus on improving those parts. This makes for good science.

Graduate students get a baptism by fire into this process when they "defend" their master's thesis or PhD dissertation to their advisory committee. The students must explain their starting hypothesis, the data they collected to build it up into a theory, the procedures they used to test the theory, their findings, results and conclusions. It can be brutal, but this is how they become battle-hardened for the professional world. (I also try to let them know that it is in everyone's interest for them to succeed and pass. This usually calms the nerves.)

Remember, in science, you are never proven to be right, only wrong. Challenging everyone's findings in science keeps it honest and avoids self-delusion. Once you get through this process you can be confident in your conclusions. That is until someone else comes up with some new data that prove you wrong. Scientists need a thick skin because numerous peers will be taking multiple shots at you. If you are confident in your data and evidence, you use it to convince the critics, just like the Los Alamos scientists convinced me. It is also incumbent upon the critics to be open to viewing the evidence placed before them and be willing to have their minds changed. Stubbornly refusing to even look at someone's data helps no one.

In a 2004 analysis of the peer-reviewed literature on climate change, science historian Naomi Oreskes found three quarters of the 928 papers she surveyed supported anthropogenic causes, while the remainder took no position on the subject.[26] Interestingly, none of the papers supported purely natural explanations for the changing climate. The important consensus was not so much among the scientists themselves, but within the body of research. Completely different studies kept finding a preponderance of evidence that climate change is being driven by global warming caused by greenhouse gases emitted from the combustion of fossil fuel.

The 99% consensus among scientists on climate change is far from a popularity contest. Anthropogenic climate change has been through the wringer and came out the other end intact. It is the equivalent of the target that survived passage through a shooting gallery. It's the last place team that somehow rallied and won

---

[26] Oreskes, N., 2004, The Scientific Consensus on Climate Change: *Science*, v. 306, no. 5702, p. 1686; DOI: 10.1126/science.1103618

the championship. It's the ugly painting that got named best of show. Or, to cite an earlier example, it's the cold fusion research that was actually reproducible and worked as advertised.

Scientists looked at the data and tried to prove it wrong. Almost everyone ended up agreeing that anthropogenic climate change is real and is the correct interpretation of the data. The agreement extends to the official positions of a number of scientific societies, including the U.S. National Academy of Sciences, the Royal Society in the U.K., the American Association for the Advancement of Science, the American Geophysical Union, the Geological Society of America, the American Institute of Physics, and the American Meteorological Society. Even the American Association of Petroleum Geologists recognizes that carbon dioxide is an issue (over the objections of some members). Given the inherent disagreeability of most scientists, it is hard to see how a supposed conspiracy could co-opt the official positions of these and other societies.

Scientists mostly shake their heads at accusations that they are hiding data or the details of their models. If the climate skeptics bothered to look, they would find climate data widely available in public data bases. After all, much of it is collected by NASA, NOAA, DOE, EPA, the U.S. Geological Survey, the National Weather Service, and other government agencies, all of whom dump their data into public databases to make it available to the taxpayers who paid for it. Most of this can be queried online these days. Articles in scientific journals typically contain links to the data for readers to view and test for themselves. Because science is open to other interpretations, people are free to analyze the data and if warranted, come up with different conclusions. If climate contrarians really wanted to disprove global warming theories, they could download the public data and develop their own credible climate models to show why current theories are wrong and why their alternatives are not. This has simply not happened.

8. "Climatologists are only raising alarms to get money and prestige." This is probably the most laughable of the misleading statements and false claims of the climate skeptics. Hardly anyone is getting rich off climate. The billionaires who are trying to address climate change, like Bill Gates, Richard Branson, and Warren Buffett all made their money elsewhere. The possible exception is Elon Musk, who is making some money from electric vehicles. Still, there are far more wealthy oil and coal executives than climate scientists. Federal spending on climate studies rose from $2.4 billion in 1993 to $11.6 billion in 2014 primarily to fund energy conservation projects and technology development programs, while actual climatology funding remained essentially flat. DOE funding for CCS research has been traditionally focused on emission reductions from coal-fired power plants so electricity could continue to be generated from coal. The first significant government funding for direct carbon dioxide removal (CDR) from the atmosphere is a program called the Carbon Negative Earthshot launched in 2021 by President Biden under Executive Order 14008 and reportedly funded at $10 billion. There is also funding in the new Inflation Reduction Act to focus on both research and implementation of climate-friendly infrastructure. Despite these initiatives, funding for research in zero-carbon energy technology areas,

such as engineered geothermal or new nuclear technology projects is measured in the millions, not billions. Although these are critical for replacing fossil fuels, they are not receiving funding at levels appropriate to an important national program, especially when compared to private sector multi-billion-dollar oil drilling projects.

As for prestige, many prominent climate scientists who have raised alarms have been savagely attacked on social media and threatened personally. In today's society, threats and harassment by internet trolls against scientists who speak up for climate and against fossil fuels are unfortunately common. The recent harassment and threats directed at Dr. Anthony Fauci and others over COVID-19 public health measures sounded depressingly familiar to many climate scientists, who invited Fauci to "join the club." If this is the "prestige" that comes from fame, who needs it?

****

How does the human influence on climate fit into the overall geologic history of the Earth? Although humans are changing the climate in ways that have not been seen for millions of years, the Earth has been through much bigger changes in the ancient past. In the long run, humanity is really not that big of a deal. We've only been around for three million years, less than 0.1% of the history of the Earth. If we were to disappear tomorrow, most traces of us would be gone in a few hundred thousand years, and almost certainly in a million years. Another advance of the Ice Age glaciers would grind most northern cities into what geologists call "rock flour," a typical product of ice sheets. Weathering and erosion would take care of the rest. There would be some remnant artifacts and chemicals in the sediments of our time and on the moon and Mars, but searchers would have to know where to look, and where to dig.

Humans do hold the distinction of being the first species in the history of the Earth capable of causing a self-inflicted environmental catastrophe and an accompanying mass extinction event. So that's at least something to brag about and puts us up there with supervolcano eruptions and asteroid impacts. Climate disruption is not our only option; we could also disappear from dangerous chemicals in the air and water, the collapse of critical ecosystems, and full-scale nuclear war. Furthermore, we know that we are capable of causing such a disaster and yet we have done little or nothing to stop it.

Most eighteenth century scholars considered the age of the Earth to be around 6 Ka, based on dates mentioned in scripture. Archbishop James Ussher of Ireland announced in 1650 after a detailed analysis of the Bible that the Earth had been created on the evening of October 22, 4004 BCE. His disturbingly precise age date was accepted for nearly two centuries.

Geological deposits were originally defined based on their relationships to biblical tales of Noah's Ark and the notion of a worldwide Great Flood. The rocks that formed when the Earth was created were called "Primary." Sediments that were deposited after the creation of the Earth but before the Great Flood were called "Secondary." Sediments from the flood itself were deemed "Tertiary," and sediments deposited after the floodwaters receded were labeled "Quaternary." Fossils were considered to be the remains of animals that had perished in the flood. Despite the complete lack of evidence for a world-wide flood, the terms Tertiary and Quaternary are still used in the geologic time scale to this day, preserving a bit of this history.

As an aside, my thoughts on the "flood" are that it may have been an abrupt rise in sea level – possibly even over 40 days – caused by the rapid release of impounded glacial meltwater at the end of the last Ice Age. There is evidence in the channeled scablands of eastern Washington State for a massive release of water from a huge ice-dammed lake. People living on the flat continental shelves in coastal areas during this prehistoric time would have viewed the resulting inundation by the sea as a "world-wide" flood. By the time the Bible was written several millennia later, the passed-down flood story had morphed into Noah and the Ark.

Archbishop Ussher's age date of 6 Ka for the Earth and the simplistic classification of sedimentary rocks around Noah's flood were not borne out by the evidence in the rocks. Dr. James Hutton, a physician, chemist, and farmer in eighteenth century Scotland became convinced that the Earth had to be much older than 6 Ka to produce the features he saw in the rocks.

Hutton essentially invented the science of geology by observing an angular unconformity at a place called Siccar Point, located south of Edinburgh, Scotland (Fig. 5.5). An angular unconformity is a geological feature where horizontal sedimentary rocks lie on top of older strata that are tilted at an angle.

With amazing insight, Hutton reasoned that the angled strata at Siccar Point must have been deposited originally as horizontal sediments, and after turning into rock, the beds were tilted upward by a later episode of mountain building. The surface was then eroded, and the upper horizontal layers were deposited on top of the tilted beds, after which they, too turned into rock. If the angular unconformity had been created through natural processes of erosion and deposition, and the rates of those processes were similar to the rates Hutton could observe around him (and there was no evidence to suggest they had been any different), then it had taken far longer than 6 Ka to create the geological features he was observing at Siccar Point.

Hutton calculated how much time was needed for natural geological processes to form the angular unconformity at Siccar Point and found that the Earth had to be at least hundreds of millions of years old. We now know that even this age was far too young; radiometric age dating of the oldest terrestrial rocks, corroborated by the ages of slightly younger moon rocks indicate that the Earth formed about 4.6 billion years ago (4.6 Ga).

The history of the Earth is divided up into different time periods, based on rock types, localities, specific groups of fossils, or mass extinction events that have occurred throughout Earth's history. The names have a variety of origins. Some, like the Devonian Period, are named for places where the typical rocks are found, in this case Devonshire, England. Other place names include the Permian Period, named after the Russian city of Perm, and the Jurassic Period, named for the Jura Mountains

Fig. 5.5 The angular unconformity at Siccar Point that inspired James Hutton to calculate the age of the Earth. (*Source: Photographed in 2019 by Dr. Brennan Jordan, University of South Dakota; used with permission*)

on the border of France and Switzerland. Other time periods are named after the dominant rock of the age, such as the Carboniferous for a period in Earth's history when the landscape was dominated by carbon-rich coal swamps, and the Cretaceous Period where the dominant rock type was chalk.

Geologic time periods are subdivided into epochs, and there tend to be more of these in the more recent periods because the details can be better deciphered. The Tertiary, for example, contains two geological periods, the Paleogene and Neogene, and these are subdivided into a number of epochs such as the Paleocene, Eocene, Oligocene, Miocene and Pliocene. The Quaternary is subdivided into the Pleistocene and Holocene epochs. Humans or at least hominids have been around since the beginning of the Pleistocene some three million years ago. The Pleistocene is the time of the Ice Ages, and the Holocene began when the last ice sheets melted about 11,000 years ago. Many geoscientists now advocate that a new epoch be added after the Holocene to describe the influence of humans on planet Earth. The name for this is the Anthropocene.

The root of terms like anthropology, anthropogenic, and Anthropocene is the Greek word ánthrōpos (ἄνθρωπος), which means "human." Human influence on the geology of the planet is widespread and substantial. We have changed the flow of rivers with dams and canals, altering natural streamflow and sediment transport. We have sculpted coastlines into harbors and waterfronts, changed longshore sediment transport, and drained wetlands, affecting organic matter accumulation rates. We have dumped trash, chemicals, plastics, and debris in a vast number of locations where it will be incorporated into the sedimentary record. Nearly half of the species on Earth that were present during the Holocene have become extinct or are threatened because of Anthropocene human influence, starting with the passenger pigeon and the dodo, and continuing with others too numerous to mention. We have changed the composition of the atmosphere and altered the climate.

A number of scientists have suggested that the distinctive, long-term marker beds left in sediments by radioactive fallout from the 1950s atmospheric nuclear tests be used to define the base of the Anthropocene. The upper boundary of the Anthropocene will be defined by the extinction of humanity, whenever that happens to occur (hopefully later rather than sooner). Some geologists have suggested, only half in jest, that the next geologic period following the Anthropocene should be called Weleftthescene.

Mass extinctions have happened in the geologic past, and they can happen again. Most people are aware that the dinosaurs and about 70% of other life went extinct at the end of the Cretaceous Period, around 66 Ma. The cause of this extinction was mysterious for many years, but evidence in the sediments and the remnants of a large, buried crater in the Yucatan Peninsula indicate that an asteroid some 10–15 kilometers (6–9 miles) across slammed into the Earth at about 20 km/second.[27] The impact caused a massive tidal wave and splashed molten rock for hundreds of miles, igniting fires everywhere. It heated up the atmosphere by several hundred degrees, killing many animals quickly. Hundreds of cubic kilometers of material were injected high into the stratosphere, darkening the Earth for months, chilling it severely and killing many plants. Dinosaurs and other animals that weren't killed by the impact, fires, heat, or cold weather ended up starving to death as the food chain collapsed. It was indeed a bad day on Planet Earth.

Fewer people know that an even larger mass extinction occurred earlier, at the end of the Permian Period about 252 Ma. The continents were joined together at the time in a gigantic supercontinent called Pangea. It included the ancestral landmasses that became the present-day Americas, Europe, Africa, Australia, Antarctica, and Asia. This single large continent disrupted plate tectonic boundaries and acted like a blanket on the Earth, causing heat to build-up in the mantle. Pangea split apart into our current continents during the early Triassic Period, but not before cooking up some massive volcanic eruptions in Australia and Siberia. The amount of basalt lava erupted in Siberia is the single largest continental deposit in the world. The volumes of CO_2 injected into the atmosphere severely disrupted climate, acidified the oceans and resulted in the mass extinction of about 80% of the species living at the time.[28]

At the dawn of the Weleftthescene, evidence of people on the surface of the planet will disappear quickly through erosion and weathering. However, a record of

[27] Collins, G.S., Patel, N., Davison, T.M., Rae, A.S.P., Morgan, J.V., and Gulick, S.P.S., 2020, A steeply-inclined trajectory for the Chicxulub impact: *Nature Communications*, v. 11, no. 1, p. 1480 (doi:https://doi.org/10.1038/s41467-020-15269-x. ISSN 2041-1723)

[28] Chapman, T., Milan, L.A., Metcalfe, I., Blevin, P.L. and Crowley, J., 2022, Pulses in silicic arc magmatism initiate end-Permian climate instability and extinction: *Nature Geoscience*, v. 15, p. 411–416 (https://doi.org/10.1038/s41561-022-00934-1)

the human destruction of the environment will be retained for hundreds of millions of years in the geology and geochemistry of the sedimentary deposits being laid down today. Layers containing contaminants, radioactivity, microplastics, debris washed into the oceans, heavy metals, and dozens of other "markers" including the extinction of at least half and maybe more of the Holocene life forms on Earth will clearly identify the sedimentary deposits associated with humans. There is no hiding it, just as there is no hiding the evidence of the extinction events that ended the Cretaceous and Permian. As threatened by many a high school principal, no matter how well we might shape-up in the future, this period of bad behavior will remain on our "permanent record" forever in the geologic history of Planet Earth.

If humanity eventually does go extinct because of our foolishness, it is likely that some forms of life will survive, and new species will evolve in the future to replace us. The Earth has been through extinctions as bad or worse than anything humans are capable of causing and life made a comeback every time. New creatures will evolve to fill the niches left by those that came before. If any of these new life forms become intelligent, perhaps one day they will figure out our existence. Some of our artifacts preserved in the rocks will no doubt be puzzling and their view of us may be as alien as if they came from another star. But if they learn from our mistakes and avoid repeating them, it will be a worthwhile lesson.

Will humanity survive into a utopian future? I certainly hope so, because as the high school teachers also used to say, we have a lot of "potential." However, we'd better clean up our act and stop abusing the Earth. We are running up a bill that is becoming increasingly difficult to pay. Treating the planet like a giant garbage dump, burning through energy, mineral, and water resources, and ignoring the physics and chemistry of the atmosphere and the oceans will one day wreak havoc upon us, and probably sooner rather than later. As one of my friends has said, "Mother Nature is a good deal worse than the mafia when it comes to collecting on debts."

Chapter 6
Replacing Fossil Fuels

Keywords Paris Agreement · Net zero · Carbon tax

****

It would be nice if the Anthropocene could continue for a while longer because humanity has done many great and wonderful things in the past and can do many more in the future. But if we want to avoid crossing over into the Weleftthescene Epoch, then we'd better get our act together. Again, this is not rocket science, and according to the IPCC, only two things are required to fix the climate crisis: (1) stop burning fossil fuels, and (2) remove some of the GHG from the atmosphere. This chapter will focus on replacing fossil fuels. I'll discuss GHG removal in Chap. 9.

Replacing fossil fuels will be difficult. Not necessarily technically difficult, because carbon-neutral and sustainable energy alternatives already exist to replace every type of fossil fuel use out there, and more are on the way. Not necessarily economically difficult either, because policies could be adopted to ensure a closer cost parity between fossil fuels and renewables, for example by making petroleum-based fuels more expensive than biofuels through a carbon tax. The main difficulty in replacing fossil fuels is political, because to keep their profitable products on the market the global fossil fuel industry sends campaign contributions and hordes of lobbyists to friendly politicians who block and discourage policy changes that would replace fossil fuels with carbon-neutral alternatives.

Other difficulties come from human nature. People don't like change, resist disruption, are happy with the status quo, and want to go through life with everything calm, steady, quiet, and predictable. Some of those opposed to replacing fossil fuels are the "hard-to-argue-with" types who distrust government, disbelieve science, see everything as a conspiracy, deny that climate change is real and also presumably think that professional wrestling is an actual sport. There is a subset of this group that consists of teenage boys who express their rebellion against climate concerns by "rolling coal," which consists of spewing clouds of black smoke out of their diesel pickup trucks at bicyclists and people driving hybrids or electric vehicles. (This is one reason why I no longer bicycle on highways, only bike trails.) None of these folks are ever likely to change their ideas, but they are also a relatively small minority.

D. Soeder, *Energy Futures*, https://doi.org/10.1007/978-3-031-15381-5_6

On the other hand, the majority of average, everyday people do generally understand that we must stop using fossil fuels if we want a habitable planet. Some wealthy individuals have installed solar panels on their homes and moved "off the grid," where they recharge their electric cars on carbon-free power. That's terrific for rich environmentalists (who are often insufferable about it), but few ordinary people can afford to trade in their Toyota for a Tesla. Most remain unsure about how to replace fossil fuel with carbon neutral energy on their existing budgets. Everyone is waiting for governments to provide some direction and leadership for transitioning to affordable energy alternatives that won't destroy the planet, but such leadership has been in short supply. People see wind turbines in the countryside and can read on the pump that their gasoline contains 10% ethanol biofuel, but is this progress? Wind turbines and ethanol are symbols, but they don't seem to have made much of a difference when scientists are calling the climate crisis a global emergency. With no other options, most of us guiltily fill up our minivan with gasoline and buy a tank of propane for the backyard grill because we need to get on with life. In the meantime, the fossil fuel industry continues to rake in profits.

The lack of a clear game plan to switch the global energy economy from fossil fuels to more sustainable, carbon-neutral energy has been a colossal failure of government in my opinion, and not just the U.S. government, but governments worldwide. Nearly all nations have pledged to eliminate fossil fuels, but none has actually done so, or even seriously gotten started on it. Humanity needs a pathway to achieve the goals of carbon-neutral energy. Virtually every nation on Earth has signed the Paris Agreement and national leaders proclaim that their nation will be at zero net carbon by 2040 or 2050. China says 2060; India says 2070. Yet no one explains how they intend to actually get there. I am totally sympathetic to young climate activists like Sweden's Greta Thunberg, who are understandably frustrated that all there is to show for the fancy talk is a stunning lack of progress.

There is a cartoon popular among scientists showing two researchers standing in front of a blackboard filled with equations on both sides. To balance the equations, in the middle between them is written "and then a miracle occurs." More often than not, this feels like the standard government approach for dealing with climate change.

Many scientists including me are growing concerned that net-zero has become a new corporate and government buzzword. Nebulous pledges to achieve it appear to be based on some miraculous, future technological breakthrough instead of specific, near-term GHG emission cuts. According to some estimates, about 90% of the global gross domestic product (GDP) is now covered by some sort of net-zero plan, but the majority of these plans are poorly defined. Even for the 74 countries that have developed detailed decarbonization plans, analysis by the United Nations concluded that their emissions would drop 70–80% by 2050, which is impressive but not net zero. The fear among climate watchdogs is that governments are using net-zero pledges as a message to their citizens that they are taking climate change seriously but then avoiding any significant and painful actions that might cost votes by postponing them to some fuzzy, mid-century target date. Vague promises about installing more charging stations for electric vehicles are no substitute for an actual action plan and a climate roadmap complete with milestones and dates.

They had better come up with something soon. The Paris Agreement is a legally binding international treaty on climate change signed by 196 parties in 2015 with the goal of limiting global warming to less than 2 °C. The IPCC found that since 1850, human activity has emitted around 2400 gigatons of CO_2 and equivalent GHG into the atmosphere. The emission of 460 billion tons more will push global temperatures across the 1.5 °C threshold. At current emission rates, this will take about 11 years.

The Paris Agreement requires individual countries to submit their plans for climate action by 2020, which has presumably been delayed because of the COVID pandemic. The plans are called nationally determined contributions (NDCs) and are intended to describe both the actions that will be taken to reduce GHG emissions and the steps necessary to build resilience for adapting to rising temperatures. Most climate scientists think that any action plan at this point, no matter how good, will end up with an overshoot of the 1.5 °C goal due to inertia in the climate system. Even our best efforts will likely result in global temperature increases of at least 3 or 3.5 °C, followed by a recovery and a temperature decline a few decades later.

The Paris Agreement is a great international effort to deal with climate, but intense pushback by the coal and petroleum industries may limit its effectiveness. The U.S. pulled out of the accord under former President Trump, who called climate change a "Chinese hoax." Although the U.S. has re-joined the Paris Agreement under President Biden, other nations in the world are now uncertain about the U.S. commitment to fighting climate change and wonder if we will waffle again under a future Republican administration. The 5 year delay on not dealing with climate issues during the Trump administration has also put the U.S. significantly behind other nations seeking climate solutions.

The difficulty of gaining support for climate action in Congress prompted President Biden to sign an executive order in 2021 to reduce the federal government's greenhouse gas emissions to net-zero by 2050. He said it will transform "how we build, buy, and manage electricity, vehicles, buildings, and other operations to be clean and sustainable." The order sets standards for reducing emissions in three key areas: electricity generation, buildings, and transportation. Because the President is the head of the executive branch, such orders have the force of law to federal agencies. However, it is not an actual law passed by Congress, and any long-term implementation would have to be made by future administrations. This is certainly not guaranteed and is in fact often subjected to politics. For example, former President Obama signed an executive order in 2015 to reduce the federal government's carbon emissions by 40% over 10 years, which was promptly rescinded by former President Trump.

Trump's claim that climate change is a hoax feeds a conservative Republican narrative that U.S. environmentalists have been duped by a Chinese diversionary tactic to convince the energy sector to cut back on coal and oil while capturing and sequestering carbon. Along with the construction of expensive wind turbines and solar farms, all of this will lead to expensive U.S. electricity and make U.S. products more costly. In the meantime, China supposedly laughs at us while they merrily burn coal to generate inexpensive power and blow us out of the water economically with low-priced manufactured goods thanks to their cheap energy.

The numbers don't actually support this. The World Bank reported the 2019 GDP in U.S. dollars for the three largest national economies as $21.43 trillion for the United States, $14.28 trillion for China, and $5.06 trillion for Japan. Thus, the U.S. economy eclipses China by more than the entire GDP of Japan. It is also important to keep in mind that China has more than four times the population of the U.S., so their lower GDP is divided among many more people resulting in a per capita income quite a bit less than ours. The Chinese economic juggernaut is driven not by cheap energy, as claimed by Trump, but by cheap labor.

I spent some time in Wuhan, China in the spring of 2019 and found it to be loaded with new infrastructure, including stylish commercial buildings, high-rise residential towers, roomy parks, new bridges and roads, a brand-new airport, and other facilities. Many structures remaining from the Chairman Mao days, such as some of the old government buildings were shoddy and run down, but these are steadily being replaced. My guide took me up into Yellow Crane Tower, a famous 13th Century landmark on the bank of the Yangtze River, for an overview of the city. I saw plants growing on green roofs on many buildings, an orderly and clean city, and a strong degree of respect for the country's ancient cultural heritage. China is so much older than the United States as a nation that the provincial museum had Chinese bronze tea kettles on exhibit that dated from the actual Bronze Age. Their display of porcelain went back nearly two millennia, and many ancient pieces had been donated by families that had passed them down for centuries. The imperial dynasties like the Ming that fiscally supported the porcelain industry received much higher quality pieces than the emperors who weren't interested.

There are also more people packed into this nation than I've ever seen anywhere. Wuhan has a population of over eight million, just a bit less than New York City, and it is not even considered an exceptionally large town in China. In contrast, the population of Beijing is 21 million, and Shanghai has 26 million. The Chinese scientist who invited me to visit, one of my former interns who received his PhD in civil engineering from Carnegie Mellon University in Pittsburgh, has his own intern now, a young man who came from a "small" town of only three million. (My former intern is also the only guy in Wuhan, China driving around with a Pittsburgh Steelers bumper sticker on his car, much to the mystery of the locals.)

China has every incentive to move away from coal and into carbon-neutral energy. The air quality in Wuhan was awful the entire week I was there from coal smoke, which hangs over everything in a constant haze, burns the lungs and throat with an acrid aftertaste and waters the eyes. Because of my history with gas shale, I visited Wuhan to meet with researchers from the Chinese Academy of Sciences who are trying to develop shale gas in the southern part of the country to replace coal.

Their target unit, the Longmaxi Shale, very much resembles the Marcellus Shale in the Appalachian basin although it is a bit older. Fracking in the Longmaxi Shale has been problematic, however, because of complex stress fields in southern China and abundant natural fractures in the rock. It would seem that natural fractures might be helpful for gathering the gas, but instead they cause the hydraulic fractures to terminate prematurely or zig-zag like cracks in a brick wall. Neither of these is very efficient for producing shale gas.

Still, because of their rapidly growing economy the demand for electricity in China has been outstripping the supply for years, keeping energy prices high as the government frantically tries to build more power plants. For the past decade or so, China has been commissioning a new power plant roughly every 2 weeks. Many of these are coal-fired to be sure (Fig. 6.1), but China has also made major investments in wind, solar and hydropower.

China has the second-largest economy in the world and emits the greatest amount of greenhouse gas. Tellingly, the other major project at the Chinese Academy of Science in Wuhan was investigating the integrity of old wells to sequester carbon dioxide for CCS. They very much understand climate is an issue and seem to be doing a lot more than the U.S. to address it.

I realize there are political differences between the U.S. and China over certain labor practices and relationships with Taiwan. Nevertheless, we must cooperate with the Chinese on ending the use of fossil fuels and removing some of the existing CO_2 from the atmosphere. We may have our differences, but we share the same planet. And one thing I took home from my visit after a researcher's 7-year old daughter gave me an origami paper frog as a gift was the realization that people are people, no matter where they live.

****

So let's get down to brass tacks. How are we going to do this?

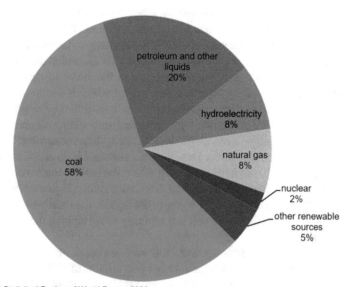

Source: BP Statistical Review of World Energy 2020
Note: Total may not equal 100% because of independent rounding. Includes only commercial fuel sources and does not account for biomass used outside of power generation.

Fig. 6.1 The energy mix in China as of 2019. (Source: U.S. Energy Information Agency (EIA), public domain)

I hate to burst any bubbles, but net zero will not be achieved by individuals recycling plastic bottles or grilling over charcoal instead of propane, although there are other good reasons for doing these things. Environmental groups often exhort people to take action individually to fight climate change, but despite the "feel good" aspect of individual actions, going vegan or drying laundry outside on a clothesline is not enough to save the planet. Even if everyone did it. A few rich people going off the grid for what seems like bragging rights is less than a drop in the bucket. A much larger and more systematic approach is needed to eliminate fossil fuels, and that requires both smart economics and government policy.

Here is what I suggest as an approach: first, focus on completely decarbonizing the electric grid. We have to get GHG emissions under control here first, because net zero electricity can be used to decarbonize other sectors of the economy, like replacing gas furnaces with electric heat pumps, for example. After electricity, the second goal should be to decarbonize the transportation sector with both electric vehicles and carbon-neutral internal combustion engines (i.e., biofuel or hydrogen). The reason for doing both is explored in Chap. 8. Once electric power and vehicles are covered, decarbonization can focus on other sectors of the economy that contribute significant amounts of GHG. These include industries like concrete, steel, and chemicals, residential and commercial use of fossil fuels, and the agricultural sector. A tally of U.S. GHG emission sources from the EPA by various economic sectors is shown in Fig. 6.2. These are where efforts ought to be focused to replace fossil fuels.

So far, we haven't made much progress. But don't take it just from me. A number of climate scientists interviewed by Yahoo News[1] prior to the 2021 COP26 Climate Summit in Glasgow, Scotland were not optimistic that world governments would finally get serious about addressing climate change. As Pacific Institute climate scientist Peter Gleick said, "You know, this is COP26, which means there have been 25 of these things already. We're way behind the curve in acting on what we have known for many, many years to be the reality, which is that humans are changing the climate, that those changes are going to be bad, that they're going to accelerate as we move forward if we don't get emissions under control, and that we're running out of time to prevent the worst-case scenarios from occurring."

UCLA climate scientist Daniel Swain summed it up as, "Climate change is like being on a train, not a runaway train where the brakes don't work, but a train where the brakes are perfectly functional, but the conductor is just actively choosing not to apply them. So if we choose to apply the brakes, the train will slow down and come to a halt. But so far, we're still just thinking about tapping the brakes lightly. It's not enough."

The climate crisis is upon us, and we must break humanity's addiction to fossil fuel. The United States is the number two emitter of GHG, but we are the country that developed petroleum as a replacement for whale lamp oil back in the nineteenth

---

[1] https://www.yahoo.com/news/scientists-express-doubt-that-glasgow-climate-change-conference-will-be-successful-090020695.html

Fig. 6.2 U.S. greenhouse gas emissions by economic sector in 2019. (Source: U.S. Environmental Protection Agency (USEPA) public domain)

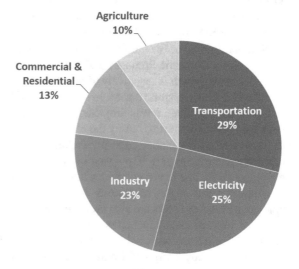

U.S. Environmental Protection Agency (2021). Inventory of U.S. Greenhouse Gas Emissions and Sinks: 1990-2019

century. Thus, the U.S. has a special obligation to lead the way out of the climate crisis that Colonel Drake's oil well eventually precipitated.

American leadership in technology and innovation in energy is a critical element of this. Zero net carbon energy sources like biofuels, wind, and solar are already commercial, but all can use improvements in energy efficiency and reduced manufacturing costs to expand their use. Advanced new energy sources, such as engineered geothermal and thorium-based nuclear are on the horizon at moderate to high technological readiness levels. (These are discussed in later chapters.) If U.S. researchers and labs were provided with serious funding to commercialize new, sustainable energy resources, these could be as big a game-changer as Samuel Kier's invention of the kerosene lantern. We all share the same atmosphere, and all nations of the world must work together to end our dependence on fossil fuels.

Somebody has to lead the way, and it should be the United States. Because if we don't, others will. It happened with the USSR and Sputnik, it happened with the Japanese and fuel-efficient vehicles, and it happened with the Chinese and consumer electronics. Replacing fossil fuels with sustainable, clean, carbon-free energy will be what economists call a disruptive technology. The thing about disruptive technologies, like smart phones for example, is that the people who are on the correct side can make a ton of money while those who are on the wrong side lose a ton of money. Apple made a fortune off the I-phone. Because it can make phone calls, view the internet, and especially take pictures and videos, people stopped using film for photography and Eastman Kodak nearly went bankrupt.

Replacing fossil fuels with new energy technology could be an enormous market worth trillions. Removing CO_2 from the atmosphere for use as chemical feedstock could also be a trillion dollar industry. We really should not be sitting on our hands here.

Those who criticize the cost of dealing with climate change should understand that the cost of NOT dealing with climate change is far steeper. A 2019 study by the U.S. National Bureau of Economic Research[2] used data from 174 countries to track the world Gross Domestic Product (GDP) on a per-person basis between 1960 and 2014. Economic growth is negatively affected by persistent temperature anomalies either above or below historic norms. In the absence of GHG mitigation policies, the average global temperature was projected to steadily rise by 0.04 °C per year. The effects vary across countries, but by the end of the century the higher temperatures will have reduced the global GDP per capita by 7.22%. In contrast, limiting the temperature increase to 0.01 °C per year by adhering to the Paris Agreement reduces the loss to 1.07%. The report concludes that climate change has long-lasting, negative effects on the output of various economic sectors, labor productivity, and employment. Disaster costs are also projected to rise as the frequency and intensity of extreme weather events increases. The U.S. General Accounting Office reports that federal disaster assistance since 2005 has topped $450 billion.[3]

People who study energy and climate change mostly agree that the following four simultaneous and non-exclusive approaches are needed to switch energy from fossil fuel to zero carbon resources: (1) implement a global carbon dioxide tax or "carbon pricing" on fossil fuels, (2) decarbonize the economy and focus research on sectors that are more difficult to decarbonize, (3) develop affordable, non-fossil energy alternatives for emerging economies, and (4) continue to innovate the technology for sustainable, non-carbon energy sources to improve efficiency and reduce costs. These are addressed separately below.

Carbon Tax

A carbon tax has been discussed in the U.S. for decades. Predictably, it has drawn strenuous objections from the fossil fuel industry. Since 1990, about 50 different proposals for carbon pricing have been introduced in the U.S. Congress.[4] Only one of those came close to passing – a "cap and trade" bill in 2009 that made it through the House but was then defeated in the Senate. Intense fossil energy industry lobbying against such efforts and resistance from Members of Congress who represent

---

[2] Kahn, M.E., Mohaddes, K., Ng, R.N.C., Pesaran, M.H., Raissi, M., and Yang, J-C, 2019, Long-Term Macroeconomic Effects of Climate Change: A Cross-Country Analysis: National Bureau of Economic Research, Working Paper 26,167, DOI 10.3386/w26167; Issue Date: August 2019; https://www.nber.org/papers/w26167

[3] https://www.gao.gov/products/gao-20-183t

[4] https://priceoncarbon.org/business-society/history-of-federal-legislation-2/

gas, oil, and coal producing states have made each proposed law a steep, uphill battle. The main excuse is that a carbon tax will kill jobs, and that the so-called "social cost of carbon" will ruin the economy. Neither has been shown to be true.

The main reason for a carbon tax is to raise the cost of fossil fuels to levels that are competitive with the cost of renewables. Electric power has two types of expenses. Capital expense or "CAPEX" is the cost of physically constructing a power plant. Operating expenses or "OPEX" are the day-to-day costs of running the plant, including funds for such things as fuel, monitoring, and maintenance. The "levelized cost of electricity" is used to compare electric prices per kilowatt/hour from various generating sources and includes both CAPEX and OPEX.[5]

Because of the pricing structure for energy, fossil fuel has always been cheaper than just about anything else. After the 1973–1974 energy crisis, a great deal of government research focused on developing more efficient renewable energy like wind, solar, and biofuels with the goal of making these energy sources more cost-competitive with fossil fuels to expand their use. Substantial efforts attempted to lower costs by improving efficiencies to reduce both the CAPEX and OPEX of renewables. This succeeded up to a point, but commercial solar and wind power in the United States still requires tax credits to be cost-competitive with coal and gas-fired electricity. Biofuel ethanol that is widely used as a gasoline additive is subsidized as well. Not many people realize that fossil fuel was also subsidized after the energy crisis in an effort to expand domestic oil and gas production. This subsidy has stubbornly hung on despite numerous attempts to kill it. Coal, petroleum, and natural gas remain cheap and abundant sources of energy and their costs undercut renewables.

Instead of struggling to bring down the cost of renewable power to compete with cheap fossil fuels, raising the cost of fossil fuels with a carbon tax is another way to achieve cost parity. A carbon tax places the environmental costs of GHG directly on the people who use fossil fuel. Higher fossil energy prices will persuade utilities and consumers to use more renewables. A carbon tax will also boost the conservation of fossil fuel, encouraging people to stop wasting it and use less because of the cost, which can only be helpful for the climate.

Funds provided by a carbon tax can be used to pay for tax credits for CCS. Thus, a power company that pays a steep carbon tax to burn coal will recover some of that money if GHG emissions are captured at the stack and stored deep underground away from the atmosphere. Under a properly constructed carbon tax/CCS credit scenario, companies that insist on continuing to generate electricity with fossil fuels will find that obtaining CCS tax credits from capturing their emissions is more cost effective than just allowing the GHG to waft away into the atmosphere.

The ultimate goal for addressing climate change is to have utilities and consumers switch from fossil energy to carbon-neutral or carbon-zero energy. One of the criticisms of CCS by the environmental community is that it will be used to support

---

[5] U.S. Energy Information Administration, 2021, Annual Energy Outlook 2021 with projections to 2050: U.S. Department of Energy, Washington, DC, 33 p. (https://www.eia.gov/outlooks/aeo/).

the continued combustion of coal and natural gas to produce electricity when we should be weaning ourselves off these fuels. If the economics of a carbon tax and credit program are structured such that coal-fired utilities receive some payment for every ton of carbon they capture but still suffer a net revenue loss from a carbon tax on coal purchases, a financial incentive will remain for switching from coal to sustainable zero-carbon energy. Tax and credit incentives have been very effective in other sectors of the economy.

The tax credit at present for each metric ton of carbon captured and sequestered as specified in Section 45Q of the Internal Revenue Code varies from less than $12 up to $50. Carbon capture costs using current technology range between $100 and $200 per ton, so only getting $12–$50 for it is not much of an incentive. There are proposals in Congress to raise the 45Q tax credit to encourage more carbon capture and sequestration. New technology under development may reduce capture costs below $100 per ton, further improving the incentive.

Climate change skeptics argue that a carbon tax or any other attempts to reduce the use of fossil fuels will disproportionately affect the poor. This wokewashing is a bald faced attempt to frighten poor and minority communities into being opposed to policies that address climate change. The goal is to threaten the most economically vulnerable with sky high electric bills and astronomical gasoline prices. There are several ways to address this, including means-tested subsidies from carbon tax revenues for power and fuel.

Oil refineries, coal processing plants, coal-fired power plants and other fossil energy infrastructure are not generally located out among the country clubs and equestrian estates of the rich folks, but in poor communities that already face major environmental justice issues. Coal mines strip the land in small, poor Appalachian communities and leave devastation in their wake. Refineries vent a host of foul smelling and hazardous chemicals into the air. Old oil wells leak methane gas, volatile organic compounds, and sometimes even hazardous gases like hydrogen sulfide into the air. Not all of these wells are located out on the Oklahoma prairie, either. There are literally hundreds of abandoned oil wells within the City of Los Angeles that were drilled in the 1920s and are now leaking hazardous gases into people's yards, under streets and driveways, and next to houses in poorer, older neighborhoods.

People living in these various enclaves suffer from a host of ailments including asthma, skin rashes, heart and lung problems, and higher cancer rates caused by the pollution of air, water, and landscapes. These poor communities will also bear the brunt of climate catastrophes such as heat waves, monster hurricanes, and sea level rise. Economically-disadvantaged places where people are already living close to the edge have little built-in resilience to overcome disasters from the climate crisis. It is in their best interests to support the energy transition away from fossil fuels.

My preferred option for a carbon tax on petroleum is to place it on the price of new gasoline and diesel-powered vehicles instead of the pump price of fuel. This would only affect those buying a new vehicle and make electric and other alternative vehicles more cost competitive. It would not change the prices at the gasoline pump, and average people filling up their vehicles to get to work would never see it.

But over time, the number of fossil-fuel vehicles on the road will decrease through attrition as old vehicles wear out and are replaced with carbon neutral or zero emission vehicles. These will be much cheaper than diesel and gasoline vehicles carrying high sticker prices from a carbon tax. Anyone who really wants a diesel truck can still get one, but it will be costly. In many cases though, people will opt for a cheaper electric truck if abundant recharging options are available, and advanced batteries give it a similar range and power. The electric version may not "roll coal," but it will help to make the climate crisis less bad. When fossil fuel vehicles fade away, so will fossil fuel.

Carbon pricing schemes that force utilities to cut carbon dioxide emissions are currently in place in only a few countries or regions. The rules for taxing fossil fuel carbon must be made uniform across global markets and enforced worldwide. Otherwise, situations like the Standard Oil Trust being banned in Ohio and reconstituting itself a few years later in New Jersey will be repeated on an international scale. Carbon-intensive industries will just move to the place with the most lenient rules and set up shop there. We all share the same atmosphere and there can be no safe haven on Earth for GHG polluters.

Another unresolved issue is the value of carbon offsets and the rules for using these. Carbon offsets are "climate credits" purchased by individuals or companies to invest in atmospheric carbon dioxide removal (CDR) projects like tree planting that are supposed to help balance out their own emissions. Carbon offsets are typically sold, for example, by travel agents that allow passengers to offset the GHG emissions of a long airplane flight. The traveler has no proof that the money was spent on CDR as promised, but simply trusts that a tree or two will be planted somewhere in the world to suck up an equivalent amount of carbon dioxide emitted by their presence on a jet aircraft. (The airline industry, by the way, only contributes about 3% of the annual global emissions of GHG.) The sale and value of carbon offsets are largely unregulated and often not very transparent, and unsurprisingly there is a potential for fraud. A lot of money goes into offsets with little accounting for how and where it is spent. Some offsets claim to be verified by "international carbon standards," but there is no official international standard for carbon and the companies that provide this verification are private businesses. Carbon offsets are a contentious global issue that requires international regulation and public oversight.

The world must come together and agree on the need for carbon pricing, the cost of a carbon tax, and the value of a carbon sequestration credit. The carbon tax must be steep enough to incentivize the substitution of renewables for fossil fuel. Sequestration credits must be substantial enough to encourage both CCS and CDR to either stop GHG from entering the atmosphere or remove what is there, but not so high that they make it profitable to continue to use fossil fuels. The carbon offset market must be regulated and monitored. Net zero climate goals have been announced by many countries from Japan and South Korea to the U.S., China, India, and the European Union. The world seems primed to agree on a carbon tax if the details can be worked out.

Decarbonize

Replacing fossil fuel with non-carbon sources of energy, known as decarbonization, is easier in some parts of the economy than others. Based on the emissions chart shown previously in Fig. 6.2, decarbonization of electric power and vehicles should be the top priorities, followed by industry, commercial/residential, and agriculture. Approaches for decarbonizing electricity and vehicles are discussed in detail in separate chapters that follow because these are the most urgent and there are multiple options. The industrial sector is more challenging, with the cement and steel industries being especially problematic. These industrial operations emit significant amounts of CO_2 as part of the process for making their products. Addressing those emissions will require creative thinking and research. Commercial and agricultural emissions are less critical but should still be addressed.

Options for decarbonizing the economy include replacing fossil fuels with wind, solar, biofuels, hydrogen, new technology nuclear, and engineered geothermal. Each of these has advantages and drawbacks, but none of them add fossil carbon to the atmosphere as a byproduct of making energy. Most of these technologies are already available, although many could use improvements in efficiency and cost. The ones that are not available are technologically close, or as DOE would put it, their Technology Readiness Level (TRL) is high.

The tall, white, three-bladed propeller wind turbines that have become common sights in many parts of the United States produce about two megawatts (MW) of electricity each. The ones in Texas tend to be larger (of course), producing four MW each. The structure on top of the tower that holds the generator is called a nacelle, and it also contains an onboard computer that rotates the whole assembly to keep the blades facing into the wind as wind direction shifts. The computer also adjusts the angle of the three blades, so that the rotation is more-or-less constant at different wind speeds. "Feathering" the blades turns them parallel to the wind and the turbine can be stopped if power is not needed or it requires servicing. The blades rotate at 10–20 revolutions per minute, which appears slow from a distance but the outer tip of a 120 foot-long (36 m) blade is typically moving at speeds in excess of 100 mph (160 km/h). Flying creatures like birds and bats often misjudge this speed and are hit, with thousands killed every year.

Estimates of the number of birds killed by onshore wind turbines in the United States vary widely, but half a million annual fatalities is a typical number. Bird deaths from offshore wind turbines, of which the United States presently has very few, are completely unknown. No one has been able to do an accurate count because bird corpses are carried away by ocean currents or devoured by voracious fish. However, the totals can be assumed to match the fatality levels of turbines on land. Birds are killed by collisions with power lines or electrocution, so the power lines servicing these turbines are also a factor. Appropriately siting wind turbines in areas away from common bird flyways will help reduce the mortality. Although the number of wind turbine bird fatalities sounds high, it is important to note that far more birds are killed annually from collisions with radio towers and glass windows in tall buildings (reportedly 600 million) and from predation by cats (reportedly 2.5 billion) than by wind turbines.

Wind turbines have an issue with the "not in my backyard" (NIMBY) syndrome prevalent in America where people want infrastructure, just as long as it is located someplace else. Onshore wind farms are usually located in remote areas away from cities. This is done in large part because the wind is more reliable on open stretches of land but transmitting electricity over long distances results in losses. Power engineers would love to place them closer to population centers, and one idea for coastal cities is to put them offshore.

The notion of offshore wind turbines initially raised the ire of people who own coastal property and objected to a spinning turbine spoiling their expensive ocean view. In response, designers moved the facilities farther offshore to a distance of about 10 miles or 16 km out to sea, where they are hidden from view by the horizon and the curvature of the Earth. This placated the landowners but caused problems with commercial fisheries, who complained that the rows of turbines block access routes to fishing grounds and scare away fish. The fishing industry has vigorously protested both proposed and constructed offshore wind facilities at Martha's Vineyard and coastal New Jersey. Offshore wind turbines are common in Europe, however, with large wind farms off the coasts of the U.K., Denmark, and other countries.

An option not discussed much in the United States but applicable to coastal regions is obtaining energy directly from the ocean. As a maritime nation, Japan has been investigating ways to harness energy from the sea, including tidal power, ocean thermal energy conversion (OTEC), and placing submarine turbines in the path of strong ocean currents. The turbine option has actually undergone some field tests and appears to be viable. The prototype device called Kairyu resembles an airplane, with two propeller-like turbine fans and a central unit that houses a system to adjust buoyancy. It is designed to be anchored on the sea floor at 30–50 m depth (100–160 feet) and transmit the power to shore via seabed cables. This could be an invisible alternative to offshore wind in some U.S. locations. The Japanese intend to anchor the production model turbines in the Kuroshio Current, one of the world's strongest, which runs along Japan's eastern coast. If the Japanese can make this work, it could produce hundreds of GW of electricity, replacing virtually all of their fossil fuel power plants.

Solar energy has two options at present. Solar photovoltaic (PV) panels generate electricity directly from sunlight, and solar thermal systems use solar heat to generate electricity. Photovoltaic panels work by using a physical principle known as the "photoelectric effect," first defined by Albert Einstein. He actually won the Nobel Prize for this in 1921, and not for his more famous work on relativity. Einstein discovered that the photons making up the energy packages in light can knock electrons loose from their orbits when the light hits certain materials. Photovoltaic cells made from thin wafers of semiconductor material like silicon use this principle to allow light to drive the loose electrons deeper into the material, where they accumulate in a lower layer. Thus, the top layer of a solar cell becomes positively charged in sunlight because it loses electrons, and the bottom layer becomes negatively charged because it gains electrons. Connecting the top and bottom layers with a conductor like copper wire creates an electric current. Each individual solar cell only produces a small amount of current, which is why they are typically placed in large arrays on "solar farms" for commercial power generation.

Solar thermal power uses mirrors to focus sunlight onto a central pylon or tower to create heat. This heat is then used to make steam and generate electricity. Solar thermal arrays tend to be vast, with literally acres of mirrors focusing sunlight onto a single spot. Commercial solar energy has some of the same problems as wind energy. It often requires large amounts of land area, and although people want it, they generally don't want it in their backyard. In recent years, the Mojave Desert between Las Vegas and Los Angeles has seen the development of both solar PV farms and a large solar thermal power station.

Biofuels are made from plants that remove carbon dioxide from the air and use sunlight-driven process called photosynthesis to build carbohydrates and other plant structural material from CO_2 and water. Because plants take the CO_2 out of the air, they serve as a CDR system. Biofuel is considered to be carbon-neutral or "net-zero" because burning the plant material for energy simply returns the CO_2 back to the air from whence it originally came and the net amount of CO_2 in the atmosphere does not change. If the emissions are captured and sequestered in a CCS process, then the energy becomes carbon negative and there is a net removal of carbon dioxide from the atmosphere. Such a process is known as BECCS: biofuel energy with CCS.

Biologically-produced methane, or biogas, is a net-zero carbon fuel that is a direct substitute for natural gas, since natural gas is almost totally methane. Biogas methane has the same composition and heating properties as natural gas. It is the simplest biofuel to make because a class of microbes known as methanogens can generate it from nearly any kind of organic material, including municipal sewage (Fig. 6.3), agricultural waste, feedlot waste, compost, and landfills. So-called "renewable natural gas" is extracted from landfills.

Liquid biofuels can be made from marine algae, vegetable oil, or by fermentation. The ethanol currently being added to gasoline is a biofuel obtained from the fermentation of corn. A number of airlines have stated their intention to replace jet fuel, which is essentially kerosene, with a liquid biofuel in the next few years. Pilots have expressed some concerns about the performance of biofuel under the pressure and temperature regimes typically encountered during a flight, and they want assurances that the biofuel will not solidify into shortening under the extremely cold temperatures present at high altitudes. A vegetable oil-type fuel might do that, but if the biofuel is formulated properly to reproduce the exact chemical structure of kerosene, the performance should be identical.

Decarbonizing vehicles and certain industrial segments of the economy will be more challenging and require affordable replacements for fossil fuels, such as biofuels, hydrogen, or long range batteries for electric vehicles. Decarbonizing electricity is an important first step because charging an electric vehicle with power from a coal-fired power plant emits a substantial amount of CO_2. Once the electric

Fig. 6.3 Methane biogas being flared off (upper right) at a municipal wastewater treatment plant. (Source: Photographed in 2021 by Dan Soeder)

power sector is decarbonized, the decarbonization of vehicles can follow. Technology innovation in the vehicle sector can provide enormous gains in GHG reduction.

Decarbonized vehicles don't all have to be electric. Methane biogas can be run in modern gasoline-powered vehicles with only minor modifications. An existing gasoline-powered vehicle can be converted for a few thousand dollars to run on bio-methane. Unlike electric vehicles, drivers are not required to purchase something new and expensive. The details of decarbonizing the transportation sector of the economy are addressed in Chap. 8.

Two other segments of the economy need to be decarbonized along with electricity and transportation: (1) certain industrial processes and (2) the residential and commercial use of fossil fuels. Among industrial processes, the manufacture of concrete is the worst, accounting for more than 8% of global CO_2 emissions in some estimations. Second place goes to the steel industry, which may be responsible for an additional 5%. The third largest offender is the chemical industry. Decarbonizing these industries will be challenging but necessary.

Concrete emits substantial amounts of CO_2 during the manufacturing process and absorbs some of that CO_2 back when it cures. Concrete consists of a gravel and sand aggregate bound together with a material called Portland cement. Cement is made by heating crushed limestone, clay and several other ingredients in a kiln to a temperature of more than 1425 °C (2600 °F). The calcium carbonate ($CaCO_3$) making up the limestone dissociates or "slakes" under this high temperature into CO_2 gas and solid calcium oxide (CaO) through a process called calcination. Calcium oxide is used as a component of the cement, but the CO_2 generated during calcination is considered a waste product and is typically vented to the atmosphere. The kiln itself also can be the source of additional CO_2 emissions if it is heated using fossil fuel (as most are).

Concrete is a versatile, durable, and popular construction material. The world-wide production of concrete is responsible for approximately 8.6% of all anthropogenic CO_2 emissions because of increased global urbanization and subsequent infrastructure construction.[6] As concrete cures, atmospheric CO_2 reacts with the CaO in the Portland cement to re-generate calcite or similar minerals that make concrete as strong as a rock. Although fresh concrete will harden in hours, a full cure to completely mineralize the cement takes weeks to months. If the cure is done properly, essentially all of the CO_2 released from the cement calcination process will be re-absorbed by the concrete. In this sense it is carbon neutral, except for the emissions from the kiln used to heat the cement ingredients during the production process.

Firing the kiln with biofuel or installing CCS to capture CO_2 emissions from fossil fuel combustion will make concrete carbon-neutral. If the CO_2 emissions from the limestone slaking process are captured and sequestered as well, the concrete will become carbon negative as it absorbs CO_2 from the air while curing. This process could help reduce atmospheric GHG levels.

The cement curing reaction can be enhanced by introducing CO_2. This can be achieved by dissolving carbon dioxide into the water that is used to make the concrete flow like a liquid when poured, leave channels in the concrete that allow air to enter, or by forcing CO_2 into the hardened but not yet cured concrete under pressure. Adding CO_2 helps to improve the cure and produces a stronger, more durable material.

The production of steel requires that iron ore, which is composed of iron oxides, have the oxygen removed to produce pure iron. That iron then has a small amount of carbon added to give it strength and resiliency and create the alloy known as steel.

Removing oxygen from the iron ore is an energy-intensive process called reduction. Iron binds tightly with oxygen (which is why steel rusts) and removing it requires adding a material called a reductant that will bind to the oxygen more tightly than the iron and carry it away. In standard steelmaking, the reductant of choice is carbon in the form of coke, which is basically purified coal with most of the volatiles driven off. The iron ore, reductant, limestone and a few other ingredients are heated until molten, usually over a hot coal fire, and then placed into a device called a blast furnace. This is a vertical smelter that blows pressurized air up through the molten iron ore to ignite the coke and encourage it to scavenge oxygen away from the iron ore and carry it off as carbon dioxide or carbon monoxide. These are vented to the atmosphere.

The more-or-less pure iron left behind in the blast furnace is called pig iron and is put through a subsequent process to make steel. The blast furnace was invented in

[6] Miller, S.A., Horvath, A., and Monteiro, P.J.M., 2016, Readily implementable techniques can cut annual CO_2 emissions from the production of concrete by over 20%: *Environmental Research Letters*, v. 11, p. 074029.

China in the fifth century BC and was widely adopted in Europe during the Middle Ages. A modern version known as the integrated blast furnace/basic oxygen furnace is used in present-day steel mills to make both pig iron and then steel in the same vessel.

The original process for mass producing steel from pig iron was patented by Henry Bessemer in England in 1856. The Bessemer process works to purify the pig iron by burning off virtually all of the carbon and other impurities. Then a precise amount of carbon, manganese, and silicon are added using an alloy material called spiegel. The finished product is a strong but malleable steel. Modern steelmaking takes a shortcut by using the basic oxygen furnace, which is the same vessel as the blast furnace but just configured somewhat differently. The pig iron is melted a second time without any additives or reductants like coke, and oxygen is blown over the surface to oxidize the remaining carbon and decrease the carbon content from about four per cent to less than one per cent. At this point, it becomes steel.

Another, less GHG-intensive process for making steel uses a device called an electric arc furnace. These make steel from scrap metal or direct-reduced iron. The direct reduction of iron removes oxygen chemically from iron ore in the solid state, without melting. The reducing agents are gases, usually carbon monoxide and hydrogen that scavenge the oxygen away from iron ore pellets. The pure iron remaining behind can be melted down and converted to steel in the electric arc furnace. This furnace uses electricity to produce the heat needed to melt iron, and the electric power can be provided by renewables like wind turbines or solar.

The industry currently produces around two billion tons of steel each year, along with more than three billion tons of CO_2, mostly from blast furnaces. There are a number of improvements to the steelmaking process that can reduce GHG emissions. The first is obviously that the CO and CO_2 carrying off the oxygen from the iron ore should be captured and sequestered instead of vented to the atmosphere. A second improvement would be to use a carbon source that is not derived from fossil fuel like coke. Biofuel carbon such as charcoal or biogas may waft up into the atmosphere from a steel mill, but at least it is not fossil fuel carbon adding to the GHG burden. A third improvement would be to increase the use of electric arc furnaces over blast furnaces, recycle more scrap steel, and use more directly reduced iron. All phases of the steel industry could benefit from efficiency improvements in the use of energy and materials.

Probably the most significant improvement to the climate-damaging aspects of steel production would be to forgo the use of carbon-based reductants altogether in favor of hydrogen-based reductants. Hydrogen readily combines with oxygen and forms water. Steelmakers in Sweden have been working on a venture using green hydrogen to make steel at a pilot plant. Substituting hydrogen for carbon reductants could cut the steel industry's emissions by around 90% and make a significant dent in global GHG emissions.

The chemical industry is the largest industrial consumer of oil and gas, about half of which is used for feedstock. It is also the largest overall consumer of energy. The consumption of both energy and feedstocks are propelled by the increasing demand for a vast array of chemical products.[7] These include plastics for packaging, construction, and automotive applications and ammonia as a base ingredient for chemical fertilizers. Recycling of plastics counterbalances only a small share of the global demand. The manufacture of plastics requires the use of petrochemicals such as ethane to make polyethylene plastic. Known as "high value chemicals" plastic feedstock includes benzene, toluene, and mixed xylenes. The United States, the Middle East, and the Asia Pacific region produce the bulk of high-value chemicals.

The production of methyl alcohol (methanol) is rising the fastest of all primary chemicals. Reasons for this include its use to make formaldehyde for the production of specialized plastics and coatings and applications as a fuel. Methanol is also used as an intermediary to produce high-value chemicals when petroleum is not available. The bulk of growth in methanol production capacity is concentrated in the Asia-Pacific region.

The chemical industry needs to reduce energy consumption to reduce GHG emissions. More efficient processes in the manufacture of chemicals are required to cut back on CO_2 emissions and also to meet more stringent air quality standards. The use of petrochemicals to manufacture plastics does not affect the climate *per se*, as long as they don't end up as GHG in the atmosphere. Plastics have plenty of other environmental problems, however, such as single use items filling landfills and great floating mats of rubbish in the oceans created by plastic trash that has washed into the sea.

Decarbonizing industry has an economic component that needs to be considered. If GHG regulations make the products too expensive, there is a real concern that companies might move production to locations with more relaxed regulations. This has been an issue of concern for environmental regulators since the beginnings of the EPA in 1970. To avoid stringent U.S. air and water pollution regulations, some companies have moved production to other countries with much more relaxed laws. This has occasionally resulted in disasters, like the methyl isocyanate gas leak from a Union Carbide pesticide plant in the town of Bhopal, India in 1984 that killed thousands. The release was blamed on insufficient environmental and safety oversight from poorly trained and lax management.

The standards for embodied emissions of GHG must be globally uniform so that no location has an economic advantage in this area. The European Union has proposed a carbon emissions tax on imports, collecting money at the border to equalize production costs and remove the incentive for more cheaply manufacturing products with high embodied emissions in other countries to undercut European producers. The U.S. is also considering such a policy.

Along with electrical power, transportation, and industry, another sector of the economy that needs to be decarbonized is the domestic use of fossil fuels. This is a bit less critical in terms of GHG volume than the others, but it still contributes a significant amount to the atmosphere (refer back to the pie chart in Fig. 6.2). Home use of coal, natural gas, propane, and fuel oil for heating, hot water, and cooking

---

[7] https://www.iea.org/reports/chemicals

may not emit much GHG on a per-household basis, but when these are added up across an entire population the numbers are significant.

The EPA greenhouse gas inventory considers the residential and commercial sectors to encompass all homes and commercial (non-industrial) businesses. GHG sources from this sector include scope 1 emissions from fossil fuel burned for heating (including hot water) and cooking needs, management of solid waste and wastewater, and leaks from refrigerants. Scope 2 emissions are primarily associated with electric power generation from fossil fuel that is consumed by homes and businesses.

Some other businesses may also produce and sell products that have scope 3 emissions, but these are considered minor compared to the transportation sector. One example that stands out is bottled soda on the East Coast, where most of the carbon dioxide used to carbonate the beverages is obtained by capturing and purifying the CO_2 emissions from a small, coal-fired power plant in western Maryland called Warrior Run. I've actually visited Warrior Run on a tour and the process for turning coal combustion flue gases into food-grade CO_2 is fascinating and represents the economic utilization of carbon dioxide. This will be discussed in more detail later but any source of income that can be found for CO_2 helps to change it from a waste product into a commodity and encourages carbon capture. Nevertheless, the carbon source at Warrior Run is coal and every time someone on the East Coast pops open a can of soda fizzing with CO_2 from there, it is technically a scope 3 emission.

The combustion of fossil fuels for heating and cooking emits CO_2, methane (CH_4), and nitrous oxide (N_2O). Natural gas combustion was responsible for about 80% of the direct CO_2 emissions from fossil fuel in the residential and commercial sectors in 2019. Home heating oil comes in second. Coal consumption is a minor component of energy use in both of these sectors in the United States, but it is much more important for residential use in China, India, and other nations.

Solid waste from residences and businesses generates CH_4 in landfills, and wastewater treatment plants emit methane as well. If this is flared off (refer back to Fig. 6.3), it emits CO_2 and N_2O. Neither the CO_2 nor methane from these sources are derived from fossil carbon, so GHG from waste products theoretically does not affect climate; however, it does add to the total inventory of GHG in the atmosphere.

Fluorinated gases such as hydrofluorocarbons (HFCs) in air conditioning and refrigeration systems are also powerful greenhouse gases. They can be released during servicing or from leaking equipment.

Achieving the 2050 Paris Agreement for GHG reductions cannot be met in the United States by only decarbonizing electric power. It requires broad reductions across multiple sectors of the economy. The U.S. household sector alone emits more GHG from heating, cooling, and electric power use than entire countries like Brazil or Germany.[8] Decarbonizing electricity will reduce much of this, but the domestic use of fossil fuels must also be phased out.

[8] Goldstein, B., Gounaridis, D. and Newell, J.P., 2020, The carbon footprint of household energy use in the United States: *Proceedings of the National Academy of Sciences*, v. 117, no. 32, p. 19122–19130; DOI: 10.1073/pnas.1922205117

A number of cities in fact already have banned the use of natural gas in new construction and some states are considering it as well. Berkeley, California, became the first city in the United States to do so in 2019, followed by other California cities like San Jose, Mountain View, Santa Rosa, and San Francisco. Seattle banned new natural gas hookups in 2021, as did New York City with a gradual phase-in by 2030. The gas industry is fighting back fiercely, and numerous states are pursuing legislation that restricts the power of local governments to enact energy bans that single out natural gas providers. Lawsuits and countersuits are flying back and forth, and the only people actually benefitting from all this at the moment are the attorneys.

I think banning natural gas hookups is the wrong approach. For one thing, what if people just went out and got propane tanks and used that instead? Propane is great for cooking and efficient for heating. It requires modification on the burners because it burns hotter, and it also emits more greenhouse gas per Btu. To avoid changing the burners, I could see some clever entrepreneur compressing natural gas into cylinders outside the city, and then selling the tanks in town where gas utility hookups are banned.

It would be far better if natural gas companies switched from fossil fuel to biogas. Biologically created methane has the exact same physical and chemical properties as natural gas produced from the ground and could be a direct substitute. The difference of course is that the carbon comes from the atmosphere, not from a fossil source where it's been packed away for millennia. Here again, a tax on fossil carbon that raises the price of natural gas to make biogas more cost-competitive would encourage the switch-over.

The other major approach to decarbonizing the residential and commercial sector is to convert as much energy use as possible to carbon-neutral electricity. Anything that can be done with natural gas can also be done with electricity. Heating and cooling with electric heat pumps, especially with a geothermal loop to provide constant temperatures is more efficient than heating with gas and cooling with electricity. Electric power can also be used to cook, although many cooks (including me) don't like it and prefer gas, and it will also heat hot water. For those who insist on cooking with gas, biogas could be substituted for natural gas.

The final sector that is a source of GHG is agriculture, which is the least serious emitter of all of them. The manufacturing process for fertilizer uses natural gas and emits CO_2, which technically falls under chemical industry emissions. Fossil-fuel powered agricultural machinery also contributes CO_2, but this is orders of magnitude lower than the transportation sector. Other GHG contributions from agriculture do not involve fossil fuels, and therefore cannot be decarbonized. These include methane emitted by cattle from a process in their digestion called enteric fermentation. This adds to the total GHG load in the atmosphere, but it does not contain fossil carbon, so it is actually biogas. Another source of methane is manure management operations that allow the manure to go anoxic, which provides conditions where methanogens thrive. Some soil management practices can release significant amounts of N_2O, and nitrous oxide accounts for about half of agricultural GHG emissions according to the EPA.

Decarbonizing the electricity, transportation, industrial, and commercial/residential sectors of the economy is important if we are to eliminate our dependence on fossil fuels. It will not be easy, and sadly it will not be enough to avoid climate change. Humanity has already added enough GHG to the atmosphere to put us across a threshold that will raise planetary temperatures past the 1.5 °C goal of the Paris Agreement and probably above 3 °C before any climate mitigation measures begin to have any effect. Other actions will have to be taken to avoid the worst impacts of climate change. These are discussed in Chap. 9.

Emerging Economies

The most carbon-intensive fossil fuel is coal, and it needs to be replaced quickly with zero net carbon alternatives to meet the Paris Agreement and other U.N. climate goals. Unfortunately, many newly-emerging economies, particularly China and India, are relying on coal-generated electricity to drive their burgeoning manufacturing sectors. More than half of global coal production is consumed in China, and over the past two decades the Chinese contribution to global greenhouse gas concentrations has soared (Fig. 6.4). In 2020 alone, China built three times more coal-fired power capacity than all other countries in the world combined.[9] India is poised to quickly ramp up into second place as that country steadily industrializes. India is the world's second largest coal producer and consumer after China.

The Asian dependence on coal will be challenging to overcome. More than 80% of the new coal-fired power plants planned in the world are located in China, India, Indonesia, Japan and Vietnam. China is not just building power plants at home but has also financed 13% of coal-fired power capacity abroad between 2013 and 2019. China announced in 2021 that it was getting out of the foreign coal power plant financing business, as are Japan and South Korea. While this is good news, it should also be noted that in 2020 China added as much new coal-fired electricity at home as it potentially canceled abroad. However, the long-term outlook for coal-fired power is that it is an old technology on the way out.

China is also a leader in renewable energy, developing about half of the world's renewable energy capacity in 2020. The country also leads in green technologies such as electric vehicles, batteries and solar PV power. Chinese-manufactured solar panels dominate the market in terms of cost and performance. The key for China to move away from coal fired power is to replace the heat sources in their thousands of thermoelectric power plants with non-fossil energy like engineered geothermal or new technology nuclear, described in detail in the next chapter. These will allow the massive Chinese investments in coal-fired generating and transmission infrastructure to remain intact by simply replacing the heat from burning coal with a carbon free heat source to boil water and make the steam that drives the turbines.

---

[9] https://carbontracker.org/

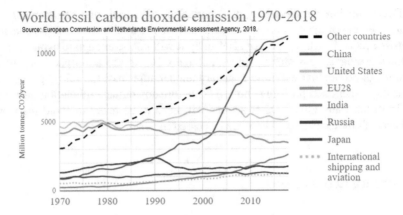

Fig. 6.4 Global carbon dioxide emissions since 1970 estimated by major contributing countries. (Source: European Commission and Netherlands Environmental Assessment Agency, 2018; public domain)

China and the United States have begun treating the climate crisis as a high-stakes competition for influence and dominance over the global energy transition away from fossil fuels. The battle lines have been drawn as each nation attempts to target competitive arenas such as solar, wind, geothermal, biofuels, and nuclear. Energy storage and zero emission vehicles are also key areas for dominance. U.S. Secretary of State Antony Blinken said in April 2021 that it was "difficult to imagine the United States winning the long-term strategic competition with China if we cannot lead the renewable energy revolution." Competition breeds innovation, and innovation breeds improvement and efficiency. The political push by the two largest economies in the world to dominate the energy transition may finally be the key to moving humanity into a global energy economy that ends the use of fossil fuels and replaces them with sustainable, carbon-neutral energy instead.

There is an environmental justice issue between developed and underdeveloped nations when it comes to the causes and effects of climate change. The bulk of anthropogenic GHG in the atmosphere was primarily emitted decades ago by the wealthy, industrialized nations. Developing nations are often asked to forgo the use of fossil fuels and curb GHG emissions without help from wealthy nations. They rightfully point out that curbing their meager emissions is a drop in the bucket compared to emissions from China, the United States and the European Union, who are "the guys that created the problem" in the words of former British Prime Minister Boris Johnson. Not addressing the climate crisis will make matters worse for everyone, but especially the poorer parts of the planet.

To add insult to injury, because of minimal infrastructure and limited social safety nets the less-developed nations often suffer the worst climate impacts from GHG they didn't emit. Industrialized nations are better suited to handling the effects thanks to resilient, widespread infrastructure. Developed countries also have disaster resilience built in through insurance, excess capital, and agencies like FEMA that are able to swoop in and offer emergency support to affected populations after an event. Much of this infrastructure does not exist in less-developed nations.

For example, heavy rainfall produced floods and some fatalities in places like New York City and northern Germany in the summer of 2021. The emergency response was swift, with many stranded people being rescued and the damage cleaned up and restored soon afterward. If those same storms had targeted underdeveloped countries like Somalia or Mongolia, the effects would have been devastating and cost countless more lives.

Third World nations are characterized by unpaved dirt roads, shallow fords or rickety bridges for crossing streams, an electric power grid that is unreliable even at the best of times, poor communications infrastructure, and widely scattered and often inaccessible medical facilities. Poor countries have far less resilience for handling climate change than wealthy nations and are disproportionately impacted.

Climate resilience can be a problem even in the less-developed parts of wealthy nations. On October 4, 2013, a winter storm dubbed "Atlas" by The Weather Channel and known locally as the "Cattlemen's Blizzard" struck western South Dakota, unexpectedly dumping 2–4 feet (0.6–1.2 m) of snow on the Black Hills and surrounding prairie. Because the storm hit so early in the season, cattle and other animals were still in summer pastures, and people had only minimal supplies of heating oil, propane, and firewood on hand. Tens of thousands of animals were lost to hypothermia and frostbite, devastating the South Dakota beef industry with a blow that would require years for recovery. Remote tribal reservations like Pine Ridge and Rosebud were isolated and inaccessible after the snow blocked roads for many days. People went without electrical power and had almost no fuel for heating. Fortunately, no humans died but more than a few came close.

Resilience is defined by the American military as self-sufficiency in the event of an external supply disruption. U.S. military bases are under orders to become "energy islands" that are capable of 60 days of independent operations off the grid and are investigating various renewable power options to achieve this. The municipal definition of resilience in cities and towns is the ability to maintain essential services to citizens in the event of a natural or human caused disaster. The services classified as "essential" are primarily life safety-related functions such as fire protection, ambulance service and medical care, access to drinking water and food, electrical power in critical areas like hospitals, and the ability to set up evacuation centers and shelters as warming or cooling centers in response to temperature extremes.

For many years, wealthy nations have promised to spend $100 billion a year to build climate resilience in emerging economies, but aid has consistently been about $20 billion a year short. In emerging economies or developing countries, establishing essential services after a climate disaster can be extremely difficult. Trying to access distressed areas after a major disaster with supplies and aid can be a logistical and financial nightmare when airports are flooded, roads impassable, and bridges washed out.

A more resilient plan for international aid would be to focus on setting up critical infrastructure at strategic locations in-country before a climate disaster strikes so that many of the key components are already in position. There are issues with the security of supplies and equipment stored in potentially corrupt and kleptocratic Third World countries, but there are ways to overcome this. According to Samantha

Power, the current administrator of the United States Agency for International Development, "You have to have a lot of auditing, a lot of vetting, you have to work with trusted partners. There is a sense of where money can be well spent, and where there's risk of waste, fraud and other forms of abuse."

An important step is to engage regional governments in contingency planning and to let local populations know that the relief supplies exist. In the event of a hurricane, heat wave, blizzard, or drought, the authorities know how, where, and when to respond, and the people know that such a response is possible and expect it.

Apart from episodic weather disasters, many areas of the planet that will be affected by sea level rise, killer heat waves, and extended drought are some of the poorest nations on Earth. Known as the "global south" these regions are likely to generate huge numbers of climate refugees as things go from bad to worse. Wealthy northern nations ought to be prepared to deal with this future crisis in a humane and fair manner. This may just be my opinion, but if our reckless use of fossil fuels has caused climate disruptions that end up displacing people from their homes, we have a moral responsibility to address the problem. Period.

Fossil fuel remains one of the cheapest sources of energy, and less developed countries in the world are clamoring to use more of it. Their argument is that the developed countries of the world were able to use fossil fuel over the past century to become wealthy (refer back to Chap. 3), and now they want their turn. Umm, yes. Well, this is awkward.

The Earth's overheated climate can't tolerate the fossil fuel we are already burning and using more would just make everything much worse. We need to gently explain to the Third World that fossil fuel is not sustainable for climate, but alternative, inexpensive, reliable, and GHG-free energy sources are being developed and will soon be distributed in the poorer regions of the world. And then we had better follow through and actually do this.

Small, hybrid solar photovoltaic (PV) and wind turbine systems can be designed for individual villages, backed up with batteries, compressed air energy storage, or pumped water energy storage for continuous energy production. In some areas, geothermal energy might be available, or solar-assisted shallow geothermal might make more sense. Biogas can be generated locally to provide a clean, carbon-neutral fuel for cooking or other uses, and this is already being done in some places. Utilizing indigenous energy resources like sunlight, wind, biomass, and geothermal in places where roads are poor, or transmission lines are long can provide a more robust and resilient source of energy. If people in these places can successfully use the available resources on hand, their energy systems will be far more reliable than depending on petroleum transported long distances over bad roads.

The Third World actually has a major advantage over industrialized nations in that there are few if any legacy systems in place that must be incorporated into a new energy paradigm. Poor countries generally don't have large infrastructure investments tied up in coal-fired power plants that will require a substitute fuel to decarbonize. They can skip the fossil fuel stage altogether and start from scratch, building smart energy systems from the ground up.

This is what happened with telecommunications. Third World villages receiving telephone service for the first time went directly to mobile cell phone systems, bypassing the hard-wired infrastructure requirements of old-fashioned landline phones and central telephone exchanges, saving a lot of money and trouble in the process. They can do the same with energy.

Technical Innovation

Technology development will be the ultimate solution to replacing fossil fuels with sustainable, zero-carbon sources of energy. The current climate crisis is largely a problem of the relatively low-level technology of fossil fuels, and more advanced technology can resolve it.

Just because fossil fuel energy technology works, it's not necessarily the best or most efficient option for energy production. This is similar to a problem faced by the early aviation industry. Propeller-equipped airplanes driven by piston engines were a workable enough technology that served from the Wright Brothers up through World War II. Propeller aircraft were limited in speed and altitude but were functional. Pre-war passenger aircraft, like the Ford Trimotor or the Douglas DC-3, were known for flying "low and slow." They flew at altitudes in the troposphere that were subjected to bad weather including turbulence, lightning, and icing and were too slow to go around or avoid severe storms in their path. There were many crashes and early aviation had a terrible safety record.

The invention of the jet engine by Englishman Frank Whittle in 1930 and its practical application on an actual flying aircraft in 1939 by Dr. Hans von Ohain in Germany changed aviation technology forever. During the war, aircraft designer Anselm Franz at Messerschmitt developed the first jet fighter, the German Me-262. Allied pilots soon learned that the Me-262 could outperform any of their aircraft in level flight, and the fastest prop-driven U.S. fighter, the P-51 Mustang, could only catch it in a dive. Fortunately for the allies, because the Me-262 jet engines consumed huge amounts of fuel, it spent most of its time on the ground in energy-strapped wartime Germany.

Nevertheless, jet engine technology advanced considerably after the war because everyone recognized that it was far superior to piston-driven propellors. New engine designs by Pratt & Whitney and General Electric in the United States and Rolls Royce in the U.K. produced high thrust jets with multiple sets of turbines and compressors that were much more efficient and used considerably less fuel. Jets have far fewer moving parts than a piston engine, which means they require less maintenance, produce considerably more horsepower, and operate efficiently at high altitudes. Modern commercial jet aircraft fly high in the stratosphere, well above most active weather, and can easily skirt around thunderstorms or other weather concerns. They also fly about four times faster than the Ford Trimotor and the DC-3. Aviation has become the safest form of travel. Similar leaps in advanced energy technology can be expected to greatly exceed the performance of fossil fuels.

Technical innovation has moved renewable electricity and electric vehicles forward much faster than predicted in the aftermath of the 1973–1974 energy crisis. Entrepreneurs like Tesla's Elon Musk, with his Silicon Valley-based development mantra of "move fast and break things" have made enormous strides toward the commercialization of electric vehicles that rival the performance and range of gasoline-powered cars. Tesla continues to innovate in battery, motor, and charging technologies and is expected to produce vehicles soon that will exceed the performance of fossil fuel automobiles. Imitation is the sincerest form of flattery, and big automakers like Ford, General Motors, and Nissan are competing with their own electric vehicles. This market is expected to grow substantially in the future as performance goes up and costs come down. Through his other company, SpaceX, Elon Musk is also advancing the technology for space travel. Vehicles are discussed in more detail in Chap. 8 and space travel in Chap. 11.

New energy technology in the form of offshore wind, engineered geothermal, new nuclear technology, carbon capture and storage, green hydrogen, biofuels, and others can fully decarbonize the electricity sector in years to come. The companies that lead in developing these new technologies will enjoy the economic benefits that may include millions of new jobs and substantial economic growth. Forward-thinking investors are putting money into developing, testing, and deploying these technologies, including a number of venture capital firms dedicated to the energy transition and the mitigation of climate change. Companies that remain stubbornly committed to fossil fuels will be left behind in the dust, hanging on to an old technology akin to videotapes that no one uses any more.

Increased government funding for research and development is needed to bring some of the more cutting edge technologies into commercialization. Engineered geothermal is one of the more obvious. This is discussed in more detail in Chap. 7, but essentially involves extracting geothermal energy at any location on Earth by drilling boreholes deep enough to reach hot rocks. The research is ongoing but woefully underfunded. The Geothermal Technology Office (GTO) at the U.S. Department of Energy had enough budget to support one field test site in Utah when it originally wanted to support several, and a few small ancillary studies are investigating the nature of fractures in deep rocks. Given the potential for the technology to decarbonize coal-fired electricity by simply drilling some geothermal wells out behind existing power plants and directly replacing fossil fuel heat sources, this research should be receiving Apollo moonshot levels of funding and attention.

The replacement of fossil fuels with carbon-neutral, sustainable energy sources will also require a systemic approach to address social issues. Acceptance of energy systems is connected to factors such as market design, convenience, cost, performance, safety, social acceptance, environmental justice, regulatory frameworks, and business models. Attempts to change energy systems with solutions that focus solely on technology will not be successful. The social and economic components are critical and must be included.

****

I will rant a bit here that climate change could have and should have been addressed back in the 1980s, when alarms first started to be raised. Some naïve environmentalists at the time assumed that if extremely wealthy and established fossil energy interests truly understood that abandoning fossil fuels was good for the planet, they would voluntarily stop selling their major products. Not surprisingly, this didn't work.

Instead, the desire for oil profits and simple greed drove political resistance against reforms to reduce GHG emissions in an earlier time window where the effect on mitigating climate change would have been greater. Forty years later, GHG is higher than ever and as far as I'm concerned the whole concept of a moral, innovative, and responsible energy industry is a sham. The fossil fuel industry has continued to do what they do best: maximize profits for shareholders. This remains their primary purpose and goal for the continued production of fossil energy. The fate of the planet? Not their department.

However, it is not entirely the energy industry that is at fault here. The climate crisis we are currently facing came about because utilities continued to generate electricity from coal and gas, the cement and steel industries never changed practices away from using fossil fuel, and nearly every vehicle produced by the automotive industry runs on petroleum. Industry apologists and anti-regulatory capitalists declared that American industry would find a way, that they would innovate and develop new technology to make products that used sustainable energy and would not harm the climate. Again, not surprisingly, this has not happened.

The U.S. government wasted the momentum from the 1973–1974 energy crisis to develop non-fossil energy sources to replace imported oil. There was some level of effort to develop renewables to be sure, but the real emphasis on improved energy security was to replace imported energy with domestic energy and the focus was on fossil. In comparison, funding for developing renewables, geothermal, and nuclear was a fraction of what was spent on fossil.

This fossil energy focus at DOE translated into millions of dollars spent on pilot plants for poorly-conceived chemical engineering experiments on coal-to-liquids, coal-to-gas, and gas-to-liquids conversions. These all worked but made expensive fuel that was unable to compete economically with real oil or gas. Additional millions of tax dollars went to develop gas and oil energy resources from coalbed methane, tar sands, shale gas, oil shale, tight gas sands, methane hydrates on the seafloor, and deep geopressured aquifers. I worked on many of these, and there were a lot of critical economic questions that never got asked.

These programs were strongly supported by the fossil fuel industry, and because DOE has a technology transfer requirement to get research results out to industry, the work was given a priority. Even the CCS work carried out more recently has been funded primarily by the coal program with support from the electric power and coal industries. Geothermal and nuclear technologies that did not have a major industry supporter were shortchanged in this push to develop fossil fuels. The technology on both of these, discussed in the next chapter, has gotten to the point where they now appear to be the best hope for economically replacing fossil fuel as zero emission heat sources in existing thermoelectric power plants. It would have been nice if we had gotten a 40 year head start on this.

Shale gas was eventually commercialized by George Mitchell with private sector money after DOE had formally given up on it. Mitchell was persistent and it helped that he owned his own oil and gas production company so he could run experiments at will. He eventually hit on the right combination of drilling and fracking to produce economical amounts of gas from shale. Although Mitchell was always careful to give credit to the earlier government research for helping him achieve his goal, it is not clear that he ever needed it and he probably would have achieved success without it.

Presidents, prime ministers, and other government officials are now expected to craft policies that will force the decarbonization of the world economy. I'm not optimistic. It remains an uphill battle against aggressive fossil energy lobbyists and vast sums of fossil energy money directly supporting politicians. No real reforms have happened to date; there has only been talk and "pledges" to decarbonize by mid-century. It is hard not to feel betrayed by governments that are supposed to protect us, the planet, and future generations.

In fact, such betrayals are negatively affecting young people throughout the world. A recent study on climate anxiety among children and young people found that government inaction on climate change has resulted in widespread psychological distress among the youngest members of society.[10] The survey of 10,000 young people, ages 16–25, across ten countries, found that increased anxiety over the fate of the planet was strongly related to perceived government inaction. Over half (56%) of young people think humanity is doomed. Nearly 60% are very worried or extremely worried about the future and more than 45% said concerns about the climate affected their daily lives. Three-quarters of respondents see the future as frightening, while two-thirds report feeling sad, afraid and anxious. They feel abandoned, betrayed, or ignored by politicians and adults, and more than half said governments are not doing enough. People are holding off on getting married, having kids, buying houses, and just living life. A little anxiety is a good motivator. Too much of it pushes people into despair and feelings of hopelessness. There is a future, but if we want it sustainable and livable, we have to work for it. All of us.

Professional climate scientists and adult activists often face similar anxieties. It is frustrating to see the world going to H-E-double-hockey-sticks (as Radar O'Reilly used to say on MASH) from a self-induced problem that we know can be fixed if people would only act on it. Many scientists cope with what is called a sense of "agency." That means seeing a path forward, even if it is narrow and dimly-lit, and doing what you can to move along it. For example, my sense of agency was to write this book and try to explain the science behind the climate crisis to the general public. For others, it might be community organizing, advocating for renewable energy

---

[10] Marks, E., Hickman, C., Pihkala, P., et al., 2021, Young People's Voices on Climate Anxiety, Government Betrayal and Moral Injury: A Global Phenomenon. *The Lancet*, preprint, 23 p., posted: 7 Sep 2021; available at SSRN: https://ssrn.com/abstract=3918955 or https://doi.org/10.2139/ssrn.3918955

or attending protests. One of my friends started a company that builds small wind turbines to independently power remote villages. Another has been working on developing geothermal resources in Africa to help the emerging economies there use carbon-free energy sources from the start. Still another invented an inexpensive process to remove carbon dioxide from the air. This is agency.

As stated earlier, individual "feel good" actions like recycling plastic or going vegan may be a coping strategy for some people, but in the overall big picture they are not going to make much of a dent in the climate crisis. What will make a dent are collective actions, such as a message to national governments worldwide that they are not doing enough. If that message comes through loud and clear from a lot of people, and is repeated on a frequent basis, it can make a difference. Especially if it is accompanied by the companion message that governments who refuse to address the climate crisis can and will be replaced. Nothing prods a politician into action more readily than the fear of losing their job.

I also want young people to understand that although addressing climate change may seem impossible, we've already done it at least once. The Montreal Protocol was signed in 1987 by all 198 United Nations member states – essentially every country on Earth – to end the use of chemicals that were destroying the ozone layer.

Ozone high in the atmosphere is critical for protecting the Earth from solar ultra-violet radiation. Some UV obviously gets through, or we wouldn't get sunburned after a day at the beach, but the ozone layer blocks most of it. A number of anthropogenic chemicals including chlorofluorocarbons (CFCs) used in air conditioning and halon used in fire extinguishers are known as ozone-depleting substances (ODS). These chemicals break down into their components high in the stratosphere and release chlorine or bromine atoms that destroy ozone. Losing the ozone layer would have had profoundly negative effects on plant life and certainly would have affected the climate.

Not incidentally, these ozone-depleting chemicals are also powerful greenhouse gases with global warming potentials thousands of times greater than similar amounts of CO_2. Thus, failing to get control of ODS emissions would have led directly to increased heat being trapped in the lower atmosphere. Higher levels of UV radiation from depleted ozone would also have reduced the ability of plants to utilize CO_2, resulting in higher concentrations of the gas. The feedback loop could have triggered a runaway greenhouse effect where the Earth would have just gotten hotter and hotter. This frightening possibility led the nations of the world to ban ODS, primarily CFC compounds. The manufacturing of nearly 100 chemicals was discontinued under the Montreal Protocol.

We clearly still have air conditioners and fire extinguishers, but these now use different chemicals that are much more friendly to ozone. The replacement chemicals are often more expensive than the ODS used previously, but the world decided it was worth the cost because all of humanity, and possibly all of nature were at risk.

The Montreal Protocol was challenging, contentious, and the devil was in the details. Despite all this, the nations of the world came together to protect the planet that we all share. Admittedly, unlike oil and gas, there wasn't a powerful, well-financed lobby of air conditioning manufacturers donating money to politicians, creating artificial uncertainty, and opposing the reforms. The refrigerant industry in fact agreed that CFCs were a hazard to ozone and went along with policies that prohibited their use. They adapted and adjusted and are still a viable industry today. We need to use similar reasoning to replace fossil fuels with carbon-neutral energy.

In terms of the total number of scientific papers published, the effects of GHG on climate are now much better documented than the effects of CFCs on ozone. There are literally no logical arguments left to support the ongoing use of fossil fuels. Despite this, some politicians continue to resist giving them up. Florida governor Ron DeSantis signed a bill in 2021 that prohibits the City of Miami from banning natural gas in new buildings. A number of Republican-led states have warned the Biden administration that if it continues to encourage major U.S. banks to invest "responsibly" with an eye to the climate crisis, they will withhold public funds from any institutions that refuse to lend to fossil fuel industries.

The reason, as usual, is money. Senator John Barrasso (R-WY) has stated in Interior Department hearings that Wyoming collects more than a billion dollars a year in royalties and taxes from oil, gas, and coal produced on federal lands. In recent years, oil produced 70% of the state revenue in Alaska, 52% of state revenue in Wyoming, and 45% of state revenue in North Dakota. The dependence of these states on oil and gas royalties is not a valid reason to delay the energy transition and ruin the climate for everybody else. There are other sources of revenue. Wyoming, for example, has the best wind energy potential of any U.S. state. I'm not sure what is motivating Florida since their income from oil and gas production is essentially zero.

And of course there is the tired argument that a clean energy transition will cost jobs. This is true of any transition, and I suspect the same argument was made by the people who manufactured whale oil lamps back when Samuel Kier started selling kerosene lanterns. All technological transitions bring an end to many of the old jobs but in turn create a whole lot of new jobs. People adapt.

There are historical examples too numerous to recount, but here is one: back in the day, teamsters drove a "team" of horses pulling a cargo wagon, hence the name. Teamsters now possess a commercial driver's license (CDL) to drive tractor-trailer rigs and other big trucks. They have successfully transitioned from horse-drawn wagons to diesel-powered semis and adapted to changing technology. They have done it so well that I think you'd be hard pressed today to find a teamster who actually knows anything about handling a team of horses.

Coal miners are one of the groups resisting an energy transition. Even though there are only about 50,000 working coal miners in the United States, their protests are heard above many others. The miners fear that "new energy jobs" such as constructing wind turbines or building solar farms won't offer the same salaries, benefits, and insurance of union coal mining jobs. There is a feeling that once the turbines are put in place, the jobs will evaporate. While job changes happen everywhere, including the coal mining business where seams eventually become depleted and mines close, it may take several generations to mine out a coal seam and the myth persists about the "good coal jobs that Daddy and Grandpop had."

Coal miners have heard talk at least as far back as the Kennedy Administration about job training programs that will give miners new skill sets to transition away from coal. Yet whenever there is a new energy initiative they seem to be forgotten, and by now they are more than a bit wary of such schemes. Hillary Clinton famously stated at a town hall meeting during her presidential campaign that she was going to "put a lot of coal miners and coal companies out of business." She proposed doing so by spending $30 billion on economic aid for coal country to help miners re-train and transition into new jobs, but all anyone heard was that Hillary planned to kill good mining jobs as part of a "war on coal." Donald Trump, on the other hand said he would "bring back coal" and not surprisingly won West Virginia in a landslide. However, the reality was that coal couldn't compete against abundant shale gas despite Trump's promises and the jobs never materialized.

Still, even in traditional coal mining areas such as West Virginia, many if not most citizens recognize the need to diversify the economy beyond coal. Being dependent on a single industry is never good, whether it is automobiles in Detroit, tourists in Hawaii, or computer software in Seattle. Coal companies have exploited West Virginia for decades, taking billions of dollars' worth of minerals out of the state and leaving behind devastated moonscapes, rampant occupational health diseases like black lung, and an impoverished population that doesn't have many skills beyond digging coal. This is a shameful legacy.

The bottom line as far as I'm concerned is that we must stop burning coal to avert the worst of the climate crisis, and to do so we must figure out how to help coal-dependent states like Wyoming and West Virginia transition into a more diversified economy. I have been living in West Virginia now for more than a decade since I began working at and retired from the DOE lab in Morgantown, and I lived here for a few years back in the 1980s as a DOE contractor. The state consists of numerous small towns in stream valleys isolated from one another by intervening ridges and mountains. The biggest city is the capital, Charleston in the Kanawha River valley with a population of less than 50,000. In most other states, a town this size would barely rate an airport. The remaining 1.7 million people in West Virginia are spread out across the countryside or live in more than 300 small towns and villages with an average population of about 5000 each. Dangerous travel on narrow, winding mountain roads has kept many businesses from investing in the state. Likewise, the lack of decent internet service and generally poor landline and cell phone communications has not helped either. As a resident of West Virginia, I have two suggestions: improve the roads and get broadband internet access to everybody.

The backwardness of West Virginia is often blamed on the mountainous terrain, but there is at least one outstanding example that shows this is a false excuse. Switzerland is a mountainous, small European country with numerous, isolated towns and villages, similar to West Virginia although it has more than four times the population. Switzerland also has a thriving economy, with a mean per capita income of $86,673 giving it the second highest GDP per person in the world.[11] (This is a modest self-ranking by the Swiss government. The World Bank actually ranks it

---

[11] https://www.eda.admin.ch/aboutswitzerland/en/home/wirtschaft/uebersicht/wirtschaft%2D%2D-fakten-und-zahlen.html

first.) In contrast, West Virginia has the lowest mean income in the United States, at just $43,469 per capita.

Switzerland is physically smaller than West Virginia, with a land area of about 16,000 square miles (41,000 sq. km) while WV covers a land area about 50% larger of roughly 24,000 square miles (62,000 sq. km). About two thirds of Switzerland consists of the mountainous terrain of the Alps and the Jura, while the remaining third is on an elevated plateau. The Swiss mountains are higher and steeper than the hills of West Virginia, with 48 peaks in the Alps standing higher than 13,000 feet (4 km). Many Swiss mountain valleys have been widened by past glaciers, and much of the population lives on the elevated plateau. In contrast, West Virginia has narrow stream valleys that were cut into an eroded and dissected plateau as ravines, and most of the population lives in isolated, small towns. Although it calls itself the "Mountain State," West Virginia does not actually have any high peaks; Spruce Knob, the highest point in the state is 4863 feet (1.5 km) above sea level.

Nevertheless, despite these differences West Virginia should consider emulating the Swiss economic model by expanding into broader markets with specialized goods and services, developing small manufacturing, and moving away from coal. Nearly three quarters of the Swiss GDP is produced by the service industry, which includes banking, finance, and publishing (Springer Nature, the publisher of this book, is based in Switzerland). An additional quarter of the GDP is generated from the small manufacturing of specialized items, such as watches. Over 99% of Swiss firms are diversified small businesses with fewer than 250 employees each.

West Virginia could develop service-based and cottage industries to make specialized goods like Switzerland. Service-based industries are much easier with good internet connections in this age of remote learning and video communications. People in isolated small towns could interact with the world over high-speed internet and various service industries like consulting, accounting, publishing, website design, software development, legal and others may flock to quaint, inexpensive WV towns to set up shop. It could bring a lot of people back to the state who were raised here, got educated, and left for the big cities because of better opportunities.

As just one example of a potential cottage industry, hardwood timber harvested from Appalachian forests is currently shipped to China to be made into furniture that is then shipped back to the United States for sale. Premium furniture could be made in West Virginia from locally-cut hardwood instead. If there were skilled artisans properly trained in the craft, decent roads to ensure that supplies and products could move more easily between towns, and a readily accessible, high-speed internet to sell to global markets, such businesses could thrive.

The widespread presence of broadband internet might also entice some technology people to move into the state. Working remotely from a beautiful (and relatively inexpensive) mountain retreat surrounded by abundant recreational opportunities could be very appealing to many tech workers stuck in crowded, expensive, polluted cities. West Virginia's poor internet is holding this back. The telecom companies in the state should replace the old-fashioned copper telephone wire (often stolen by criminals to sell for scrap, which adds to the communications problems) with optical glass fiber to improve the speed and clarity of transmission. Glass fiber is also not worth much as scrap. The telecom companies have been adding optical fiber, but progress has been slow.

The narrow, winding, hilly two-lane roads in West Virginia need reconstruction for safety purposes. I recently found a state road map from 1940 inside the preserved Esso service station in Arthurdale (refer back to Fig. 3.5). This town was built during the Great Depression in the 1930s by the Works Progress Administration (WPA) as a "hand-up, not a handout" for the unemployed, and became a pet project of First Lady Eleanor Roosevelt. She visited frequently from Washington, D.C., driving over in her own car. Except for the few interstate highways added since the Second World War, the 1940 road map was spot-on accurate. For all I know, it could have been left behind by Mrs. Roosevelt herself. Most of the roads currently being used in West Virginia were built and paved by the WPA in the 1930s and have seen little improvement since then.

The roads need to be wider with shoulders and breakdown lanes for people to pull off safely. Curves could be straightened and hills flattened. Such features on two-lane highways are common in other states, but not in West Virginia. The second thing needed is a passing lane on hills, of which West Virginia has many. Getting stuck behind a gravel or logging truck laboring up a long hill can be tedious, and sometimes impatient people will pass on blind curves or under other dangerous circumstances. It only takes one poor sap coming the other way for there to be a deadly collision. We don't need four lane highways everywhere throughout the state but adding that third lane on long hills would be extremely beneficial. They have it in western Maryland and other mountain states, and there is no reason not to in West Virginia. My mantra for the roads is SWiFT: Straighter, Wider, Flatter Transit. The politicians will whine about the cost, but last year the state budget for West Virginia had a record surplus of $1.3 billion. At the same time, the state is ranked 50th in the nation for infrastructure. I think the issue is priorities, not cost.

The bottom line is that West Virginia and other coal-dependent states have options for transitioning away from a coal-based economy. Some infrastructure investments are needed, but new infrastructure is required for virtually every transition (the interstate highway system, for example, was vitally important to the teamsters for the development of the trucking industry). Coal-based economies in a GHG-constrained world are going to collapse sooner or later, and the earlier a transition is started to broaden the economic base, the better.

Energy job transitions in the oil and gas business are different than coal mining because these workers are forced to be some of the most flexible employees on the planet. The normal petroleum boom and bust cycles and the wide geographic spread of the various plays have made workers good at changing jobs, adapting to different circumstances, relocating, and learning new skills. Very few oil and gas workers expect to hold a lifetime position with a single fossil energy company.

Many oil and gas wells in remote locations or offshore require workers to stay at the rig site for weeks. They typically work 12 h shifts on "tours" that last 2 weeks or more before returning home for a couple of weeks off. "Home" is often in some central location like Houston or Tulsa with easy access to the oilfields and a good airport. Over the course of a year, they may report to work at a dozen different drill sites from Alaska to Indonesia.

People were sleeping in their trucks when the number of workers on the Bakken shale oil boom in 2014 far exceeded the hotel room capacity of northwestern North Dakota. "Man camps" made from re-purposed marine cargo containers were still common up there when I was working on environmental issues in 2018–2019. A decade earlier during the boom years of the Marcellus Shale, I remember seeing hotel parking lots in northern West Virginia and southwestern Pennsylvania filled to capacity with white company trucks bearing Texas or Oklahoma license plates. Oilfield workers go where the work is.

These workers will be able to readily transition into drilling geothermal wells, capturing and sequestering carbon dioxide, maintaining wind turbines, and performing field work on solar panels. Operating construction equipment is pretty much the same everywhere, whether you are building a pad for an oil well or excavating a foundation for a wind turbine. The mobility of oilfield workers suggest that they will be the first ones to show up at remote field sites to take on renewable energy projects. Even if they have to sleep in their trucks.

****

Fossil fuel advocates contribute substantial amounts of money to both Republican and Democratic campaigns. Energy industry lobbyists are abundant in Washington, D.C. and even around state legislatures. In Florida, the utility companies employ an average of one lobbyist for every two legislators. Overcoming institutional and political resistance to decarbonizing the energy economy is an enormous challenge. Nevertheless, contending that we should continue to burn fossil fuels as a cheap source of energy while the climate crisis worsens is irresponsible and wrong.

Senator Joe Manchin (D-WV) has said we need a "fuel neutral" federal energy policy that doesn't favor renewables over fossil fuels. I disagree. We need a "carbon neutral" federal energy policy. Take another look at Figs. 2.1, 2.2, and 2.3 and tell me how we can continue to burn fossil fuels. Manchin was a coal broker before he became the West Virginia governor and later senator, and he still has connections in the industry. "Fuel neutral" is a term favored by the coal and oil industries to claim that fossil fuel has to remain in the mix for the next 30 or 50 years to meet the energy needs of the public (at a profit, of course) because these transitions take time. Going fuel neutral and not identifying fossil energy as the problem plays to the notions of "fairness," "balance," and "the science is not settled" arguments of climate skeptics.

As one of my professors used to say, "Bull gravy!" All the evidence we've explored in the previous chapters indicates that fossil fuels are definitely the problem. Requiring 30–50 years to transition electric power and vehicles to carbon neutral energy is nothing but a stall tactic. The technology exists, as explained in the next two chapters, and it can be applied to current gasoline powered vehicles and existing electrical generating infrastructure at a modest cost. The transition could be done in less than a decade if we really wanted it to happen. By this point in the book, however, readers can probably figure out why it has not.

The West Virginia state treasurer, Riley Moore, has said that the state will divest from companies in the financial sector that prioritize environmental, social and governance issues (ESG) over profits. Moore said coal and gas provide substantial tax revenue and jobs in the state, and ESG companies are "telling us that these industries are bad, and we have to fight back." Divesting state funds from ESG companies may be a feel-good response that excites the political base, but it is not going to stop killer heat waves, major wildfires, monster hurricanes, melting glaciers, sea level rise, and climate refugees as the climate crisis grinds on.

The state treasurer should also understand that ESG companies are not pulling some kind of public relations stunt when they say fossil fuels are bad. They are bad. Fossil fuels are bad for the air and water, they are bad for landscapes, they are bad for human health, they are bad for ecosystems and the environment, and they are very bad for climate. In my opinion, these kinds of statements from West Virginia state and federal officials indicate very clearly that the state must diversify the economy. Holding on to coal and gas for dear life in the face of the climate crisis because you have nothing else suggests that you very much need something else.

Foreign policy issues with nations that rely on energy exports to support their economies are another challenging problem. Like oil or coal-dependent U.S. states, many of these countries are not willing to give up a major source of income for the sake of the climate. Yet, some of these nations are likely to be the most adversely affected by climate change. For example, it is already hot in the Middle East and adding heat waves onto normal summer heat there could be deadly, even with air conditioning. Like West Virginia, these nations and other energy exporters would all be better off if their economies had wider diversity than just fossil fuel.

On the other side of the coin, few nations seem to have learned from the experience of the United States during 1973–1974 OPEC oil embargo. Many countries are still dependent on fossil energy imports that may come with significant political strings attached. China obtains a substantial amount of the coal used to produce electricity to run their economy from Australia. Ongoing tensions in the South China Sea may or may not affect this supply. Japan continues to import most of the energy needed to run their economy, despite the lessons of the Second World War, waged primarily by Japan to gain access to supplies of oil, coal and other raw materials in China and Indonesia. We all know how that turned out. Even the United States, with it's vaunted "all of the above" energy strategy and number one oil and gas production status is still cautious about offending certain oil-exporting countries like Saudi Arabia over politics, even in the face of obvious human rights abuses.

The most recent example of fossil energy import dependence is the reliance of western Europe on Russian oil and natural gas. After the Russian army invaded Ukraine in February 2022, the European Union and associated countries declared that they would immediately impose harsh economic sanctions against Russia in retaliation. Except, oops, they still needed Russian gas and oil to get through the winter. How embarrassing. The Baltic states, Poland, Germany, and other European nations eventually figured out work-arounds and the U.S. began shipping more LNG to Europe.

Nevertheless, the continued dependence on Russian oil by EU members Bulgaria, the Czech Republic, Hungary, and Slovakia remains an obstacle to a proposed across-the-board EU embargo on Russian energy. Purchases of Russian oil put money in the treasury and finance the war against Ukraine. The Ukrainian dilemma shows that replacing fossil fuels with indigenous, sustainable, resilient, renewable energy resources is as much of a national security issue as a climate concern.

The bottom line is that coal has to go. Oil and natural gas have to go. And the sooner the better.

Chapter 7
Decarbonizing Electricity

Keywords Thermoelectric · Geothermal · Nuclear

****

A brief description of how the electrical generating system works in the U.S., Canada, and Mexico might be helpful to many readers before we get into the details of decarbonization. Most electrical systems in other developed parts of the world operate on similar principles, although voltages, frequencies, and current are often different than those in North America. Power systems on a particular continent, such as North America or Europe tend to be standardized to allow the electricity supply to be networked and transferred as needed.

Many people ran experiments in elementary school science class where an iron nail with an insulated copper wire coiled some 20 or 30 times around it became magnetic when attached to a battery and could lift paperclips. The magnetism disappeared as soon the electricity was cut off. Massive examples of such "electromagnets" can be found on cranes in recycling yards where they are activated and deactivated at will to pick up and drop scrap steel.

Just as electricity can produce magnetism, magnetism can produce electricity. Electricity is nothing more than the flow of electrons pushed by a magnetic field through a conductor like copper wire. A "primary" energy source is required to power the magnet to move past the coils of conductive wire and cause the electrons to flow. The primary energy source can be any number of different things: water flowing through the penstocks of a dam and pushing on a water turbine to turn the magnet, wind blowing across the blades of a wind turbine to rotate it and turn a magnet, or steam generated by burning coal, natural gas, or nuclear heat spinning yet another turbine and rotating a magnet. The exception to moving magnets is solar photovoltaic, described earlier as sunlight on a solar cell creating an electrical potential between the front and back of the cell, causing electricity to flow with no moving parts.

The electric power system is an energy transfer system. Coal burned in a power plant creates steam that generates electricity, which then travels through power lines to a person's house where it heats a coil of wire on their electric stove and boils

D. Soeder, *Energy Futures*, https://doi.org/10.1007/978-3-031-15381-5_7

water for tea. Of course it would be more efficient to just burn the coal under the tea kettle directly, but the advantage with electricity is that it can heat the kettle with primary energy sources that may include oil, nuclear, hydroelectric, biomass, wind, solar, geothermal, and natural gas, as well as coal. Electricity is also much cleaner than burning coal in the house.

Utility-scale electric power generation requires a delicate and dynamic balance between supply and demand known as "dispatch." Demand for electricity is constantly shifting as air conditioners and heat pumps kick on and off, streetlights cycle on at dusk and off at dawn, people come home from work in the evening and turn on stoves, TVs, and other appliances that had been turned off all day, and commercial buildings ramp up power use during the day and shut it down at night. Electric power suppliers must anticipate and meet these demand swings.

My father worked for an electrical utility back in the days when neighborhood substations were staffed around the clock and dispatch was done manually to adjust local supply and demand. He worked all shifts, weekends and holidays because "people always need electricity." He was especially busy during storms. These substations are now fully automated and monitored remotely. His last job at the power utility was to dispatch excess electricity out to the national grid or bring it in as needed.

The utilities keep their electrical systems steadily charged with a modest level of power known as the "base load," and then add or remove "peak" power to meet load changes caused by demand spikes. These spikes occur daily, as power usage varies between day and night, weekly as power varies between the workday and the weekend, and also seasonally as power usage varies between winter and summer.

Base load power is the cheapest type of electricity to be generated, and typically comes from large nuclear, coal-fired, or hydro power plants. These plants are difficult to start or stop quickly in response to shifting demand but can supply constant levels of electricity into the system once they reach a steady state of power generation. Peaks in power demand, in contrast, are met with a combination of power sources that can be rapidly brought online as needed and quickly shut down when not. These may include natural gas, wind, solar, and some smaller hydroelectric facilities. Balancing power supplies around demand spikes is known as "peak shaving."

The most common type of power generation is called "thermoelectric" because it uses heat to create steam or other hot gases to turn a turbine and generator. Thermoelectric power plants include coal-fired, natural gas-fired, and nuclear powered generators. Solar thermal plants, which focus sunlight from hundreds of mirrors onto a central tower also use heat energy to run a turbine and generator and are considered thermoelectric as well. Non-thermoelectric power generation includes wind turbines, solar photovoltaic (PV) panels, and hydroelectric power plants.

Thermoelectric power plants generally operate on one of two thermal cycles. The Brayton Cycle is an open system where gas is compressed, heated by combustion, and flows across a turbine to produce power, exiting the system as exhaust. It is characterized by the absence of a phase change; the working fluid throughout is in the gas phase. A jet engine is an example of the Brayton Cycle.

In contrast, the Rankine Cycle is a closed system that does utilize a phase change. Liquid water is boiled into steam and the steam is used to drive a turbine blade. Instead of exhausting the steam afterward, it runs through a condenser loop that cools it down enough to convert it back into liquid water where it is re-boiled and used again. The condensed water is typically hot and requires less energy to generate steam than an outside water supply. The clouds that can be observed coming from the cooling towers on nuclear and coal plants are condensation from the cooling water used to convert the spent steam back into liquid water for reuse.

In terms of fossil fuel use, coal-fired power plants are Rankine Cycle designs where the coal combustion boils water into steam, and the steam is used to turn a turbine. Nuclear power plants also use the Rankine Cycle because it is closed, and no radioactivity can escape into the environment. The Rankine Cycle is not limited to water and can work at various temperatures with many different phase-changing materials such as volatile hydrocarbons like butane or propane, alcohol, and even with carbon dioxide.

Natural gas-fired power plants typically employ both the Brayton Cycle and the Rankine Cycle to generate electricity, and hence are known as "combined cycle" (CC) power plants. A Brayton Cycle is used on the front end to turn a gas turbine and generate electricity. Instead of venting the exhaust to the atmosphere however, the hot gases are diverted to a boiler to create steam, which drives a Rankine Cycle turbine and a second generator. The gas CC power plants generate about 50% more electricity with the same amount of fossil fuel as single cycle gas or coal plants.

These favorable economics have led to the dominance of natural gas-fired electricity in the United States. The CC plants also happen to be cleaner, better for the environment, and emit substantially less GHG than coal plants, but industry is driven by costs, not by environmental altruism. In this case, it is fortunate that both the economics and environmental advantages happen to align. This may suggest some approaches that might be successful at decarbonizing electricity.

Government policies such as a carbon tax on GHG emissions or simply outlawing them and requiring CCS technology on smokestacks are two approaches that could be taken to force utilities to replace fossil fuels with cleaner, alternative generating technologies. The gas and coal industries along with the electric utilities that use these fuels predictably oppose such policies, and this resistance is thwarting global efforts to address the climate crisis.

Nevertheless, although government policies and leadership would be helpful, they may not be critical for decarbonizing the electric power generation sector. This industry is obsessed with cost and economics. As described above in the example of gas-fired CC power plants, any reduction in the cost to generate electricity will be adopted quickly by utilities. If that cost reduction can be achieved with non-carbon, sustainable primary power sources, utilities will quickly adopt these with or without government incentives.

The switch from coal to natural gas in the past decade for power generation in the United States is even more startling given the history. Utilities traditionally had been wary of using gas for power generation because of uncertain supplies. In fact, they were prohibited by law under the Fuel Use Act from using gas to generate electricity after the winter gas shortages in 1977 and 1978. The Fuel Use Act expired in 1987 with gas deregulation under the Reagan administration, which also brought a large amount of new gas production online, resulting in a natural gas supply surplus in the 1990s.

Several hundred gigawatts of new natural gas generating capacity were built between 1997 and 2003, only to have the price of gas climb steeply after conventional production peaked in 2003–2004. Much of the new generating capacity was idled, because although gas was available, it was expensive. Ironically, these same high gas prices helped George Mitchell profitably develop natural gas from the Barnett Shale and led to the shale gas revolution. The resulting glut of shale gas drove prices down, but the drilling industry rapidly developed new efficiencies in drilling and fracking, keeping production high. The abundant shale gas supplies that became available during the first decade of the twenty-first century along with the efficient CC technology convinced many utilities to switch from coal to gas-fired electricity.

The economics of natural gas-fired electricity are superior to coal on two fronts. First, coal has transport and handling costs, including barges or rail cars that require unloading, preparation of the coal through a process called comminution to produce a uniform particle size and remove non-combustible mineral matter, transport of the processed fuel to the burner through a series of belts and hoppers, controlling sulfur, mercury and other toxic emissions while burning the coal, and then finally trucking away the solid combustion products such as fly ash for disposal. Natural gas, on the other hand, can be brought right up to the burner tip by a pipeline with no processing or handling costs. The only combustion products from natural gas are water vapor and CO_2, both of which are vented to the air and do not need to be hauled away.

The second advantage gas has over coal is in the configuration of the power plants for CC generation as described earlier. These have much higher efficiencies and superior economics compared to coal. The loss of coal mining jobs and the closure of mines with some companies going bankrupt was almost exclusively due to abundant natural gas from the shale gas revolution displacing coal in the power sector, and not because of any additional environmental regulations or an imagined "War on Coal." It was always about cost.

I suggested back in Chap. 1 that the proposed ban on fracking in the "Green New Deal" legislation sponsored by Representative Alexandria Ocasio-Cortez (D-NY) and Senator Ed Markey (D-MA) is misguided. Senator Bernie Sanders (I-VT) has also been an advocate for this ban. To elaborate on my concerns, because 70% of our natural gas now comes from shale, a national fracking ban will lead to immediate and substantial gas shortages. I can only presume these legislators think that a fracking ban will force electric utilities to replace natural gas-fired electricity with greener alternatives like wind and solar. As explained above, the utilities are always focused on costs, and this idea ignores economic reality.

Wind and solar generate electricity directly without producing heat. They require brand new infrastructure and cannot be used with existing thermoelectric power plants fired by natural gas. Replacing natural gas plants with these renewables would require huge capital expenditures to build new wind farms and solar farms and run miles of new power lines. Replacing a single 500 MW gas plant with 2 MW wind turbines would require at least 250 new turbines at $3 to $4 million each for a total cost of as much as a billion dollars.

But there is a much cheaper alternative. Simply switching natural gas plants back to coal can run the Rankine Cycle generators in existing power plants with little to no modification. Many CC power plants in fact already have a standby option for coal on the Rankine Cycle in case extreme cold weather or other issues result in a high gas demand. Living spaces like residences and critical facilities like hospitals will receive priority for natural gas supplies in a shortage and cut into the allocation for utilities. So coal is their back-up.

This was demonstrated during the COVID pandemic, or more precisely as we were coming out of the pandemic lockdown. The economy dipped in 2020, and then rebounded sharply in 2021. Natural gas was among the industries disrupted by COVID, and when energy demand came back faster than expected, the gas industry was unable to fully supply fuel to the electric power sector. As such, many power plants switched to their back-up coal generation, increasing the amount of coal-fired electricity by 17% in 2021, the first such increase since 2014.[1] This also resulted in an increase in U.S. emissions of GHG.

It is important to note that coal generates nearly twice as much carbon dioxide per Btu as natural gas when burned (Fig. 7.1). Coal is mined, not fracked so any ban on fracking will have no effect whatsoever on coal. It should come as no surprise that the coal industry is one of the major supporters of a national fracking ban. Natural gas shortages resulting from a ban on fracking will almost certainly send the electric utilities straight back to coal.

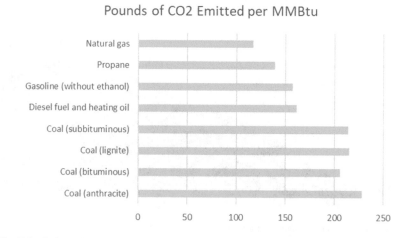

Fig. 7.1 Carbon dioxide emissions for equivalent energy values from different fossil fuels. (*Source: U.S. Department of Energy; public records*)

---

[1] h t t p s : / / w w w . n b c n e w s . c o m / s c i e n c e / s c i e n c e - n e w s / us-greenhouse-emissions-rose-2021-making-unlikely-climate-targets-will-rcna11431

No one should be under the illusion that decarbonizing the electric power sector will be easy. According to the U.S. Energy Information Administration (EIA), total U.S. electric generation in August 2019 was 400 GW. This electricity was produced from a mix of sources, including 273 GW (68%) from fossil fuels (primarily coal and natural gas), 72 GW (18%) from nuclear; 34 GW (9%) from renewables (primarily wind and solar), and 21 GW (5%) from hydroelectric (Fig. 7.2). Decarbonizing the 273 GW of electric power produced from fossil fuels will be a major challenge.

Some prominent environmentalists have stated that fossil fuel power plants should be completely replaced by renewables. The numbers suggest that this is not feasible from a cost standpoint, and cost plays into the narrative of climate skeptics who claim we need to keep using fossil fuels because renewables are too expensive. This is probably true for wind and solar because these require completely new electrical generating infrastructure. However, there are alternatives that can produce carbon-free heat under a boiler in an existing power plant.

Utilities have invested billions of dollars in thermoelectric power plant infrastructure and are probably not willing to simply walk away from it all and replace it with new wind turbines or solar panels. In addition to abandoning substantial capital assets and taking on a huge new CAPEX to build wind or solar installations, renewable power generation is often located in remote areas in order to gain access to reliable wind or unobstructed sunlight. In contrast to compact fossil fuel power plants, wind and solar both have what is known as a low energy density. They require significant amounts of land and supporting infrastructure such as new, high voltage transmission lines to generate an equivalent amount of power.

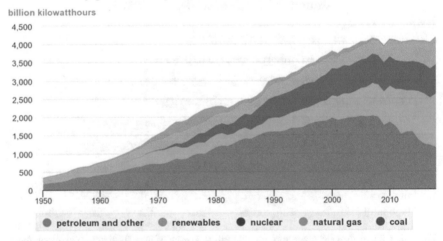

Fig. 7.2 Electricity generation in the United States by primary energy source. (*Source: U.S. Energy Information Administration, 2019; public domain*)

Individual, land-based wind turbines typically generate about 2 MW of electricity under normal operations. For the purposes of this discussion, we can assume a one-to-one replacement of fossil with wind based on rated power output. At present, there are about 70,000 wind turbines operating in 41 states in the U.S. To replace 273 GW of fossil fuel electricity with standard two-MW wind turbines, about 136,500 additional turbines would be needed. Using wind industry cost estimates for installed turbines of $3 to $4 million each, the additional turbines could cost more than half a trillion dollars.[2] The reality is even worse than that.

The numbers above assume that wind power can directly replace fossil fuel on a one-to-one basis. In the real world, wind turbines have a "capacity factor" of only about 40%, meaning that they produce electricity 40% of the time, and are shut down 60% of the time. This may be due to a lack of wind on calm days, or the inability to store wind power efficiently, so the turbines are only used for peak shaving during times of high demand and shut down otherwise. In comparison, the capacity factor of coal-fired power plants is 85%, because they typically supply base load power, and for natural gas CC power plants it is 87% because they supply both base load and peak power.

Given the low capacity factor of wind, power utilities assume that it takes 3 MW of wind power to replace 1 MW of fossil. Thus, in the example given earlier, instead of 250 two-MW wind turbines replacing a 500 MW fossil fuel power plant, the actual number needed would be more like 750. The capacity factor of solar is even worse, ranging between 25% and 29%.

Even with larger wind turbines the numbers needed to replace fossil energy are still formidable. As mentioned in the previous chapter, the largest onshore wind turbines in the United States are in Texas, which generate about 4 MW each. Assuming that a capacity factor similar to fossil fuels could somehow be gained, some 68,250 more of these would be needed to generate 273 GW of electricity. If these were spaced in a line at the recommended industry cross-wind distance of 500 m (1640 feet) apart, the linear extent of these new turbines would be some 34,125 km (21,204 miles). This is 85% of the distance around the Earth.

Solar has similar issues. The U.S. Department of Energy has been pursuing solar thermal power in addition to solar PV. Solar thermal uses acres of mirrors to direct beams of sunlight onto a focus point in centralized towers to generate heat that drives a steam turbine. The largest existing solar thermal power plant in the U.S. is currently the Ivanpah Solar Power Facility in the California desert that generates 392 MW (Fig. 7.3).

This facility cost $2.2 billion and nearly 700 more just like it would be required to produce 273 GW of electricity. These power plants occupy enormous amounts of land that can't be used for anything else. The two existing U.S. solar thermal facilities, Ivanpah in California and the smaller Crescent Dunes project near Tonopah in Nevada are located in deserts, which gives them abundant sunlight and keeps them off usable farmland, but also often places them far from where the power is needed.

---

[2] https://www.windustry.org/how_much_do_wind_turbines_cost

Fig. 7.3 The solar thermal power plant located on Ivanpah Dry Lake in the Mojave Desert of California. Arrays of mirrors focus sunlight onto the central towers, creating heat. The plant has a peak output of 392 MW. (*Source: U.S. Department of Energy photograph; public domain*)

Like wind turbines, solar thermal facilities exact a toll on birds, in this case instantly incinerating any unfortunate avian that happens to fly into a concentrated beam of sunlight. The beams are invisible in the clear desert air until they hit the central towers, and the birds fly into them unknowingly while chasing insects. The number of birds killed at Ivanpah in southern California and Crescent Dunes in Nevada is estimated to be greater than 16,000 per year.[3] The acres of mirrors also displace habitat for birds and other wildlife.

Commercial solar PV also requires land and infrastructure. As the cost of PV panels continues to fall, it has become a "distributed" energy resource with many homeowners adding PV panels to their rooftops and selling excess electricity back to the power grid. On the Hawaiian island of Oahu, for example, rooftop solar panels have become so numerous that the electric grid could not handle all the extra power flowing into it, and the utility company temporarily suspended solar panel hookups. Computer-controlled dispatching on a so-called "smart grid" is needed to deal with all of the distributed energy sources from rooftop solar panels. Although it has been discussed for decades, the national smart grid always seems to be just a few years away from becoming a reality.

In addition to being mounted on rooftops by individual homeowners, solar panels are also being deployed industrially on "solar farms" that occupy large tracts of land. A guideline for mid-latitude locations is that each MW of solar generating

[3] Walston, L.J., Rollins, K.E., LaGory, K.E., Smith, K.P., and Meyers, S.A., 2016, A preliminary assessment of avian mortality at utility-scale solar energy facilities in the United States: *Renewable Energy*, v. 92, p. 405–414

capacity requires about 10 acres of land (4 hectares), but some additional space is needed for utilities and access. A 30 MW solar farm that opened in Texas in 2011 required 380 acres of land, or about 12 acres per MW.[4]

The land requirements can be a problem. Some people object to valuable agricultural land being converted to solar energy use. Placing solar arrays in deserts (where they get more sunlight anyway) and other places that are not farmland can help to resolve this land use issue, and indeed, some large tracts of desert south of Las Vegas are hosting solar farms. However, plans to place solar farms on Pennsylvania farmland are drawing protests from local farmers.

Another objection to solar arrays is that they create large amounts of impervious surface area in watersheds, increasing runoff, causing flash flooding, reducing groundwater recharge, and damaging biota in streams. Studies have shown that adding impervious surfaces to as little as 10% of the catchment area in a watershed will greatly increase runoff and harm aquatic ecology. The solution to this problem is to place solar panels on roofs, or above paved parking lots. These are already impervious surfaces and adding solar panels will not make any difference. Providing shade for hot cars is not a bad idea either.

Of perhaps greater concern is the number of toxic substances involved in the manufacture and composition of PV panels. The manufacturing process used to reduce silicon dioxide (silica) to the elemental silicon needed for the monocrystalline and polycrystalline wafers produces silicon tetrachloride as a byproduct. This chemical is toxic if not properly handled, and can result in chemical burns on skin, lung damage if inhaled, and releases hydrochloric acid if exposed to water. Toxic chemicals within the solar panels include cadmium telluride, copper indium selenide, cadmium gallium (di)selenide, copper indium gallium (di)selenide, lead, hexafluoroethane, and polyvinyl fluoride.[5] These substances are not an immediate risk for users, but as PV panels wear out and are disposed of in landfills, they could lead to environmental problems. It is important that the manufacture, use, recycling, and disposal of PV panels be carefully monitored and regulated.

Wind and solar have a major problem by not being "on" all the time. There is no wind power on calm days and no solar power at night. These renewables can be used intermittently for peakshaving but not for baseload power unless there is some kind of energy storage to get them through the off periods. Electric power is a tricky thing to store. Current technology relies on banks of lithium batteries that can only supply the grid for a few hours at most. Other energy storage options include gravity such as pumped hydro where water is stored in an elevated lake and then used to generate electricity as needed. Another mechanical energy storage injects compressed air into underground rock formations and releases it to operate a generator as needed. Most storage options have significant efficiency losses, requiring renewables to be overbuilt to make up for this. Until the energy storage issue is resolved, wind and

[4] https://www.ecmag.com/section/green-building/austin-energy-activates-30-mw-solar-farm

[5] https://sciencing.com/effects-chlorofluorocarbons-humans-7053.html

solar remain intermittent power sources and by themselves are not the sustainable solution needed.

New battery technology is constantly under development. One new system is called "aqueous air" and claims to be capable of delivering 1 MW of power for 150 hours. These and other advances in battery technology suggest that storage will be less of a technical issue in the future, but it will remain an economic concern as an additional cost.

It is also important to note than the massive construction of renewables will produce substantial embodied emissions. Some LCAs suggest that the steel, concrete and glass needed for tens of thousands of solar panels and wind turbines may emit more GHG during manufacture than the amount saved over the lifetime of electrical production from the unit.

It is not impossible to replace 273 GW of fossil fuel electricity with wind and solar. But readers should be aware that it would be a huge and very expensive undertaking. There are better options.

****

Removing fossil fuels from electric power generation is the important first step to decarbonizing everything else. But how can we do that if renewables are so expensive? I've thought about this for some time, and I believe the way to decarbonize the electric power sector in a manner that is fast, cost-effective, and efficient is by finding carbon-neutral or carbon zero sources of heat that can replace fossil fuels in existing thermoelectric power plants. While we should continue to develop wind and solar power to supplement the grid as part of an "all of the above" energy strategy, any technology that can be integrated into already operating and grid-connected generating facilities by simply substituting a carbon-free heat source for a fossil fuel burner would be a quick and economical way to decarbonize. The boilers at a power plant don't care where the heat comes from as long as it is hot enough. This modification would allow existing power generation, transmission, and distribution infrastructure to be retained and used, saving considerable amounts of both time and CAPEX.

There are three climate-friendly sources of heat that I know of, and if anyone has ideas for others, I would love to hear them. Here are mine: (1) biogas, which is carbon neutral, (2) engineered geothermal, and (3) new technology nuclear, both of which are carbon zero. Details about the technologies are discussed separately below.

There are three requirements for climate-friendly heat sources. First, of course, is that they must be carbon neutral or carbon zero, and second, they must be sustainable, so we are not going through this exercise again 50 years down the road when we run out of some critical material. Third, they must also be resilient because future storms, sea level rise, and wildfires are not likely to be gentle on delicate or exposed infrastructure.

The amount of heat required to make live steam is generally between 200 and 400 °C (390–750 °F) and could be considered the minimum replacement heat values for coal or gas burners. All three climate-friendly heat sources are capable of meeting this temperature range. However, this only applies when using water in the Rankine Cycle. If other fluids like alcohol or hydrocarbons with lower boiling points are used, less heat is needed, but different turbine designs are then required, and power production may not be the same. Research and development on biogas, geothermal, and nuclear are woefully underfunded and proceeding slowly. In my opinion, these should be receiving Manhattan Project-like priority and Apollo Moon Mission levels of funding.

Geothermal research is struggling to understand a host of engineering issues while trying to develop and test new technology with only one available field site. There was some money set aside for it in the 2021 bipartisan infrastructure bill, but much more is needed. Three or four additional field test sites placed in a variety of different geologic environments and geographic locations would help answer some of the technical questions a lot faster.

Biofuel research has been focused primarily on liquids to replace diesel fuel and gasoline in vehicles, not to decarbonize electricity. The easiest and simplest biofuel to make is methane biogas, which can directly replace natural gas in the fossil fuel pipeline. Biogas is carbon neutral and can be made from nearly any available organic matter. Microbes do all the work, and some clever genetic engineering could produce microbes that generate substantially greater amounts of methane that could meet the needs of power plants and towns. Yet research in this area is lacking because microbiologists see it as boring and uninteresting compared to engineering bacteria to create liquid fuels, plastics, and other durable goods. Putting some funding behind biogas would increase their enthusiasm levels.

New nuclear technology is facing challenges at even getting permission to build reactors. Bill Gates and Warren Buffett have invested in the development of a new technology liquid sodium-cooled reactor in Wyoming that is running into all kinds of permit problems. Even with a boatload of money and the backing of billionaires, the decarbonization of the electric power sector will not happen overnight. On a shoestring budget, it may take forever.

1. <u>Biofuels</u>. As described in the previous chapter, these are made from plants structured with carbon dioxide taken from the atmosphere. When combusted they simply return this CO_2 back to the carbon cycle; thus they are carbon-neutral. Biofuels can be made from microbes, algae, grasses, trees, or other plants.

Configuring the biofuel combustion process as BECCS to capture and sequester the GHG emissions is carbon negative. The CO_2 taken from the air by plants is captured after the biofuel is burned, and the captured CO_2 is then sequestered from the atmosphere. The BECCS process actively performs CDR from the atmosphere and obtains "negative emissions" while also providing energy to run the capture and storage process along with income to the power company. The economic incentive behind BECCS is that it generates electricity revenue in the process of capturing carbon. BECCS has been promoted as a way to achieve profitable negative emissions, although so far it has not been widely adopted.

The easiest and simplest biofuel to make is methane, which is created from waste biomass through a fermentation-like process as a metabolic byproduct by a group of anaerobic bacteria called methanogens. These microbes occur almost anywhere there is an accumulation of organic matter in the absence of oxygen (refer back to Fig. 6.3). Since the microbes do all the work, the only things humans need to supply are some large, sealed fermentation tanks that can provide the proper anoxic environment, a supply of organic waste material, and a system to collect the gas. The bio-methane can be burned exactly like natural gas to provide energy.

Other biofuels such as liquid ethanol made from corn or solid wood harvested from trees can also be burned to provide energy, of course. The beauty of biogas is that it is pure methane, and natural gas is composed of about 99% methane. Therefore, biogas methane can be put directly into the natural gas system as a net zero substitute fuel. They are exactly the same substance except for their origins, and any CC power plant running on natural gas can run on biogas.

There are significant economic advantages here. Biogas can use existing natural gas transmission and distribution pipelines, gas storage fields, gas burners, and gas appliances without requiring any modifications, refits, or new construction, unlike hydrogen. Hydrogen gas is often touted as a carbon-free natural gas replacement, but it has some serious economic, physical, technical, and logistical issues that are discussed in more detail in Chap. 8.

Biogas will allow existing electrical generating facilities to retain their highly efficient combined cycle power plants. Biogas in the natural gas distribution system will allow residential and commercial customers to retain their gas furnaces, hot water heaters, and stoves. It is the most economic, realistic, and achievable alternative to fossil fuel as the gas industry navigates into the carbon-constrained future.

Instead of spending tens of millions of dollars to drill and frack new shale gas wells, gas producers would be better served by spending that capital to build facilities for large-scale biogas methane production. Besides being carbon neutral, biogas is also much more sustainable than natural gas from wells. Even shale gas wells will eventually decline according to the Hubbert peak oil concept. However, as long as organic waste matter is available, and I seriously doubt we will ever run out of sewage, a biogas facility can continue to manufacture methane indefinitely.

Fossil fuel natural gas is often transmitted long distances through limited-diameter pipelines, which can subject it to short-term supply disruptions during times of high demand. We are also seeing in Europe the potential for the political disruption of natural gas supplies transmitted from one country to another if there are disagreements over policy. In addition, long distance transmission of natural gas risks fugitive emissions at every pipe joint, every valve stem packing, and every compressor station. Over long distances, these small leaks can add up to significant amounts of fugitive methane emissions, which are a GHG concern. Building a biogas plant close to the powerplant and the city that will use the fuel greatly minimizes the potential for both transmission disruptions and fugitive emissions.

The cost of natural gas can vary widely with supply availability. Natural gas is abundant at present thanks to shale gas and fracking, but sooner or later, gas supplies will become short, and prices will rise. Since the electric power sector has been burned in the past by the gas industry with price spikes and shortages, I think they would welcome a secure source of fuel that can be made onsite as needed.

Properly constructed and scaled, microbes in a biogas facility can produce sufficient methane to generate electricity. It may be possible to genetically engineer or selectively breed the methanogens to be more efficient. The system may end up needing to be quite large depending on the requirements of the power plant. Nevertheless, the advantages of cost, security, sustainability, resiliency, and carbon neutrality from on-site biogas production are reasons for electrical utilities to seriously consider this as a means to fuel a CC power plant.

Animal waste, agricultural waste, and just about any other source of organic carbon can be turned into biogas. For example, confined animal feeding operations, a common method for producing meat animals in the U.S., generates prodigious amounts of manure. The waste products from poultry, hogs, and cattle are an environmental hazard, contribute to water quality problems, and have little use as fertilizers, soil amendments, or other products. Many livestock farmers are stuck with the waste and not sure what to do with it. Organic waste washed into streams has created anoxic "dead zones" in the Chesapeake Bay and the Gulf of Mexico. Funneling all of this manure into biogas tanks to convert it into usable methane fuel will reduce the volume of the waste and provide a clean energy source. Farmers could turn a financial liability into an income-producing asset. Municipal wastewater treatment plants could also benefit from biogasification to help reduce the volume of waste material and make some income off energy production. Biogas facilities are being constructed in some locations by industry and government, but this needs to scale up much, much larger.

According to the EIA, the United States burned 30.5 trillion cubic feet (TCF) of natural gas in 2020.[6] The bulk of this (nearly 40%) went for electric power generation, and another significant volume (a third) went to industrial uses. The use residential of natural gas in homes for heating, hot water, and cooking only makes up 15% of the total.

Is it possible to make 30 TCF of methane from biogas to replace natural gas? What kinds of facilities would this require? How large? Is there enough organic material available? Do we even need 30 TCF of gas if we use other renewables to generate more electricity and substitute it for gas? I don't know, but these issues could really benefit from some well-funded research and definitive answers.

The vast overproduction of shale gas through fracking has produced a glut of cheap natural gas that has made it difficult for biogas to compete economically except in some rare instances. The best way for the government to assist biofuel development is to put a tax or tariff on fossil carbon at the mine or wellhead, while providing tax credits for carbon-neutral fuels. Instead of spending their capital drilling expensive new wells to produce taxed natural gas, the gas producers might consider instead putting their millions into tax-credited biofuel facilities.

There are pitfalls to biofuels, and well-intentioned plans can have unexpected results. In the pre-Brexit days of 2009, the European Union put out a directive for member countries to use more renewable energy to generate electricity. In

---

[6] U.S. Energy Information Administration, Natural Gas Annual, September 2021

a response that would have startled King James I who required in 1615 that English wood be reserved for shipbuilding, a number of coal-fired power plants in the U.K. were converted to burning wood to achieve carbon-neutral emissions. The wood was required to be in pellet form for ease of handling and uniform flame intensity. Technically, the king's edict was still being upheld because much of the wood for these pellets was harvested from forests in the southeastern United States and shipped across the Atlantic. Half of the wood pellets were going to one large power plant in England.

This is where an LCA is important. The lifecycle of this "renewable" electricity included diesel-powered machinery to cut, trim, and transport the wood to pellet plants. The energy used to turn the wood into pellets at the plants was supplied in large part by coal-fired or natural gas fired electricity. Then the pellets were transported to a harbor and placed aboard a ship, which carried them across the Atlantic Ocean burning bunker oil in the engines. At the other end, they were unloaded by more diesel powered equipment and transported to the power plant. By the time all the emission sources were added up, the British wood-fired power plant still managed to come out somewhat better on emissions than the coal-fired power plant it replaced but it was far from carbon neutral.

The EU rescinded the renewable energy directive in 2021. It is important in the biofuel business to be aware of all of the other scope emissions that might be associated with the production and use of the fuel. If any of these are from fossil energy, they will act to offset the goal of carbon neutrality for the biologically-derived fuel.

Liquid biofuels are not well-suited to electrical generation. Fuels like alcohol are created as the byproduct of a fermentation process, but so is CO_2, which is typically vented. Although this isn't fossil carbon, it is still carbon, and it seems to me that it is wasteful to add it to the atmosphere without really gaining any energy from it. Liquid biofuels are mainly of interest to the transportation sector and are discussed in the next chapter.

2. Geothermal. Geothermal has been limited in the past by the need to be near natural geothermal heat sources like volcanic features, geysers, hot springs, etc. Much of the current geothermal development has been around volcanic "hot spots" in places like Iceland, on the North Island of New Zealand, Hawai'i, and "The Geysers" near Santa Rosa, California. However, the Earth is hot everywhere if you go deep enough, with the temperature of the core over 9000 °F (5200 °C). This is actually hotter than the surface of the sun. New geothermal technology seeks to drill wells that can tap this deep heat from the planet. We don't need 9000 °F, but it should be possible to obtain the temperatures needed for a power plant (200–400 °C or 390–750 °F) in wells at depths that can be reached with existing drilling technology.

Geothermal heat comes from three main sources: (a) primordial heat from the initial accretion and differentiation of the Earth's interior that has been retained for billions of years, (b) heat produced by the decay of radioactive elements trapped inside the Earth, and (c) frictional heating as parts of the semi-liquid outer core and mantle slide past one another, probably pushed and pulled in part by tidal forces from the moon.

The temperature inside the Earth increases with depth along what is known as the "geothermal gradient." This gradient varies from place to place but averages 25–30 °C/km (15 °F/1000 feet). The temperature at any particular depth can be estimated from surface temperature plus the geothermal gradient, where the average surface temperature of the Earth is about 15 °C or 60 °F. Using these values as guidelines, rocks at the temperature of boiling water (100 °C or 212 °F) should be found at depths of slightly more than 10,000 feet (3 km). This is easily within a typical drilling depth for oil and gas wells.

I learned about geothermal gradients first hand while spending a hot Texas summer in Galveston County back in the 1980s. We were monitoring watered-out gas wells on a secondary gas recovery project that were producing saltwater brine from the Frio Formation at depths of about 10,000 feet and bringing natural gas with it to the surface as bubbles in the brine. The fizzy water went into a tall vertical tank next to the well where the gas and brine were separated by gravity. The gas was sent to a pipeline and the brine was stored in a large holding pond at the wellsite for later disposal by injection into a shallower formation. This "pond" was simply a lined and bermed open pit with no guardrails or fence and filled to a depth of perhaps 10 feet (3 m) with saltwater at a temperature of 88 °C (190 °F) according to our readings at the inlet. It was still hot enough to brew tea. An itinerant young geologist falling in would have twitched a few times and then cooked up like a boiled lobster. I was terrified to go near it.

A new geothermal technology under development by the Department of Energy called engineered or enhanced geothermal systems (EGS) employs a pair of vertical wells that deviate into parallel laterals after reaching deep, hot rocks. Hydraulic fractures are created to connect the laterals with flow paths through the hot rock so fluids can move from one borehole to the other. A working fluid such as brine is then injected into one well, collects heat from the rock as it moves through the fractures, and brings that heat to the surface in the second well. A heat exchanger is used to boil water, make steam and run a generator.

EGS is a modification of an earlier technology called "hot, dry rock" that sought to employ vertical wells connected with fractures at depth (Fig. 7.4). However, intercepting vertical well bores with vertical fractures is very challenging and this technology never actually worked.

Using laterals in deep, hot rock instead of vertical wells places the boreholes in an orientation where they can be more easily connected by hydraulic fractures. Horizontal boreholes also greatly increase the volume of rock from which heat can be extracted. Staged hydraulic fracturing is used to create multiple flow paths through this large volume of rock, further increasing the heat recovery. This technology sounds a lot like George Mitchell's design for extracting gas from large volumes of shale, as described back in Chap. 4, and it should because it was adopted from the shale gas industry.

The EGS technology has not been without problems, and the DOE Geothermal Technology Office (GTO) is funding a field research site in Milford, Utah called

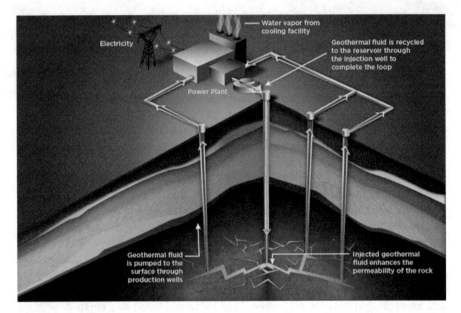

Fig. 7.4 Schematic of a hot, dry rock geothermal system with a single vertical injection well and multiple production wells connected by fractures. The Enhanced Geothermal System uses horizontal wells in the hot rock for better interception by vertical fractures and a greater volume of hot rock that can be reached by the circulating fluid. (*Source: U.S. Department of Energy, Geothermal Technology Office; public domain*)

FORGE (Frontier Observatory for Research in Geothermal Energy) to test concepts and work out difficulties.[7] Some of the problems include keeping the hydraulic fractures to a uniform size and getting them to break in the proper direction to intercept the other well.

Hydraulic fractures tend to be of different widths, or apertures. Fluids will preferentially flow through the wider aperture fractures and not as much through the narrower ones. It's like traffic on a six lane freeway versus a two lane highway. This difference in flow ends up removing more heat from the rocks that contain the wider aperture fractures but tends to leave it behind in the other rocks. The efficiency of the system would be higher if the flow was distributed more evenly, such as through a sedimentary rock like a sandstone with a uniform pore network. Unfortunately, most formations at the depths necessary to provide sufficient heat for electrical power generation are igneous or metamorphic rocks without much inherent porosity that must be fractured to create flow paths.

Another problem with pushing fluids through fractures is the potential to trigger earthquakes. This so-called "induced seismicity" has been a headache for the oil industry for the past decade. While there have been instances of fracking-induced earthquakes, most notably in England, much of the induced seismicity in the U.S. is

---

[7] https://www.energy.gov/eere/forge/forge-sites

related to the disposal of oilfield waste fluids deep underground. Some of these are flowback and produced water from hydraulic fracturing, but the bulk of the volume comes from enhanced oil recovery (EOR) operations.

EOR typically uses an oilfield brine-injection process called a waterflood to push the residual oil in a nearly-depleted field toward production wells. The oil is recovered along with a lot of wastewater that must then be disposed of by deep injection down Class II Underground Injection Control (UIC) wells. Putting large volumes of water down disposal wells without giving it time to spread out into the formation causes pore pressures to increase. This pressure can "unlock" the walls of an existing fault and if there is any stress built-up across it, the two sides will slip past one another and generate an earthquake. Places that are not usually considered seismic hotspots, like southern Oklahoma or northeastern Ohio suddenly found themselves subjected to frequent earthquakes from oilfield wastewater disposal operations.

Induced seismicity may also be a problem with EGS, but data are sparse at present, consisting mostly of modeling studies. However, we do know that fluids moving through underground fractures will increase the pore pressure, and any increase in pore pressure always has the potential to find its way into a pre-existing fault and trigger slip. Induced seismicity tends to quiet down over time as residual, built-up strains on faults are relieved by the action of multiple earthquakes. Still, the local effects can be quite damaging in some places that are not built to withstand earthquakes. Masonry structures that would never be built in places like California without major seismic reinforcement came tumbling down in Oklahoma.

Some ancillary work on EGS is also being done by a national lab research consortium in the Sanford Underground Research Facility (SURF) in Lead, South Dakota (the former Homestake gold mine), which provides access to the deep subsurface. It is named for Denny Sanford, a South Dakota philanthropist who supported the solar neutrino measurements in the mine during the uncertain transition period from the Homestake Mining Company to the U.S. DOE. The SURF geothermal group is drilling horizontal boreholes into tunnel walls at a depth of 4100 feet (1250 m) to test rock mechanics, measure stress fields, and learn how to control the direction of hydraulic fractures in deep rocks.

This is a follow-on to some earlier studies done on the 4850 feet level. The geothermal experiments had to move to the 4100 feet level because the cost of operating the facility is now being covered by Fermilab in Chicago, and they are in the process of developing the deeper level for some upcoming (and really expensive) solar neutrino physics experiments. Once the physics people found out that the geothermal work was going to involve fracking, they demanded that the geologists and their drill rig move elsewhere. I have visited both of these levels in the mine, and it is quite an experience to be that deep underground. The ride down in the lift cage takes more than 10 minutes.

One wouldn't think that intercepting a horizontal well with a vertical fracture would be all that difficult, but it has proven to be a challenge. The stress fields underground are hard to predict, and as the fracture pries apart the walls of rock and opens up space, the stress direction changes, altering the direction of the breaking fracture. Lawrence Berkeley Lab and several other organizations are leading the study of these fractures and learning some interesting insights that will be applied later to the FORGE site.

Another issue with geothermal energy is that hot fluids moving upward from below tend to contain dissolved minerals that precipitate out as the liquid cools. Anyone who has ever visited the hot springs at Yellowstone National Park has seen the large, terraced deposits of hydrothermal minerals that precipitated out of the spring water as it cooled. In a geothermal power plant these minerals can cause the rapid buildup of deposits called scale on the inside of pipes and clog up the plumbing. Fluids in natural geothermal systems typically have a long residence time in the ground that allows them to incorporate more minerals. Scale deposits may be less of a problem with the rapid flow through hydraulic fractures in an EGS.

The Rotokawa Power Plant in New Zealand taps into a natural geothermal system and is famous in the industry for the rapid build-up of hydrothermal mineral deposits in the pipes. The surface plumbing must be replaced and cleaned out on an almost annual basis. Fortunately, the mineral deposits contain significant amounts of gold, which usually covers most of the cost.

To obtain the preferred steam temperatures between 200 to 400 °C (390 to 750 °F) for a thermoelectric power plant, EGS wells in the 20,000 feet (6 km) depth range are needed. While technically more challenging than shallower wells, this is not an unusual borehole depth for the oil and gas industry. As mentioned previously, this temperature range is only for water and lower temperatures can be used with other liquids.

If EGS can be developed to run a modified commercial power plant, coal and natural gas can be eliminated from existing thermoelectric facilities by simply drilling some deep wells on the power plant property and extracting heat from the Earth. There is a capital expense involved in drilling the wells and running a geothermal loop to the boiler, but once the system is functional, the heat is essentially free and sustainable. Modifying a coal-fired power plant to run on geothermal heat is a far cheaper way to decarbonize it than abandoning an existing facility and replacing it entirely with wind turbines or solar. EGS technology can go anywhere. The Earth gets hotter with depth no matter where you drill. Most conventional geothermal resources, like hot springs and geysers are not located where the electricity is needed, and this has severely restricted development. The ability of EGS to be "location agnostic" is a huge advantage that allows it to be applied to existing power plants.

A potentially even more sustainable improvement to EGS is a hybrid geothermal technology that includes solar heat in the system. Known as Solar Assisted Geothermal Energy or SAGE, it uses wells drilled into shallower, porous sedimentary rocks where the fluids move through an interconnected pore network instead of fractures. Flow through a porous rock is much more uniform and evenly distributed than flow through a fracture system, improving the efficiency of heat extraction and storage.

The SAGE system uses a solar heating component at the surface, such as mirrored parabolic troughs with black pipes running down the center. Groundwater is pumped up from the aquifer, flows through the solar troughs where it is heated, and is then injected back into the rock formation. Once circulation is established, the heat from the injected water is transferred to the body of the aquifer rock itself, warming it and providing hot water day or night. Depending on the size and input of the solar component, SAGE systems can reportedly get quite hot. SAGE can be used to supply a district heating system for the direct heating of buildings and other structures, or to generate electricity if the temperatures are high enough.[8]

A concept for a SAGE system in a direct heating application is shown in Fig. 7.5. Solar heat is added at the surface between the extraction well and the hot injection well, and the natural flow gradient through the aquifer carries the hot water into the rock. The thermal mass of the aquifer rock stores the heat until it is extracted for use. Afterward, the spent, cool water is injected into a third well downgradient of the hot zone to prevent it from cooling the aquifer.

The advantages of a SAGE system include shallower wells than those required for EGS, with the associated substantial savings on drilling costs. If such a system could be developed in a depleted oil field using existing wells to recover and inject fluids, even more money could be saved on drilling costs. Heating up the brine in a watered-out oil reservoir with solar energy could be an excellent way to repurpose old oilfields as useful energy resources, but in this case the energy is carbon-free and sustainable.

Because the system adds the solar heat needed for geothermal energy, there is no ambient temperature requirement for an aquifer, only that it be deep enough to contain the pressure. Once the rocks warm up, an aquifer that was previously too cold for effective geothermal use can supply abundant hot water at efficient temperatures. Aquifers and depleted, conventional oil reservoirs with high porosity and permeability can be used for SAGE without the need for hydraulic fracturing. This will

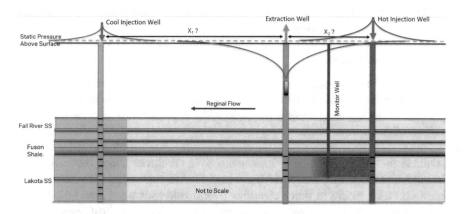

Fig. 7.5 Conceptual cross-section of solar assisted geothermal energy (SAGE) transferring solar heat to an aquifer. Warm water extracted for heating buildings is injected into a third well farther downgradient to keep the now cold water away from the hot aquifer. (*Source: South Dakota School of Mines & Tech DOE proposal team, used with permission*)

---

[8]Zhou, C., Doroodchi, E., and Moghtaderi, B., 2013, An in-depth assessment of hybrid solar–geothermal power generation: *Energy Conversion and Management,* v. 74, p. 88–101.

provide a more uniform distribution of heat through a pore network compared to preferential flow through fractures, and also greatly reduce the potential for induced earthquakes. Finally, a SAGE system doesn't actually consume any groundwater; all produced water is only used to transfer heat and then returned to the aquifer afterward.

This system stores heat and avoids the intermittent power losses of other solar-based systems like photovoltaics due to night or cloudy days. More solar heat can be injected downhole during the summertime to build up the temperature, and the thermal inertia of the rock will allow the aquifer to continuously produce hot water, day or night, rain or shine. The system could operate efficiently with what is called a 5-spot pattern where four hot injection wells surround a single extraction well. There are many engineering challenges to overcome, but if solar heat can be used to create high temperature groundwater in locations that previously had no geothermal resources, this could be a revolutionary source of energy.

Sites like FORGE in Utah should be replicated elsewhere. Research on EGS in deep sedimentary basins and in a variety of other rock types and environmental settings would be helpful for developing this promising technology. SAGE research should receive massive funding because even if an EGS site is marginal, adding heat from sunlight can improve its performance. Geothermal has an incredible amount of potential to change the way we make electricity and instead of budgeting millions, DOE should be giving it billions.

3. New technology nuclear. Nuclear reactors have a major advantage of generating copious quantities of heat for thermoelectric power without producing any GHG emissions. However, nuclear energy is not viewed by the public as cutting-edge technology, and in fact many existing nuclear power plants can be accurately described as old and cantankerous. This is understandable as many were built four to six decades ago using Cold War technology. They are literally fueled with radionuclides left over from the construction of atomic weapons. We can do much better than this.

Nuclear power has both a perception problem and a logistics problem. The perception problem has to do with fears among the public about the risks of nuclear power, and the logistics problem has to do with nonproliferation and the long-term storage of radioactive waste.

The fears are driven largely by the disasters and reactor meltdowns at Three Mile Island in the United States in 1979, Chernobyl in the Soviet Union (now Ukraine) in 1986, and Fukushima Daiichi in Japan in 2011. Although some 441 reactors have been operating safely in thirty countries[9] worldwide over more than 60 years of commercial nuclear power generation, these three accidents still terrify people. Like fracking, nuclear power suffers from a perceived risk that is substantially greater than the actual risk.

Although the probability of a nuclear accident is quite low, the consequences can be severe. Remember from the risk discussion back in Chap. 1, if the consequences

---

[9] https://www.statista.com/statistics/267158/number-of-nuclear-reactors-in-operation-by-country/

are severe enough, even generally "safe" activities like skydiving can be considered risky. The immediate consequences of a nuclear accident may include deaths from acute radiation poisoning, such as the 29 firefighters who reportedly perished at Chernobyl while attempting to put out the reactor fire and contain the radioactivity. Two people were also killed there in the initial blast. The Three Mile Island accident officially resulted in zero deaths, and no immediate fatalities from the reactor breach were recorded in the Fukushima Daiichi accident, although it was linked to a lung cancer death 7 years later.

Of perhaps greater fear to many people are the potential long-term impacts of nuclear disasters, which may show up years to decades later. Foremost among these is cancer resulting from long-term exposure to low-level, but still elevated, doses of radiation. Radioactive fallout from the Chernobyl fire, smoke, and ash cloud descended across much of Europe, and elevated radiation levels affecting ecosystem and human health remain a significant concern to this day. In fact, a 2014 report from the German government found that one third of the wild boars hunted in Saxony were too radioactive to eat. Similar radioactive pigs have been found in the Czech Republic and elsewhere in central Europe. It is thought that the wild mush-rooms the boars consume as part of their normal diet have concentrated radioactive cesium in the soil from the Chernobyl fallout and are depositing it in the pigs. Given the half-life for cesium decay, this can be expected to remain a problem for the next 50 years.

The recent Russian "special military operation" in Ukraine passed through the Chernobyl area, with tanks and other vehicles reportedly stirring up dust in the "Red Forest," one of the most radioactive regions on Earth. Some reports had infantry soldiers digging trenches and bivouacking in this highly contaminated soil. The consequences of this remain to be seen, but the invading soldiers who paid little heed to the warnings of Ukrainian nuclear engineers at the Chernobyl power plant may soon come to regret their actions.

The only radionuclide release from Three Mile Island in the U.S. was a small amount of krypton gas, and although there have been claims that the accident resulted in above-average rates of cancer and birth defects in the surrounding area, this has not been proven. In Japan, the 16,000 estimated deaths from the earthquake and tsunami that triggered the Fukushima reactor malfunction were far higher than any deaths attributed to the reactor leak itself. Along with the lung cancer case men-tioned above, the only other deaths that the World Health Organization was able to directly link to the reactor accident are 573 fatalities from road accidents or stress-related illnesses such as heart attack or stroke that occurred during the panicky evacuation of the area around the nuclear plant.

Nevertheless, decades of chronic radiation exposure from reactor leaks can add up to a significant toll. The United Nations estimates that 4000 to 9000 people may eventually die of cancer due to radiation from the Chernobyl incident.[10] Radiation

---

[10] Bennett, B., Repacholi, M., and Carr, Z. (eds.), 2006, Heath Effects of the Chernobyl Accident and Special Health Care Programmes: Report of the UN Chernobyl Forum Expert Group "Health": Geneva, Switzerland, World Health Organization, 167 p.

exposures were high because of the widespread area covered by fallout and the reluctance of the Soviet government to report the true scope of the accident. As such, some researchers have claimed that Europe may actually suffer between 30,000 and 60,000 cancer fatalities because of the widespread, low-level radioactive fallout from Chernobyl.

Overheating of the reactor cores due to a loss of coolant was the ultimate cause of the disasters at all three plants. Without coolant, the reactors melted down into a molten substance called "corium," composed of radioactive fuel, the zirconium cladding of the fuel rods, and concrete, sand, steel, and other materials from the reactor structure. At Three Mile Island and Fukushima Daiichi the corium stayed within the containment domes, but at Chernobyl it melted through the floor and flowed like lava into the basement, creating a structure called the "elephant's foot" (Fig. 7.6).

Chernobyl was obviously the worst of the three nuclear power plant disasters because its obsolete design used a reactor core made from combustible graphite as a moderator. When it overheated, the graphite burned like charcoal in a massive fire that lofted radioactive fallout high into the sky and distributed it far and wide. The high pressure water reactors at Three Mile Island and Fukushima Daiichi suffered meltdowns from coolant leaks or pump failures, but at least they didn't catch on fire. It is important to note that each of these incidents resulted in the recognition of root

Fig. 7.6 The glowing "elephant's foot" made of corium at Chernobyl photographed in 1996 ten years after the accident. "Lightning streaks" are traces of the inspectors' flashlights and speckles are radiation damage to the film. (*Source: U.S. Department of Energy, public domain*)

causes for previously unrecognized engineering problems, leading to major safety improvements in nuclear facilities world-wide.

The Three Mile Island accident produced better control systems, greater redundancy, and additional monitoring sensors in the reactor and surrounding cooling system. Fukushima showed that contingency planning is important for all natural disasters, even low probability events like a tsunami. Chernobyl demonstrated the foolishness of using old-style, graphite-moderated reactors that were first developed in the 1940s under the Manhattan Project. As a result of these lessons learned, nuclear power is considerably safer today than the electricity generated from the early reactors of the 1950s and 1960s.

Even with our current old-fashioned Cold War reactor designs, nuclear electricity is far safer for human life than any fossil-fuel based energy when compared to kilowatt-hours of power generated.[11] Burning coal to produce electricity releases particulates, toxic metals, and organic compounds into the air, and results in a much greater number of premature deaths from air pollution than any other kind of electrical generation (Fig. 7.7). This is especially true for lower Btu brown coal or lignite, which requires the combustion of significantly more material to produce the same energy output as denser black coals. Even oil, gas and biomass contribute to fatalities because they all produce smoke or vapors. Nuclear has an extremely low death rate because it emits essentially nothing.

Reducing risk on nuclear energy requires reducing both the probability of an accident and the consequences if one does occur. We can reduce the probability of an accident by improving the redundancy of safety systems, adding more backup systems, and using more sensors paired with better monitoring systems to ensure that things don't get out of hand in the first place. The consequences of an accident can be reduced by making the reactors smaller and more compact, designing them with alternative cooling systems, and using a safer nuclear fuel cycle based on thorium instead of uranium.

New nuclear reactor designs are smaller, and smaller reactors are much less prone to overheating and suffering catastrophic meltdowns. They literally do not contain enough radioactive fuel to melt into corium after a loss of coolant event. Modern designs use molten salt or liquid sodium metal for cooling, eliminating some of the problems that come with water-based cooling systems. These include plumbing ruptures from over-pressurized steam (the root cause of coolant loss at Three Mile Island) and the thermal dissociation of water into flammable hydrogen gas and oxygen (the root cause of the explosions at Fukushima Daiichi).

One of the enduring criticisms against nuclear power is that it produces high-level radioactive waste that must be handled properly. This is true, but first some background: The fuel cycle of typical reactors uses what is called "enriched uranium," a mixture of the two main uranium isotopes $_{235}U$ and $_{238}U$. Both of these are unstable and break apart or "fission," but the lighter isotope $_{235}U$ releases more

---

[11] Markandya, A. and Wilkinson, P., 2007, Electricity generation and health: *The Lancet*, v. 370, No. 9591, p. 979–990.

Death rates from energy production per TWh

Death rates from air pollution and accidents related to energy production, measured in deaths per terawatt hours (TWh)

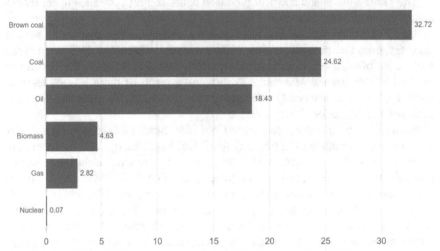

Source: Markandya and Wilkinson (2007)   OurWorldInData.org/energy-production-and-changing-energy-sources/ • CC BY
Note: Figures include deaths resulting from accidents in energy production and deaths related to air pollution impacts. Deaths related to air pollution are dominant, typically accounting for greater than 99% of the total.

Fig. 7.7 Death rates from different forms of energy production. (*Source:* https://ourworldindata. org/safest-sources-of-energy; *Open Access*   )

neutrons when it decays and supports the "chain reaction" that creates the explosion in an atomic weapon. Unfortunately for the weapon-makers, the heavier isotope $_{238}$U is far more common.

"Little Boy," the first atomic bomb dropped in Japan on Hiroshima in World War II was a uranium weapon powered by the $_{235}$U isotope, tediously separated by weight from the heavier $_{238}$U isotope. This separation process was so challenging that the bomb almost didn't get built, and there was only one of these weapons in the U.S. arsenal. You can envision the difficulty by imagining a crowd of nearly identical people in a room, some of whom weigh 235 pounds mixed in among a majority that weigh 238 pounds and your job is to separate them by weight. The Manhattan Project successfully isolated the uranium isotopes using a complicated process that involved uranium hexafluoride gas and separation of the isotopes using centrifuges. It was carried out under a project called K-25 at the Clinton Engineer Works located on a federal reservation at Oak Ridge, near Knoxville, Tennessee.[12] The K-25 operation is long gone, but the DOE Oak Ridge National Laboratory is a neighbor.

Most of the uranium in nuclear power plant fuel rods is $_{238}$U with a small percentage of $_{235}$U added to "enrich" it and create a low-level chain reaction that produces heat. Nearly pure $_{235}$U is used in atomic bombs, so processing uranium ore for

---

[12] Atomic Energy Commission, 1948, Manhattan District History, Book II - Gaseous Diffusion K-25 Project - Volume 4 Construction, 366 p. [declassified 2013]

weapons produces a substantial amount of $_{238}U$ as a leftover waste product. This so-called "spent uranium" is a very heavy element, nearly 20% heavier than lead, and some military ordinance employs it as projectiles. However, the major use is in power plants. It does not emit enough neutrons during fission for an explosive chain reaction and cannot be used in an atomic weapon.

When exposed to the particle flux inside a nuclear reactor, some of the $_{238}U$ will add an extra proton to the nucleus, gain a matching electron and become $_{239}Pu$ or the element plutonium, which can be used in atomic weapons. In fact, "Fat Man," the second nuclear bomb dropped on Nagasaki, Japan was a plutonium weapon, made with Pu created from $_{238}U$ in a reactor located at the Hanford Site in Washington state, now the location of the DOE Pacific Northwest National Laboratory. Because $_{239}Pu$ is a totally different element than $_{238}U$, it can be separated chemically from the uranium. This was much less challenging for early atomic scientists than separating the two uranium isotopes.

Which brings us to the nuclear waste problem. As uranium fissions under the flux of subatomic particles in a reactor, in addition to creating heavier elements known as "transuranics" such as plutonium, neptunium, americium, curium, and others, it also breaks apart into lighter daughter products like thorium, polonium, bismuth and lead. These build up over time in the fuel rods, absorb neutrons and slow down the nuclear reactions. Even though a considerable amount of the original uranium may still remain in the fuel rod, about a third of the nuclear fuel rods in a power plant reactor must be replaced with fresh ones on an annual basis. This means that the entire inventory is replaced every 3 years. There are two options for dealing with this "spent fuel" after it is removed from the reactor: it can be reprocessed, or it must be handled as waste.

Many countries with nuclear power plants reprocess the old fuel rods to reduce the amount of nuclear waste they must deal with. Reprocessing involves a chemical "cleaning" of the fuel rod material to remove the non-fissionable daughter products and other elements. The remaining uranium from the original fuel rod plus the other fissionable materials created in the reactor such as transuranics are then reconstituted into what is called mixed oxide fuel, or MOX, along with some fresh uranium to top it off.

Since the days of the Carter Administration, the United States has expressed concerns that MOX contains plutonium, and the more reprocessing cycles that the spent fuel goes through, the greater the Pu content becomes. The U.S. decided against reprocessing spent fuel because of fears that terrorists or other bad actors could somehow obtain the fuel rods, easily separate out the Pu chemically and build a bomb. The reasoning behind this is called "nonproliferation" and it was supposedly designed to limit the number of nuclear weapons in the world and who has access to them. As such, all spent fuel from U.S. nuclear power plants is handled as high-level radioactive waste without reprocessing, and therein lies the problem. We have nowhere to put this stuff and there is a lot of it.

I have multiple issues with nonproliferation being an excuse for not reprocessing spent nuclear fuel. First of all, the decision against reprocessing was made in the 1970s, when the nuclear club was quite small. These days, everyone from Pakistan to North Korea has nuclear weapons, which is not really surprising because the nuclear technology behind the Manhattan Project is over 75 years old. You can find instructions for building an atomic bomb on the internet. The physics are actually pretty simple, and the key secret in 1945 for initiating the nuclear chain reaction was in the design of the electronics. I think we can all agree that any 75-year-old electronics secrets are beyond obsolete. If a terrorist really wants a nuclear weapon, they can probably get one. Given the availability of nukes in the world, stealing power plant fuel rods to cobble one together is almost laughable.

Instead of actual atomic weapons, the real concern about terrorists should be over the development of "dirty bombs." These are conventional explosives loaded up with finely pulverized radionuclides so that when the bomb explodes, it spreads radioactivity over a large area and contaminates it for years to decades. Unlike atomic weapons, dirty bombs are easy to make, and they don't need any specific radioactive elements. Just about anything radioactive will do, from medical waste to spent power plant fuel rods. As such, the total amount of fissionable material in the world should be kept to a minimum. By not reprocessing spent nuclear fuel into MOX, the United States is greatly increasing the global quantities of radioactive material and putting us all at risk. This is the true terrorist danger.

The "once-through" process for powering U.S. nuclear powerplants is not only a security issue, but also environmentally wasteful. Uranium is not an abundant element and mining it can cause significant environmental impacts. Most uranium production these days uses solution mining where wells are drilled into a subsurface ore body. Fluids and chemicals are circulated down one well and up another in a process called in-situ leaching (ISL) to dissolve the uranium out of the rock and bring it to the surface.[13] Although this is less impactful than digging a mine, there are concerns about this process because the uranium is recovered in a water-soluble oxide form, and leaks or spills can be highly dangerous. Greater amounts of uranium must be mined to meet the needs of powerplants than would otherwise be necessary if the spent fuel rods were being reprocessed into MOX. This additional mining has environmental consequences that could have been avoided if uranium demand was lower.

In the end, not converting spent fuel into MOX greatly increases the volume of high-level radioactive waste that must be handled. Disposing of this waste in an environmentally acceptable manner has been an unsolved problem in the U.S. for almost 70 years.

I spent 8 years working out at Yucca Mountain in Nevada to characterize this site as a potential repository for high-level nuclear waste. The facility was huge, with a five mile (8 km) long main tunnel 25 feet (9 m) in diameter that was supposed to connect with hundreds of side tunnels to store thousands of casks of nuclear waste (Fig. 7.8). Although no physical or technical problems have ever been identified that

---

[13] https://world-nuclear.org/information-library/nuclear-fuel-cycle/mining-of-uranium/in-situ-leach-mining-of-uranium.aspx

would compromise the ability of the site to contain the waste, it had plenty of political problems.

The State of Nevada felt that they had been subjected to a certain degree of high-handedness from Washington by being forced to accept the nation's nuclear waste. This generated a massive amount of resentment, and the state took every opportunity to block, oppose, and slow down the repository. Permits were issued at glacial speed. Any potential safety or environmental violation, no matter how small, was considered grounds for shutting down site operations. Every scientific finding that was challenged, such as the origin of mineral deposits in natural fractures led to extensive investigations, long, drawn-out debates, and expensive analysis to prove or disprove. Environmental groups and native tribes that had been protesting nuclear weapons testing shifted their focus to Yucca Mountain when testing ended in 1992. As a result, the work fell years behind schedule and costs went way over budget. Former President Obama shut the program down in 2011.

However, even in Washington they eventually realize that ending a program doesn't necessarily solve the problem. The bottom line is that we still don't have anywhere to put high level nuclear waste. Even with new technology reactors, any nuclear fission process is going to produce some waste. Utility companies would be foolish to add more nuclear power when they can't dispose of the nuclear waste they already have. We must decarbonize electricity to fight the climate crisis. Nuclear power is an obvious option, but it will not expand until the radioactive waste disposal issue is resolved. So here we sit.

Where is this stuff now? It is residing at power plant sites in vertical concrete casks sitting out in the open. This so-called dry cask storage is supposed to keep

Fig. 7.8 The north ramp of the five-mile-long exploratory tunnel under Yucca Mountain in Nevada. (*Source: Photographed in 1997 by Dan Soeder*)

spent radioactive fuel rods isolated from the environment temporarily until a permanent disposal solution is found. Some of this "temporary" storage has gone on now for six decades or more.

The idea for a nuclear waste repository located under a remote mountain in the desert goes back to the Cold War. Nuclear power plants are placed relatively close to cities because electrical power loses voltage if it is transmitted over long distances. In the paranoid days of the 1950s when potential atomic destruction could be visited upon us at a moment's notice, some Dr. Strangelove-type character had a nightmare. Instead of nuking a city, what if an adversary decided they could obtain a much larger cloud of deadly radioactive fallout if they nuked the atomic power plant next to the city? All the radionuclides in the nuclear reactor would be vaporized by the atomic fireball and incorporated into the mushroom cloud, enriching it significantly. If a bunch of nuclear waste was sitting around at the powerplant and got vaporized as well, all the better! On a day when the wind was blowing in the right direction, the fallout and fatalities could be epic. Thus, removing as much radioactive material as possible from cities and burying it somewhere in a remote desert was considered critical to reducing this risk.

An underground repository for high-level radioactive waste was recommended by the National Academy of Sciences in 1957 at the very start of civilian nuclear power. The reason for a "repository" as opposed to options like dumping it into a deep ocean trench or shooting it into the sun was to monitor the waste and even retrieve it if necessary. No one knew how the storage canisters would hold up over time from corrosion or leakage. There was also a concern that the large stock of fissionable material in a repository might be needed in the future. It was impossible to predict if some new technical or engineering application would surface in a decade or a century that required these radionuclides, and they should remain an accessible resource. In the meantime, isolating the waste deep underground in a remote area would keep it secure for the time being and protect both the environment and public health.

Sadly, despite the ending of the Cold War it is still possible for atomic destruction to be visited upon us at a moment's notice. Maybe not by missiles raining down from the Soviet Union, which no longer exists, but there are plenty of other players and terrorists. Russia still has a substantial number of atomic weapons and they have been saber-rattling lately. North Korea makes routine threats against their neighbors and the United States on a regular basis. Other unfriendly nations have nuclear weapons or are trying to get them. Some guy with a homemade nuke in the back of a van might decide to make a statement someday.

Of course, we all hope that nuclear weapons are never again used in war because they will destroy both sides and make such a war unwinnable. However, there are plenty of reckless leaders and outright madmen out there who might not see the calculus that way. Nuclear power plant reactors are still located close to cities. Loading up the back parking lot with casks of spent fuel rods makes the fallout risk even worse. It seems to me that for basic societal hygiene we should get as much nuclear material as possible out of the cities and locked away in the remote countryside.

DOE initially investigated ten different potential repository locations across the country, with geology that varied from shale to bedded salt to volcanic rocks. The 1982 Nuclear Waste Policy Act directed that full technical assessments be performed on the three top candidate sites, and the best one would be selected based solely on technical merits. The minimum technical requirements for a high-level radioactive waste repository included the following:

- The site must be remote, stable, and on land either presently controlled by the government or readily acquired by the government.
- Groundwater travel times from the nuclear waste storage area to the accessible environment must be at least 10,000 years and the longer the better.
- The site must contain no existing natural resources of any consequence that might entice someone to drill or mine it in the future.

Other factors like accessibility and distance from population centers were also considered and by the mid-1980s, DOE had selected three locations to fully characterize for site suitability. These were a bedded sedimentary salt in the Texas high plains southwest of Amarillo, the Columbia River basalt flows at the Hanford Site in east-central Washington state, and the welded volcanic tuffs at Yucca Mountain, on the western border of the Nevada Test Site.

Congress intervened before any meaningful site characterization could begin at these three locations. A 1987 amendment to the Nuclear Waste Policy Act stopped the investigations at the Texas and Washington sites, designating Yucca Mountain in Nevada as the only site that would be fully characterized. This wasn't just done on a whim – it was a sincere attempt to save money and there was some sound technical reasoning behind the decision.

The Texas location was on privately held land that the federal government would have had to acquire at substantial cost. The target horizon for storing the nuclear waste was a bedded sedimentary salt about 1000 feet deep (300 m), which was considered to be a stable formation. The fact that the highly soluble salt existed in the first place indicated that there was little to no groundwater movement through it. However, to reach the salt formation, the repository shaft would have been required to penetrate the overlying Oglala aquifer, a critically important regional water resource. Despite government assurances that there would be no contamination of the Oglala aquifer, local farmers and others vigorously opposed the Texas repository site.

The Hanford Site in eastern Washington State was acquired by the government in 1943 under the Manhattan Project to manufacture plutonium for the first atomic bombs. A repository at Hanford would be complicated because individual basalt flows at the site were not thick enough to accommodate it within a single flow unit. Cutting tunnels across multiple basalt layers would potentially provide groundwater an easy flow path along the contacts, and at some locations the highly fractured basalts may also have a direct hydraulic connection to the Columbia River. The potential for extremely short groundwater travel times became a show-stopper for Hanford.

Yucca Mountain really was the best of the three sites, but the 1987 amendment to the Nuclear Waste Policy Act created a huge amount of resentment in Nevada for being singled out as the only state in the country forced to host a "nuclear waste dump." There are no nuclear power plants in Nevada, yet they were being forced to take everyone else's waste. Nevada at the time had two Senators and only one Representative in the House, and they were essentially outvoted by nearly everyone else to pass the bill. This also added to the resentment. The 1987 amendment became known locally as the "screw Nevada bill," and opponents of nuclear power used it to stoke public anger and raise state-wide political opposition to the very idea of a nuclear waste repository at Yucca Mountain.

When I lived there, I remember that a city councilman once suggested that Nevada should at least look at the potential employment benefits from the nuclear waste repository. He was nearly run out of town on a rail for such blasphemy. This opposition continued to grow until Senator Harry Reid (D-NV) finally convinced Obama to shut it down.

Okay. But we still don't have a solution for what to do with high level radioactive waste. We must develop new nuclear power technology to replace fossil energy with a carbon-free energy source for the climate crisis. We can't move forward with new nuclear technology until the waste problem from the old nuclear technology is resolved. So, what do we do?

A Blue Ribbon Commission was appointed after the shut-down of the Yucca Mountain Project to investigate other possible options for the disposal of high-level nuclear waste.[14] The commission recommended obtaining consent from state and local governments prior to siting any nuclear waste facilities, create a new organization to oversee nuclear waste given the poor track record of the now-disbanded DOE Office of Civilian Radioactive Waste Management, and improve transportation options for moving large volumes of high-level radioactive waste to a geologic repository. The commission also noted that the lack of a repository was damaging U.S. efforts to innovate in nuclear energy technology, where we once were the international leader.

Unsurprisingly, no other state has stepped up and volunteered to accept this waste. The leaders of a few Native tribes expressed an interest in the possible financial benefits of a waste facility on their reservations but were quickly overruled by their enrolled tribal members. The U.S. Department of Energy is still in charge of nuclear waste disposal with oversight by the Nuclear Regulatory Commission. Most of their focus over the past decade has been dealing with plutonium pits and weapons-grade uranium as nuclear weapons are decommissioned and disassembled in compliance with strategic arms reduction treaties. The weapons-grade waste is stored underground in a salt formation at a facility in New Mexico that has no intention of accepting any civilian waste. The storage of spent power plant fuel rods on reactor sites in dry casks is now classified as "indefinite."

---

[14]Hamilton, L.H. and Scowcroft, B., 2012, Blue Ribbon Commission on America's Nuclear Future, final report to U.S. Secretary of Energy, U.S. Department of Energy, Washington, DC, 157 p. (http://brc.gov)

Massive amounts of time, effort, and money were spent to characterize Yucca Mountain as a nuclear waste repository site. Investments include the tunnel complex, aquifer tests and groundwater studies, rock characterization, geologic mapping, seismic and other geophysical surveys, earthquake risk assessments, volcano risk assessments, future climate assessments, geochemical investigations, ecological and cultural surveys, and physical studies of how the rock responds to heat and radiation. As mentioned, no technical flaws have ever been found that would make the site unsuitable, and substantial amounts of information are available on the site's performance as a repository.

In my opinion, walking away from all this work to start over someplace else would squander huge amounts of time and money for reasons that are exclusively political. The heavy-handed approach that so alienated the people of Nevada would almost certainly have the same result elsewhere. Likewise, the gentler approach recommended by the Blue-Ribbon Commission will work just as well in Nevada as anywhere else. Based on my experience from nearly a decade of working at Yucca Mountain while living in Nevada, I offer the following three suggestions:

1. Negotiate with Nevada. As a state, Nevada is famously agreeable to almost anything, including drinking, gambling, smoking, recreational marijuana, and legal prostitution as long as it is regulated and taxed. They make so much money off vice taxes that there is no state income tax. It has easy marriage and easy divorce. You can play slot machines in grocery stores and gas stations. What happens in Vegas stays in Vegas and so on. The only casino I know of that ever went out of business there was one that became health conscious and banned indoor smoking in the 1990s. Legal brothels are present outside of the counties that host Reno and Las Vegas, including one on U.S. 95 just south of Yucca Mountain.

It turns out that no one from the federal government had actually approached state or local officials at the beginning of the Yucca Mountain Project and asked them for advice on how to make nuclear waste acceptable for their citizens. Federal agencies assumed that if it had been okay to set off nuclear weapons at the Nevada Test Site for 30 years, no one would care if nuclear waste was stored out there as well. However, Yucca Mountain became somewhat of a poster child for government over-reach, and it crossed a red line. By the time DOE figured out that there was a problem, the state had dug in and was uniformly opposed to the entire idea.

The federal government would have received support for a repository had the local population been assured of receiving something in return for accepting the waste, such as financial payouts, new roads, improved power lines, or other infrastructure but none of this was offered at the time. Nevada being Nevada, there is probably still an opportunity to revive Yucca Mountain as a repository site if the government was willing to pay for some tangible and visible benefits. However, at this point, the benefits had better be both very tangible and very visible.

My suggestion for an infrastructure project that is critically needed in Nevada is water. Southern Nevada is a serious desert, and water supplies for Las Vegas are limited. Climate change is negatively affecting the existing water supply for Las Vegas while also increasing water demand. Climate model projections show that this is likely to get worse in the future, not better. People are cutting back on water use, including ripping out their lawns, but there are limits and a minimum demand must be met for the city to survive.

The western United States is experiencing the worst drought in centuries. Winter snowpacks in the Rocky Mountains provide a significant amount of flow in the Colorado River when they melt in the spring. The drought has reduced mountain snowpacks to a fraction of their normal volumes. As a result, Lake Mead behind Hoover Dam has reached a historic lows; so low in fact that the bodies of apparent mob victims have been emerging as the water recedes. Lake Mead provides 90% of the Las Vegas water supply, with the remainder from groundwater withdrawals out of alluvial aquifers at rates far above recharge. The Las Vegas metropolitan region has a population of about 2.2 million and consumes approximately 489 million gallons of water per day.[15]

Water supplies from the Colorado River are fully allocated, and the only way to increase Las Vegas' share is to take it from someone else, which is not popular with others who are dependent on the river and facing water shortages of their own. The explosive growth experienced by the city over the past three decades is about to come to a screeching halt as developers run up against water limits.

Here's an idea: What if seawater from the Pacific Ocean was transported to Yucca Mountain for desalination? The straight-line distance is about 250 miles (400 km), which is not an outrageous length for a pipeline. In comparison, the Dakota Access Pipeline, which carries Bakken oil from North Dakota to Illinois has a length of nearly 1172 miles (1875 km), is 30 inches (0.75 m) in diameter, and moves 31.5 million gallons of oil per day. The seawater supply could use multiple pipelines or a single pipe 6 or 8 feet in diameter (2–2.5 m) to deliver a sufficient volume of water. It could be run mostly above ground with solar panels mounted on top to provide electricity for the pumps. Seawater could then be distilled into fresh water at Yucca Mountain using waste heat from the spent nuclear fuel rods. Although the spent fuel is no longer hot enough to generate steam at the 200–400 °C temperatures needed to run an electric turbine, as long as it remains above boiling it will work for distilling fresh water. There is plenty of open space in Jackass Flats next to Yucca Mountain to construct a massive desalination plant.

Packing the waste canisters close together in the underground passages could easily produce sufficient heat. Such a "hot repository" was one of the options in the Yucca Mountain design plans. Less excavation is required if the canisters in a repository are stored close together. However, there were concerns about what the concentrated heat might do to the groundwater in the surrounding rocks, so the other option was a "cold repository" where the casks are spread out to diffuse the heat. This requires a lot more space and a lot more digging. Placing the waste

---

[15] Renner, T., Las Vegas Goes All in On Water Resources: *Water & Wastes Digest*, February 2020.

canisters close to each other in a hot repository and then removing that heat for distilling fresh water solves both issues by keeping the repository cool and reducing excavation costs.

The energy cost for using this heat is zero. The fuel rods have been paid for by the electricity they generated, and the residual heat will just be wasted on rocks if not used. A heat exchanger system to pull heat from the repository would keep the distillation process isolated from the radioactivity. The resulting fresh water could then be piped 100 miles (160 km) down the valley to supply Las Vegas. Jackass Flats is about 1000 feet higher than the city of Las Vegas, so the water could actually run downhill without pumps in a properly constructed pipeline. If it is more than the city needs, the excess fresh water could be stored in Lake Mead or used to recharge the alluvial aquifers. Compared to most western water projects that involve enormous dams and hundreds of miles of canals, this one is not hugely expensive.

Many of the existing ideas for augmenting the Las Vegas water supply involve transporting groundwater hundreds of miles from other arid valleys against strong opposition from the locals. Most of the population in Nevada is concentrated in Las Vegas and increasing the water supply to the city from a completely independent source like the Pacific Ocean will ease the pressure on these other locations, none of whom has agreed so far to allow Las Vegas to access (some would say "steal") their water.

The desalination of seawater using leftover heat from the nuclear waste could provide a significant quantity of fresh water to the city, and possibly ensure its very survival in a harsh desert climate. A desalination plant built to deliver 50 million gallons of freshwater per day has been proposed for Huntington Beach, California. Building one ten times larger next to Yucca Mountain could essentially replace the Las Vegas water supply from Lake Mead.

One of the main environmental concerns about desalination plants near the ocean is the discharge of the concentrated, residual salty brine back into the sea, where it can harm marine life at the base of the food chain. Running a desalination plant in the desert ensures that this residual brine will not return to the ocean, but in fact it could be evaporated in pans under the hot Nevada sun to form crystalline sea salt. This is another valuable commodity that could be sold for revenue to further improve the economics of this proposed project.

Using waste heat from radionuclides in a Yucca Mountain repository for seawater desalination is sustainable, climate-resilient, economic, and does not impact anyone else's water supply. Developers and politicians ought to support accepting the nuclear waste with such an obvious benefit.

2. Build the facilities using local companies and contractors. Government agencies commonly write "cost plus fixed fee" contracts with the so-called "beltway bandits" in Washington, D.C. These are large companies with three-letter-acronym (TLA) names located around the Capital Beltway. Their lifeblood is Department of Defense, Department of Energy, and other government contract work. Most of the Yucca Mountain site characterization studies were done by these contractors under the leadership of a TLA Management and Operations contractor. People

were brought in from Virginia, New Jersey, Ohio, and other eastern states to work on the project in Nevada. Even most of the U.S. Geological Survey folks who worked on the project were based out of Denver. I was part of a very small contingent located in Las Vegas or at the actual site.

Local engineering and natural resource companies resented this of course, especially when hiring them could have made a real economic difference. The tunnel, for example, could have been constructed by local gold miners, who were struggling to stay afloat at the time with low commodity prices for gold. Las Vegas was loaded up with trade and construction workers who had built the big resorts and were looking for new jobs. Tapping into the local talent pool will create much more support for the repository than contracting out the work to TLA beltway bandits in D.C., who don't really need the money in any case.

The excavation of the repository, surface handling facilities for the nuclear waste, the seawater pipeline, desalination plant, and freshwater pipeline should all be constructed as much as possible by local workers from Nevada and California. There are almost three million people living in southern Nevada these days. Surely the talent is available. Engaging them will really help with acceptance of the facility.

3. Figure out workable transportation options. As much as anything, this was the Achilles heel of the Yucca Mountain Project (and in fact any potential repository site) and was specifically called out by the Blue Ribbon Commission in their report. Hauling nuclear waste on a train through Las Vegas along the Union Pacific rail line that runs adjacent to the Strip and through downtown would not have pleased the wealthy owners of fancy Las Vegas resorts. The main rail line doesn't go anywhere near Yucca Mountain, so an expensive and long spur would have been needed, either from Las Vegas or along a more northern route from Caliente.

Highways are equally problematic; the only road that passes near Yucca Mountain is U.S. 95, a two-lane highway at that point, and many of the roads around the mountain itself are unimproved dirt. Hundreds of trucks transporting hundreds of heavy nuclear waste canisters to the site would pulverize these roads in no time.

A transportation option that has not been seriously considered to my knowledge is to bring the waste in by air. Instead of reinforced roads or long, expensive railroad spurs, the only infrastructure required for air freight is a concrete runway near the repository entrance portal. Existing military cargo aircraft such as the C-17 Globemaster and the C-5 Galaxy that can handle heavy cargo like M-1 Abrams tanks or a load of M777 Howitzers should easily be able to carry multiple canisters of high-level nuclear waste. The canisters and the aircraft can be designed to shield the pilots and crew from radiation during transport, and the short travel time by air compared to days of overland travel would minimize radiation exposure in any case. Having just a runway at the site for bringing in the waste will also keep Yucca Mountain isolated in a way that a paved highway or railroad spur would not.

People obviously worry about airplanes full of nuclear waste falling from the sky, but it is important to remember that power plant waste is solid fuel and will not contaminate soils or groundwater. In the event of an accident, every single fuel pellet can be recovered. Accident probabilities for aircraft are very low anyway; trucks and trains have a much greater risk of accidents per mile compared to aircraft. The aircraft accidents that do occur tend to be weather-related, but cargo flights are not required to follow a fixed schedule, so the flights can be restricted to periods of good weather. If bad weather turns up unexpectedly, cargo aircraft can change routes to dodge it or land and wait it out.

Flight paths can also be routed to avoid population centers, keeping the waste away from cities in a way that can't be done with roads or rails. A number of cities like Flagstaff, Arizona, which sits astride several major rail and highway routes have declared themselves "nuclear free zones" and vowed to block any radioactive waste shipments that try to pass through. Flying the waste around them at a distance is a better option.

Nuclear security concerns about bad actors trying to hijack canisters of transuranic waste are minimal when the waste is flying high above in a cargo plane. Nuclear reactors tend to be located near big cities like Chicago and Philadelphia with major airports, which can be used as launch points for flights to Yucca Mountain. Transporting the waste from the power plant to the airport probably would be the riskiest part of the trip.

A final suggestion for Yucca Mountain is to change the classification of the waste. This may sound like semantics, but there are significant regulatory consequences for something designated "permanent storage" versus "monitored, retrievable storage." Yucca Mountain should be designated as a monitored, retrievable storage site, and the waste canisters should be carefully tracked for signs of corrosion or deterioration. In a century or two when space travel has become routine and reliable, the entire stock of radionuclides could be transferred to the moon if desired. There is no water on the moon to corrode the canisters, no air to spread radionuclide dust, and no ecosystem to destroy with leaks. The moon is the ultimate, safe, permanent storage site for humanity's radioactive waste.

The issues and concerns with Yucca Mountain or any other repository site must be resolved to give the nuclear power program a new lease on life. Providing a critical commodity like fresh water in return for hosting the repository places a solid value on it that can be used in negotiations. For a state like Nevada, showing that a repository is worth a major water project will be crucial to sealing a deal. If nuclear power is to play a significant role in decarbonizing the electricity sector of the economy, we must get over our fear of all things nuclear and come up with a viable solution for isolating and storing high-level radioactive waste. The low carbon energy future needs new technology nuclear reactors with an operational repository for the waste, and we need them soon.

****

New technology nuclear reactors are much smaller than traditional pressurized water reactors and use different fluids for cooling. As mentioned earlier, failure of the cooling systems was a root cause of the core meltdowns at both Three Mile Island and Fukushima. Three Mile Island lost cooling after a steam explosion and pipe rupture, while at Fukushima the coolant circulating pumps failed after the plant was swamped by a devastating tsunami. The loss of coolant at Chernobyl was also a root cause of the meltdown but the combustibility of the graphite reactor vessel and the ensuing fire made everything so much worse. The use of a cooling system other than water may be more efficient, more resilient, and avoid a lot of problems.

New nuclear technology is supported by the likes of billionaires Bill Gates and Warren Buffett, both of whom have made substantial investments in nuclear power. A new technology nuclear plant called Natrium is being built by Gates' TerraPower in conjunction with Buffett's PacifiCorp as a demonstration project for a liquid sodium-cooled reactor. It has been proceeding slowly as they work through permit issues but is expected to go online in 2028. The liquid sodium cooling should be safer and more effective than a traditional, water-cooled nuclear plant. The Natrium reactor also has the potential to recycle spent fuel, reducing the total output of high level radioactive waste.

New radioactive fuel options are also part of improved nuclear reactor technology. Old style reactors that use $_{235}U$-$_{238}U$ enriched uranium are relics of the Cold War atomic weapons manufacturing process, and even those that use reprocessed MOX fuel with $_{239}Pu$ are still reliant on weapons programs to supply their initial fuel. A different and safer fuel cycle can be based instead on the element thorium, which does not fission itself, but is known as a "fertile" element in reactors.[16] When exposed to a neutron flux, thorium undergoes a series of nuclear reactions that eventually create the light uranium isotope $_{233}U$. This isotope is fissionable but does not produce the heavy transuranic elements that come from enriched $_{238}U$. Eliminating transuranics from nuclear waste makes it less toxic and easier to deal with over long time scales, and without plutonium the spent fuel can be reprocessed to reduce waste even further without any concerns about nonproliferation. Thorium is often produced as a byproduct from rare earth element (REE) mining, but because of the limited market for it at present, it is usually treated as waste.

Modern reactor designs are small enough that they can be retrofit into existing coal fired or natural gas power plants to provide a heat source for the boilers. Some new reactor designs can provide 10% of the power of a traditional reactor in 1% of the space. A TerraPower molten salt reactor being developed at Idaho National Laboratory is reportedly about the size of a refrigerator. This technology is known as "ultra-safe nuclear," and these small reactors have the ability to shut themselves down if they begin to overheat. A safety factor in the small reactor designs is that there is simply not enough mass of fissile material in the reactor core to melt down into corium (refer back to Fig. 7.6) if there is a loss of coolant accident.

---

[16] IAEA (International Atomic Energy Agency), 2005, Thorium fuel cycle — Potential benefits and challenges: Report IAEA-TECDOC-1450, International Atomic Energy Agency, Vienna, Austria, 113 p.

Another advantage of small, modern reactors is that they can be manufactured in a factory instead of being constructed piecemeal on site like older reactors. Building precision structures with high engineering tolerances like nuclear reactors on-site requires tedious, repeat inspections for quality assurance, and are one of the main reasons the CAPEX for nuclear electricity is so high, and the construction schedules are so long. Factory construction of small reactors can include the required quality assurance steps in the manufacturing process, turning out uniform, precision units with known performance parameters. A quick test at the factory before shipping should provide all of the necessary safety and performance data needed to license the reactor for operations.

Like engineered geothermal wells drilled on the grounds of a powerplant, small nuclear reactors are location agnostic. Installing them under powerplant boilers will allow utilities to preserve billions of dollars in existing generating and transmission infrastructure while transitioning away from fossil fuels. It is an additional capital cost to be sure, but far cheaper than replacing or abandoning the entire power plant. Reactors can be made small enough to fit inside submarines, and even small enough to fit on spacecraft. Surely with some focus and ingenuity they can be constructed to fit under powerplant boilers to generate steam and produce electricity.

The future of nuclear power will be defined by new reactor technology, new fuel cycles, a reduction in the costs and commissioning times for new reactors, and an agreed-upon policy and plan for dealing with high-level nuclear waste.

****

No matter how the electric power sector is decarbonized, the cost of generating electricity will be more expensive at least initially than burning coal or natural gas and allowing the GHG emissions to vent up the stack. Capturing and sequestering these emissions (discussed in detail in Chap. 9) adds a cost. Substituting manufactured biofuels for fossil fuel adds a cost. Building all new generating capacity for wind or solar adds a significant cost. Even replacing fossil fuel heat sources with geothermal or nuclear heat in an existing thermoelectric power plant will add cost.

Climate skeptics argue that these added costs of GHG mitigation will "ruin" the economy and thus any attempts at decarbonizing electricity will provide a cure that is worse than the disease. This argument fall apart when one realizes that fossil fuel is cheap only because the damage done to the environment by burning coal, oil and gas is not included in the cost of the product. As described in previous chapters, polluted air, contaminated water, and damages resulting from climate change, such as rising sea levels, stronger storms, and more frequent wildfires are all externalized costs paid for by society, not by the fossil fuel producers. When the cost of mitigating such damage is included in the price of fossil fuel electricity, it becomes considerably more expensive.

As a case in point, back in 2009 when the U.S. Congress was seriously considering a carbon tax along with cap and trade on GHG emissions, coal-fired power plants would have been required to install CCS to reduce carbon dioxide emissions. The cost of electricity from these coal plants after adding CCS would have increased to levels comparable with nuclear electricity. As a result, the Nuclear Regulatory Commission received four new applications for old style, pressurized water nuclear reactors in 2010 for the first time in decades.

The cost of electricity is a complicated calculation that involves the economics of capital, fuel, maintenance and operations, and financing. It is affected by different utilization rates and access to fuel resources. Two measures are commonly used to make cost comparisons between different technologies: the levelized cost of electricity (LCOE) and the levelized avoided cost of electricity (LACE).[17]

The cost per-kilowatt-hour to build and operate a generating plant is the LCOE. The LCOE includes capital costs (CAPEX) and financing costs, OPEX that includes fuel, operations, and maintenance costs, and an assumed utilization rate for each power plant type. These costs vary among different regions and over time, and also depend on the type of use planned for the powerplant, such as baseload, peak shaving, or seasonal.

For example, technologies such as solar and wind have no fuel costs for electric power generation, thus their OPEX is usually small. The LCOE for these power sources is mostly determined by their fairly high CAPEX. Coal plants, on the other hand, have a more moderate CAPEX and financing costs, and also a modest OPEX even though it includes fuel costs and ash disposal. If CCS is added to the coal fired power plant either by the utility or in response to government regulations, however, both the CAPEX and OPEX increase significantly. All of these various factors must be considered when comparing the cost of electricity from different sources.

The cost of not adding a specific generating technology to the electrical system but supplying the required additional capacity using existing sources is LACE, the Levelized Avoided Cost of Electricity. LACE represents the value per-kilowatt-hour to the electric grid when generating capacity using a specific technology is added to the system. LACE reflects the cost that would be incurred to provide the same electricity supply if new capacity using that specific technology was not added. For example, if a set of wind turbines are not added to a power grid, the electricity they would have supplied has to be made up by running an existing coal plant at full capacity. Adding the wind turbines allows the coal plant to run at a lower capacity, burning less fuel and incurring lower operating costs. The cost difference per kilowatt-hour between not adding and adding the turbines is the LACE. A new generating technology is generally considered economically competitive when its LACE exceeds its LCOE, or in other words, when it is cheaper to add it than to not add it.

The attempt by the Energy Information Administration to assemble all this information for comparison is shown in Fig. 7.9. The findings are interesting. Electricity produced by coal with CCS is more expensive per kilowatt-hour than advanced nuclear power. Combined cycle natural gas is one of the cheapest, but so is traditional geothermal. A gas combustion turbine without the combined cycle option is

[17] Energy Information Administration, 2015 Annual Energy Outlook Report: Levelized Cost and Levelized Avoided Cost of New Generation Technologies.

as expensive as nuclear. The high CAPEX of the marine construction needed for offshore wind makes this an expensive power option, as is solar thermal with its acres of mirrors. Without the tax credits these costs would be even worse. Biomass is pricey due to the high OPEX, but this is probably dependent on the type of fuel used. Wood pellets are considerably more expensive than methane made from sewage.

Any discussion about decarbonizing the electric power sector of the economy needs to take these costs under consideration. The tax credits shown for the renewable technologies are helpful to the economics and could be improved. Policies that increase the tax credits for non-GHG power like geothermal and add tax credits for other carbon-free power options like nuclear and biomass would help to encourage a switch to decarbonized electricity. Imposing a substantial carbon tax on electric power generation that emits CO_2 would provide an even stronger economic incentive for utilities to decarbonize. If biogas, geothermal, or nuclear heat can be used to replace fossil fuel heat sources in existing thermoelectric power plants, a substantial

Plant type	Capacity factor (%)	Levelized capital cost	Levelized fixed O&M	Levelized variable O&M	Levelized transmission cost	Total system LCOE	Levelized tax credit[1]	Total LCOE including tax credit
Dispatchable technologies								
Coal with 30% CCS[2]	85	84.0	9.5	35.6	1.1	130.1	NA	130.1
Coal with 90% CCS[2]	85	68.5	11.0	38.5	1.1	119.1	NA	119.1
Conventional CC	87	12.6	1.5	34.9	1.1	50.1	NA	50.1
Advanced CC	87	14.4	1.3	32.2	1.1	49.0	NA	49.0
Advanced CC with CCS	87	26.9	4.4	42.5	1.1	74.9	NA	74.9
Conventional CT	30	37.2	6.7	51.6	3.2	98.7	NA	98.7
Advanced CT	30	23.6	2.6	55.7	3.2	85.1	NA	85.1
Advanced nuclear	90	69.4	12.9	9.3	1.0	92.6	NA	92.6
Geothermal	90	30.1	13.2	0.0	1.3	44.6	-3.0	41.6
Biomass	83	39.2	15.4	39.6	1.1	95.3	NA	95.3
Non-dispatchable technologies								
Wind, onshore	41	43.1	13.4	0.0	2.5	59.1	-11.1	48.0
Wind, offshore	45	115.8	19.9	0.0	2.3	138.0	-20.8	117.1
Solar PV[3]	29	51.2	8.7	0.0	3.3	63.2	-13.3	49.9
Solar thermal	25	128.4	32.6	0.0	4.1	165.1	-38.5	126.6
Hydroelectric[4]	64	48.2	9.8	1.8	1.9	61.7	NA	61.7

[1]The tax credit component is based on targeted federal tax credits such as the PTC or ITC available for some technologies. It reflects tax credits available only for plants entering service in 2022 and the substantial phase out of both the PTC and ITC as scheduled under current law. Technologies not eligible for PTC or ITC are indicated as *NA* or not available. The results are based on a regional model, and state or local incentives are not included in LCOE calculations. See text box on page 2 for details on how the tax credits are represented in the model.

[2]Because Section 111(b) of the Clean Air Act requires conventional coal plants to be built with CCS to meet specific CO2 emission standards, two levels of CCS removal are modeled: 30%, which meets the NSPS, and 90%, which exceeds the NSPS but may be seen as a build option in some scenarios. The coal plant with 30% CCS is assumed to incur a 3 percentage-point increase to its cost of capital to represent the risk associated with higher emissions.

[3]Costs are expressed in terms of net AC power available to the grid for theinstalled capacity.

[4]As modeled, hydroelectric is assumed to have seasonal storage so that it can be dispatched within a season, but overall operation is limited by resources available by site and season.

CCS=carbon capture and sequestration. CC=combined-cycle (natural gas). CT=combustion turbine. PV=photovoltaic.

Source: U.S. Energy Information Administration, *Annual Energy Outlook 2018*.

Fig. 7.9 Comparisons of the LCOE per kilowatt-hour for a variety of electrical generating options. (*Source: U.S. Energy Information Administration, Annual Energy Outlook 2018; public domain*)

part of the electric power sector of the economy could be decarbonized quickly and at a reasonable cost.

We must develop robust biogas production, engineered geothermal, solar assisted geothermal, and advanced nuclear power as soon as possible to replace fossil fuel electricity. Funding for research on these technologies is a pittance in light of their potential importance for combating climate change. We must also deal with the nuclear waste issue, possibly by re-visiting the already well-characterized site at Yucca Mountain and developing some creative ways to deal with the politics and transportation issues.

Speaking of waste, the one thing we can't waste is any more time.

Chapter 8
Zero Carbon Vehicles

Keywords Electric vehicles · Methane gas · Hydrogen

****

In America, we love our cars. After the Second World War, car culture in the United States blossomed in a major way. Part of this was due to pent-up consumer demand that had been stifled for two decades, first by the Great Depression and then by WWII. In the booming 1950s, people wanted houses, appliances, and most of all, automobiles.

The second reason for American car culture was that in the latter half of the twentieth century, many people in the U.S. moved out of the inner cities and into post-war suburban housing. The suburbs were built for the automobile and lacked transportation infrastructure like streetcars, buses and subways that existed in cities. Even pedestrian sidewalks and bicycle paths were often hard to find in suburban neighborhoods. It was an absolute necessity to have a car to get around in the suburbs. Institutions like drive-in movies, drive-up bank tellers, and drive-in restaurants were established to cater to the automobile. Las Vegas even has drive-up wedding chapels and at least one drive-up divorce lawyer.

California became (and still is) the center of American car culture. Trends that started primarily in southern California spread throughout the nation. California-based pop bands sang to us about cars in the 1960s. Jan and Dean told us about the Little Old Lady from Pasadena winning races, but then warned us about the risks of racing at Deadman's Curve. Ronnie and the Daytonas sang an ode to their Little GTO. Mustang Sally and her disregard of speed limits worried Wilson Pickett. The Beach Boys had their Little Deuce Coup along with a girlfriend who had Fun, Fun, Fun in her daddy's T-bird. The Hondells had a fast Little Honda. And so on. We were immersed in car songs.

The practice of "cruising," or randomly driving around on designated streets to show off customized, souped-up automobiles to friends and rivals became a staple of teen culture in the 1950s and 60s. If you didn't own one of these "muscle cars," you needed to catch a ride with someone who did if you wanted to be part of the scene (this was my strategy – the only car I had access to as a teenager was my

D. Soeder, *Energy Futures*, https://doi.org/10.1007/978-3-031-15381-5_8

mom's rather sedate Chevy Nova). Mayfield Road through the eastern suburbs of Cleveland was one such cruising route, and I went up and down it many times with my high school friends who had cool cars like Camaros and Mustangs and such. I even had one friend with a souped-up Chevy Nova that was way cooler than my mom's.

By the late 1960s, the United States had become almost totally dependent on gasoline-powered vehicles for travel to work, shopping, church, school, and almost everywhere else. Automobiles so dominated American culture that the only major post-war national infrastructure project to be built was the Eisenhower Interstate Highway System. Drivers loved the new freeways, and these roads quickly became the preferred travel routes from anywhere to anywhere. Railroads that used to run passenger trains between cities switched to hauling freight as people chose to drive long distances and even across the country on freeways. (Despite the introduction of jet passenger aircraft in the 1960s, flying was regulated, expensive, and rare for most people. The current popularity of air travel is a relatively recent occurrence.)

My first drive across the country was in 1973, and although the interstate highway system wasn't yet completed, most of it had been built. We made it from Ohio to California in 3 days on a drive that used to take more than a week. The interstates were so new that there often were no gas stations, hotels, or restaurants near the exits. You had to drive a couple of miles into town to find any services.

By the 1970s, families often owned multiple vehicles, especially if they had driving-age children. I remember at one point between my parents, myself, and my brothers, we had about six automobiles parked in our driveway. It looked like a used car dealer's lot. Prior to WWII, owning more than one automobile per family was practically unheard of, except for the very wealthy.

Where did the suburbs come from? Most date from after the Second World War, which is telling. Military veterans who had postponed plans to get married and start families until the war was over formed a large group of people seeking the "American dream" of single-family home ownership. Loans provided through the Veterans Administration (VA) home mortgage program, authorized in 1944 as part of the GI Bill and still active to this day provided many returning veterans with the ability to purchase houses.

The problem was that there were not a lot of available houses to buy. New home construction had fallen off significantly during the depression years of the 1930s and essentially came to a halt in the 1940s as both materials and construction workers were pulled into the war effort. In an attempt to meet postwar demand, inexpensive housing stock was built quickly in previously rural areas adjacent to city limits during the late 1940s and through the 1950s. Much of it was thrown together, made from cookie-cutter designs, and constructed using cheap materials called "ticky-tacky" in popular culture.

The classic example of postwar ticky-tacky housing is Levittown, PA, a suburb of Philadelphia. Starting with land purchased in 1951, it was planned and built by William J. Levitt. His sons William Levitt and Alfred Levitt designed the houses, which consisted of only six different models. Levitt & Sons used an assembly line process for home building with workers assigned to single tasks like pouring concrete, framing, installing wiring, etc., who moved from house to house. This technique supposedly allowed Levitt's workers to complete about four finished houses per hour. A total of 17,311 homes were built in Levittown between 1952 and 1958. It inspired political activist Malvina Reynolds to write a protest song called "Little Boxes" that became a popular recording by folk singer Pete Seeger in 1963.

I grew up in such a tract home in the Cleveland suburb of South Euclid. It was somewhat larger than the "little boxes" of Levittown and was built in the mid-1950s, right before I was born. The other suburban home my parents had been living in with my older sister was indeed a little box deemed to be too small for a second kid. So, my dad found a new house with four bedrooms that was big enough for him, my mother, my sister, myself, and eventually my three younger brothers. The area had been a rural farm locality when my dad was young, but in my day, it was a fully developed suburb.

Not all the reasons for the migration from cities to the suburbs had to do with the availability of housing stock. Some were grimmer. Large numbers of African-Americans from the south had come to northern industrial cities like Cleveland during the 1940s to take wartime manufacturing jobs. Black neighborhoods expanded as these people earned enough money from factory work to become first-time homeowners.

In order to provide a supply of housing stock, real estate agents needed to move white people out of older houses adjacent to existing Black neighborhoods so they could sell these to African-Americans. Some unscrupulous real estate agents practiced "blockbusting," which is now illegal. The agents would play to the racist fears of many urban whites, urging them to quickly sell out and move to the suburbs or risk huge financial losses and high crime rates once "the blacks" moved in and property values dropped. There was no actual reason for property values to drop when Black people moved into a neighborhood, but it was enough to panic white homeowners into selling. These people would often sell their homes to African-Americans at low prices or even at a loss just to get out, making the decline in property values a self-fulfilling prophecy. Sometimes all the homes in an entire city block would go up for sale at the same time, hence the origin of the term.

Many cities actually banned "for sale" real estate signs on front lawns because when every house on the street had one, it looked pretty bad. The real estate agents themselves weren't necessarily racist. They raked in their commissions no matter who bought the house and the more properties they turned over the more money they made. Regrettably, my grandparents and my parents were among those caught up in this so-called "white flight" from the cities to the suburbs.

Another 1950s fear factor for moving from the cities to the suburbs was that the central parts of cities were considered targets for atomic weapons in the early days of the Cold War. Many people naively felt that the suburbs would be safer in the event of a nuclear exchange. This actually has not been the case since the days of megaton-sized thermonuclear warheads, which will pretty much take out the entire county.

Southern and western cities experienced population and economic growth after the war as the widespread use of air conditioning made living in the south much more tolerable. Population booms were accommodated by rings of suburban sprawl on former farmland around cities like Atlanta, Charlotte, Dallas, Houston and others.

The dependency on cars in the suburbs soon led to their increased use within the cities themselves. Many older American cities, especially in the east have street layouts that were designed in horse and buggy days and are definitely not car-friendly. Five minutes of driving around in downtown Washington, D.C. should be enough to convince anyone that this city was not designed for automobiles. The architect, French-American military engineer Pierre Charles L'Enfant had of course never heard of automobiles when he laid out the city in 1791. Still, L'Enfant's grid-like street layout would have been navigable in an automobile, but his plan was then modified by Andrew Ellicott, who added radiating streets from traffic circles to enable directional lines of fire for cannon emplacements. Cannon emplacements were probably important in 1791, but they are a nightmare these days to negotiate in a car.

Automobiles had already been invented when newer, western cities like Los Angeles were laid out. Accommodating cars here resulted in significant urban sprawl and made Los Angeles the largest American city in terms of land area. Other sprawling newer cities include Las Vegas, Houston, and Phoenix, among others. I first visited these cities in the early 1970s, and I cannot believe how much they have grown. Northern Avenue in Phoenix used to actually run along the north boundary of the town; it is now practically downtown. Getting around any of these cities without a car is nearly impossible. The predominance of people in vehicles driving into cities for work and back out again to their homes in the suburbs has led to one additional post-war invention: the rush hour traffic jam.

****

Electric Vehicles

According to state vehicle registration data, there are presently 289.5 million cars, trucks and SUVs on the road in the United States.[1] This works out to roughly ten vehicles for every dozen American citizens. U.S. drivers burn about 200 billion gallons of motor fuel each year, and transportation accounts for nearly a third of American GHG emissions. President Biden issued an executive order in 2021 for automobile manufacturers to increase average gasoline mileage to 40 miles per gallon by 2026. This will help but it's not enough.

Fossil fuels must be completely eliminated in the transportation sector. The only options at present for zero emission vehicles are electric cars and trucks, now being offered by multiple manufacturers and growing in popularity. However, replacing

---

[1] https://hedgescompany.com/automotive-market-research-statistics/auto-mailing-lists-and-marketing/

nearly 300 million personal vehicles with electric cars will be an environmental disaster. But wait, you say; aren't electric vehicles good for the environment? They are good for climate, but not for the environment. This seemingly contradictory statement is one of the complications of decarbonizing vehicles.

Electric vehicles (EVs) require substantial amounts of exotic, expensive, and environmentally-damaging materials in their construction. Every EV requires a robust battery that can hold enough power to provide a reasonable driving range between charges. Most EV manufacturers use variations of the lithium-ion (Li) battery similar to those found in smart phones and laptop computers, although much larger of course. These are relatively lightweight, have high energy densities, and are rapidly rechargeable. Companies that make hybrid vehicles (HV) lean more toward nickel-metal hydride (NiMH) batteries. These can handle sudden power demands just as well as Li batteries, but the NiMH battery can operate and perform better than Li in extreme cold weather. The Li batteries have recharge, size, and weight advantages over NiMH. An NiMH pack weighs about 25% more and occupies about 20% more space than an Li pack with a comparable capacity and output.

Other EV battery types are lithium polymer, lead–acid, nickel-cadmium, and the less common zinc–air, iron-air, and sodium nickel chloride. A hot new design called a vanadium redox flow battery promises to hold substantial amounts of power and go through many recharge cycles witout degrading. Lithium, lead, nickel, cadmium, vanadium, and zinc are all relatively rare metals, and this brings us to the EV environmental problem. All of these must be extracted from the Earth by mining. Replacing every American fossil-fueled vehicle with an EV means that huge amounts of these metals must be mined. In addition, EVs require copper for wire and motor windings, the REEs neodymium, praseodymium, and dysprosium for the magnets in electric motors, and cobalt for the cathode structures in Li batteries.

China possesses more than a third of the REEs in the world and supplies more than 85% of global production. Chinese REE deposits are mined from weathered granites in the southern part of the country in an unusual geologic setting found nowhere else.[2] China may or may not be willing to export these critical materials to the United States. Some 70% of global cobalt comes from the Democratic Republic of Congo, much of it produced by euphemistically-named "artisanal" mining operations, generally small-scale digs that often include child labor, other human rights abuses, and significant environmental damage. The second largest cobalt producer is Russia. Cadmium is a toxic metal produced as a byproduct of zinc mining, and sometimes from lead and copper production. A large percentage of cadmium comes from Asia.

So, we seem to have a problem here. We need more EVs to help climate but building 300 million of them will simply ravage the Earth. In addition, there are questions about whether the Earth can even supply these elements, known as "critical minerals," in the quantities needed. Transitioning to 100% electric vehicles to

---

[2] Xu, C., Kynický, J., Smith, M. et al., 2017, Origin of heavy rare earth mineralization in South China: *Nature Communications*, vol. 8, article 14,598 (https://doi.org/10.1038/ncomms14598)

decarbonize the transportation sector will mean that huge amounts of these various critical minerals will have to be obtained.[3] At the moment, the supplies are simply not available.

People are exploring for new sources of critical minerals and also attempting to develop a circular economy for these by more diligent recycling. Electronic trash (i.e., old computers, cell phones, televisions, etc.) contains significant quantities of REEs and other materials that can be reused. A technology called "biomining" uses plants to extract and concentrate the desired metals, which are then obtained from the harvested plants. It is designed for use on low concentrations of metals, such as the tailings piles around mines where lower grade ore was dumped and on obsolete electronics in landfills. Plants can more efficiently collect and concentrate the metals from these low-grade resources compared to inorganic chemical processes like heap leaching.

REEs can be extracted from volcanic ash beds in coal seams called tonsteins, and from similar ash beds in shales. Lithium is typically obtained from saltwater brines pumped from deep underground. Nickel, zinc, cadmium, lead, and copper can be gleaned from old mine tailings. All of these elements also can be recovered from landfills. Electronic waste buried in landfills contains many of the critical metals and rare earths needed for EVs. Developing technologies for extracting these elements from this waste will reduce the toxic metals content of the landfills, provide a domestic source for these materials, and recycle valuable resources. Demands (and prices) for critical minerals are expected to increase sharply as the manufacture of electric vehicles ramps up, and some entrepreneurs see this as a developing market.

Tesla has made enormous strides in bringing EVs to market since its founding in 2003 (and not by Elon Musk, by the way – he joined as an investor in 2004). Big companies like Ford and General Motors now also produce electric vehicles. Nevertheless, depending on who's numbers you look at, EVs only make up 2–6% of the total U.S. automobile fleet, with 94–98% of vehicles in America still running on fossil fuel.

The requirements for EV batteries are strenuous: they must be able to store substantial amounts of energy while quickly recharging and retain their energy density over thousands of discharge and recharge cycles. They can't be too heavy, or the extra weight will reduce the driving range. They have to be robust enough to survive mud, rain, snow, road salt, and dirt, desert heat, polar cold, bad roads, and potholes. It's a tough order, and it's why companies like Tesla that are dedicated to the production of electric cars are essentially battery companies that almost manufacture vehicles as a side item.

Battery capacity is defined by electrical output in kilowatt-hours (KWH). A kilowatt is a thousand watts. For example, a 100 KWH battery like those used in the Tesla Model S can deliver 100 kilowatts of electricity for 1 hour. Think of it as being able to run a hundred 1000-watt hair dryers for an hour. Normal driving uses less

---

[3] International Energy Agency, 2021, The Role of Critical Minerals in Clean Energy Transitions: World Energy Outlook Special Report, IEA, Paris, 287 p. (https://www.iea.org/reports/the-role-of-critical-minerals-in-clean-energy-transitions)

energy than that, so the battery will actually provide power for several hours before it requires recharging. EV batteries are designed to last at least a decade under normal use and perform well through thousands of recharging cycles. The vehicle will probably quit before the battery does.

One of the other concerns with EVs is their low energy density. This reflects the amount of energy that can be supplied per mass or weight of fuel. The lithium-class chemicals used in EV batteries typically have an inherent energy density of about 0.7 KWH per kilogram, about five times higher than old fashioned lead-acid batteries. That sounds impressive until compared with the 12 KWH of energy in a kilogram of gasoline. A thousand-pound battery in an EV can provide a driving range equivalent to about ten gallons of gasoline.

The really high energy densities are in nuclear, where a kilogram of uranium has the energy density of 16,000 kilograms of coal. This was recognized early on as a significant advantage of nuclear electricity. The high energy density along with the lack of combustion products makes nuclear power useful for applications with limited space and no access to air, such as submarines and spacecraft. Although a nuclear-powered electric car would have zero emissions and could run for decades without refueling, I don't think we are going to see these anytime soon.

There are significant concerns about embodied emissions from the materials and steps that go into producing a finished electric vehicle. Lifecycle analysis suggests that although EVs have low downstream scope 3 emissions, they contribute GHG to the atmosphere from scope 1 and scope 2 emissions. These are in fact difficult to quantify.

The makeup of a typical 450 kg (1000-pound) EV battery includes about 14 kg (30 pounds) of lithium, 28 kg (60 pounds) of cobalt, 59 kg (130 pounds) of nickel, 86 kg (190 pounds) of graphite, and 41 kg (90 pounds) of copper along with steel, aluminum, and plastic. The embodied emissions from the mining, refining, processing and manufacturing of battery materials depend on the grade of ore used, how that ore was mined, where that ore was mined and how far material had to be transported. These all tend to be poorly defined because so many of the components that make up an EV are sourced globally from multiple locations. Each source will have a different emissions profile.

The scope 1 and scope 2 emissions associated with the factory that assembles the vehicle also need to be considered, especially if that factory is run on electricity generated by fossil fuel. The upstream scope 3 emissions are difficult to quantify as described above, because the sources of the EV components are quite diverse, and it really depends. Many parts are manufactured in Asia where significant amounts of electricity are generated from coal.

The downstream scope 3 emissions of an EV are thought by most people to be zero, but this actually depends on how the car battery was charged. Emissions are zero if the battery was charged by carbon-zero or carbon-neutral electricity. However, if coal-fired electricity was used to charge an EV, the GHG emissions can exceed those from driving the same distance in a gasoline-powered vehicle. This is why the electric power sector ought to be decarbonized first before focusing on vehicles.

****

Biofuel Vehicles

EVs charged up with carbon neutral electricity are one method to decarbonize the transportation sector. They are not the only option, and maybe not even the best option, but they can help. The other way to decarbonize vehicles is to use internal combustion engines operating on carbon-neutral fuels such as biofuel or carbon-zero fuels such as hydrogen.

Internal combustion engines were invented in the late 1800s and work by basically creating a small explosion inside a cylinder. The energy of the expanding gases pushes a piston out of the cylinder, where it turns a crankshaft and does work. The explosion requires a uniform mixture of fuel and air inside the cylinder for combustion. Gasoline is the preferred fuel, because it can be easily handled in liquid form, while vaporizing readily for the combustion process. However, other fuels can be used, and essentially anything that will mix with air to create an explosion will work.

Carbon dioxide emissions from biofuel combustion, as mentioned previously, simply return the gas to the atmosphere from where it was taken when photosynthetic plants made the biofuel. Liquid biofuels are favored for transportation applications, but biogas can also be used. The most common biofuels are ethyl alcohol (ethanol), vegetable oil, and methane gas.

Liquid biofuels like ethanol are considered by some to be a replacement for gasoline. One of the problems is that ethanol has a lower energy density than gasoline, about a third less per gallon.[4] Thus a vehicle running on ethanol will require more fuel to go the same distance as a gasoline vehicle.

I experienced this first hand when I worked for the DOE National Energy Technology Laboratory in Morgantown, WV. The lab had several government vehicles that were "flex fuel" and could run on gasoline, gasoline blended with 10% ethanol, and ethanol blended with 15% gasoline known as E-85. We were required whenever possible to fill up with E-85 at the motor pool before leaving the lab complex to be more carbon-neutral.

I had to drive to a university meeting in Wilkes-Barre, PA, a bit more than 300 miles (480 km) from Morgantown. I left the lab with a full tank of E-85, but by the time I got to Wilkes-Barre, I was running on fumes. I couldn't find any place in Wilkes-Barre that sold E-85, so the next day before I left town, I filled up the vehicle with regular gasoline. I followed the exact same route back to Morgantown, and when I arrived at the lab, I still had over a quarter tank of fuel left. That was vivid proof that ethanol will not take you as far as gasoline.

Ethanol has some other issues besides energy density. It is primarily made from fermented corn, and at one point nearly every ear of corn grown in the state of Iowa was destined to become biofuel. Although some Iowa farmers have become quite wealthy growing corn for ethanol on large swaths of prime agricultural land, this

---

[4] https://afdc.energy.gov/fuels/ethanol_fuel_basics.html

raises concerns in a world with limits on arable land. Corn in fact has many other uses as an animal feed and human food stock. Growing it for energy seems wasteful. The fermentation is done on a large scale at ethanol plants and the product is delivered to refineries to be blended with gasoline. The fermentation process emits carbon dioxide as a byproduct; even though it is carbon neutral CO_2, it is still a GHG. Regular engines can run on a 90% gasoline/10% ethanol mix. Vehicles designated as "flex fuel" are able to run higher ethanol concentrations.

Ethyl alcohol made from corn is basically "white lightning" or "moonshine." This clear elixir has been painstakingly hand-crafted for decades by "artisan" distilling operations in the Appalachian backwoods and carefully packaged in Mason jars. The ethanol used in vehicles is "denatured," with gasoline or methanol added to make it undrinkable, but that's really the only difference.

The addition of ethanol to gasoline comes from the Renewable Fuel Standard (RFS) established as part of the Energy Policy Act of 2005 (P.L. 109–58; EPAct) and expanded in the Energy Independence and Security Act (P.L. 110–140; EISA) in 2007. These laws called for four billion gallons of renewable fuel in the transportation sector in 2006, increasing to 36 billion gallons by 2022.[5] The EPA is supposed to determine the levels after that. Money from EPAct funded a large part of my research at the U.S. Department of Energy from 2009 until 2015, when parts of it expired.

Ethanol replaced a compound called methyl tertiary butyl ether (MTBE) as a gasoline additive. MTBE is an "oxygenate" that helps gasoline to burn more cleanly and reduces smog. It was mandated in the 1980s to meet clean air standards. MTBE is also water-soluble, and as a volatile ether compound it contaminated significant amounts of groundwater when it leaked from underground storage tanks. The EPA and state agencies assigned regulators to trace MTBE contamination back to a leaking underground storage tank (LUST) and required owners to fix or replace these. These regulators became known as LUST experts and learned to keep smiling even as they heard the same tired jokes dozens of times.

The 2005 EPAct rescinded the reformulated gasoline standard that required the use of oxygenates like MTBE because the catalytic converters on modern exhaust systems do a similar job of preventing smog-inducing chemicals from being emitted by tailpipes. The law also instituted a requirement for renewable fuels. As such, the refineries made a wholesale switch to remove MTBE from gasoline and add ethanol. The other soluble components of gasoline known as BTEX (benzene, toluene, ethylbenzene and xylenes) remain groundwater contaminants, and as far as I know, the LUST experts are still out there.

The use of renewable fuels is not actually doing all that well, where "no ethanol added" gasoline options are now available at some service stations. Originally intended for small engines like lawnmowers, chain saws, or weed eaters that don't

[5] Congressional Research Service Report R43325: The Renewable Fuel Standard: An Overview, April 14, 2020, 17 p.

do well on ethanol blends, people are now filling up their boats and pickup trucks with ethanol-free gasoline. Americans used just 20 billion gallons of renewable fuels in 2019 and it looks like we are going to fall well short of achieving the 36 billion-gallon per year goal mandated for 2022.

The other primary liquid biofuel is vegetable oil. This can be used to replace diesel fuel with slight modifications, and in fact a number of urban vehicles are operating on recycled vegetable oil. Users seek old oil from restaurant deep fryers that needs replacing and would have been discarded in the past. It turns out that the old rancid stuff works just as well as fresh oil in a diesel engine, and suddenly a waste product is now a commodity. Biodiesel is also produced directly from soybean oil for use in vehicles. Airlines are beginning to experiment with it as an alternative to kerosene for jet engines.

It may surprise some people to learn that methane biogas is a very good option for a carbon-neutral vehicle fuel. Compressed methane gas (CMG) can be added to an existing, gasoline-powered vehicle with a simple and inexpensive conversion that allows drivers to keep their current automobile. This idea is similar to replacing the heat source in a thermoelectric power plant while holding onto the rest of the infrastructure. Converting vehicles already on the road to run on carbon-neutral CMG biofuel is an efficient and effective way to decarbonize the transportation sector quickly and cheaply without forcing drivers to decarbonize by replacing their existing vehicle with a new and expensive EV or hydrogen car.

The typical methane conversion design leaves the original gasoline tank in place and adds a CMG cylinder to the vehicle (often in the trunk) as a second fuel source. The engine needs only slight modifications if any. Many of the computer-controlled fuel injection systems on modern vehicles can handle methane directly without much fuss. The conversions, known as "bi-fuel" vehicles, generally have a range of about 160 km (100 miles) or so on the CMG fuel, and then with the flip of a switch they can run on gasoline. Since most people don't drive 100 miles per day, the vehicle can run on CMG most of the time. The gasoline is there for those who suffer "range anxiety" and absolutely must have a car capable of going 400 miles, even if their daily commute is only 40. It is also possible to add larger tanks or multiple tanks of CMG to give the vehicle greater range and do away with the gasoline tank altogether.

The CMG vehicle technology is neither new nor difficult, having been developed for natural gas in Italy during the 1930s. Conversion of vehicles to compressed gas became popular in western Canada and New Zealand during natural gas surpluses in the 1980s. Biomethane has the exact same chemistry as natural gas, and it will work just as well. The technology could be phased-in immediately using natural gas and switched over to biogas when it becomes more widely available. With currently abundant shale gas and low natural gas prices, a CMG vehicle is significantly cheaper to operate than a gasoline-powered one. As natural gas transitions to carbon-neutral biogas, these vehicles could continue to be run without any additional modification.

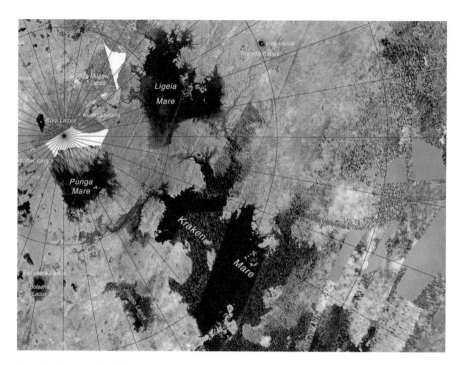

Fig. 8.1 Lakes of liquid methane showing remarkably Earth-like landforms near the north pole of Saturn's moon Titan, radar mapped by the Cassini probe. (*Source: Image credit NASA/JPL; public domain*)

Even fossil methane in natural gas is better for the climate than gasoline, producing about two thirds of the CO_2 per Btu of energy (refer back to the chart in Fig. 7.1). Methane does not emit particulates or volatile organic compounds like gasoline when burned. These are the components of smog. Since it is expected that these bi-fuel vehicles will operate largely on methane for short trips, local commutes, and within cities, the air in cities will be much cleaner.

The other option besides compression to include more fuel in less space and increase the range is to liquefy the methane gas. Liquid methane is a cryogenic fluid like liquid nitrogen or liquid oxygen, becoming a liquid at a temperature of −162 °C (−260 °F). If not kept this cold, it would have to be held under a very high pressure to remain a liquid.

Liquid methane exists naturally in the solar system in places where the temperature is very, very cold. Saturn's moon Titan, for example, contains large lakes of liquid methane on the surface. It rains methane on Titan to fill the lakes, and despite having a "hydrologic" cycle based on methane rather than water, the landforms look quite similar to those on Earth (Fig. 8.1). Liquid methane is used by SpaceX as a rocket fuel, where it can be maintained at approximately the same temperature as liquid oxygen (−183 °C or −297 °F), simplifying the cryogenic cooling system.

Despite the challenges of preparing and handling a cryogenic liquid, using liquid methane as a vehicle fuel has the benefit of rapid refueling (pouring in a liquid is usually quicker than filling a tank with compressed gas). This is important in some commercial operations like trucking, where drivers are on tight schedules and have limited time at refueling stops. The other advantage is that liquid methane is more densely concentrated than the compressed gas, allowing vehicles to carry more usable fuel in the same space.

The methanogens that make biogas don't require specific sugars or starches like the microbes that ferment ethanol from corn or certain other crops. As mentioned earlier, just about any source of organic carbon will do for methane, including waste streams such as animal manure or human sewage, allowing biogas operations to use a wide variety of organic feedstocks. Finally, as a gas, if biomethane leaks it will not pollute groundwater the way liquid fuels do. It is a mystery to me why this technology has not caught on with automobile manufacturers, motorists, natural gas companies, and environmentalists.

****

Hydrogen Vehicles

Hydrogen is the lightest element, and typically occurs as a gaseous, diatomic molecule (H_2). Hydrogen is a carbon-zero fuel because it contains no carbon at all, so there is no CO_2 to put into the atmosphere. It burns readily with oxygen, creating water (H_2O) as a combustion product. As mentioned earlier, water vapor can contribute to atmospheric warming by absorbing IR radiation from the Earth, but it also cools the planet by blocking sunlight with clouds. These two effects interact in a complex manner, and scientists think they essentially cancel each other out. Some climate skeptics like to argue this point, but the bottom line is that contributions of water vapor to the atmosphere from the human combustion of hydrogen are never likely to be anything more than minor given the vast amounts of water in the Earth's natural hydrologic cycle.

Unlike methane, hydrogen cannot be used in existing vehicles without significant modifications. For starters, it burns with a flame temperature that is approximately 500 °F or 260 °C hotter than methane, so the engine would require extensive changes or perhaps outright replacement to run on hydrogen. It would probably be cheaper and more efficient to simply design an entirely new, purpose-built car from scratch to enable the use of hydrogen as a zero-carbon fuel. Likewise, replacing natural gas with hydrogen in a utility distribution system would require modifications to the burners on every gas stove, water heater, and furnace serviced by the system to prevent overheating and creating a fire hazard. The flame speed of hydrogen is nearly ten times faster than methane, making it more explosive. When

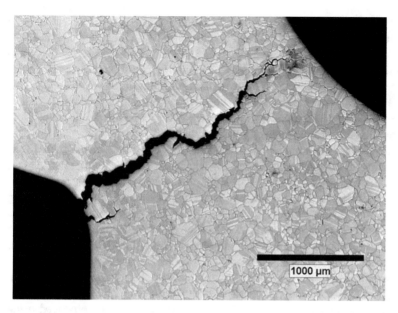

Fig. 8.2 Photomicrograph of a fracture in steel formed along grain boundaries weakened by hydrogen embrittlement. Scale bar is 1 mm in length. (*Image source: Iannuzzi, M., Barnoush, A., and Johnsen, R., 2017, npj Materials Degradation v. 1, no. 2; doi:10.1038/s41529-017-0003-4 (open access)*)

the masses or weights of hydrogen and methane are equal, the hydrogen will supply double the energy value of the methane. However, the methane molecule is eight times heavier than the hydrogen molecule, so on a per volume basis, it takes about three times as much hydrogen to equal the heating value of a given volume of methane.

One additional problem with hydrogen is that the molecules are very small, and permeate into metal along grain boundaries, where they react and form metal hydrides. These weaken bonds, separate the grains, and eventually cause brittle failure of the metal (Fig. 8.2).

Carbon steel, which is already somewhat brittle is especially susceptible to this so-called "hydrogen embrittlement." As such, using steel tanks to carry hydrogen aboard vehicles risks leakage and ruptures. Likewise, ideas about swapping out natural gas for hydrogen gas and transmitting it through the same interstate pipeline system and other natural gas infrastructure will not work without some major material upgrades to keep the gas away from the steel. The entire system would have to be protected somehow before hydrogen could be introduced into steel pipelines.

Hydrogen is carried onboard vehicles as a compressed, pressurized gas in tanks that have been specially treated to avoid hydrogen embrittlement. Like methane, hydrogen can be liquefied at cryogenic temperatures, and NASA in fact uses liquid hydrogen as the fuel on the new Space Launch System (SLS) rocket for the Artemis moon program. However, to remain in the liquid phase, hydrogen has to be kept much colder than liquid methane at −253 °C or −423 °F.

Hydrogen does not exist naturally on Earth as a stand-alone molecule. It is so chemically active that it occurs only in combination with other elements like oxygen, carbon, or nitrogen. Thus, to be available as a fuel, H_2 has to be extracted and isolated from one of these compounds, which must be considered in its economics as an energy resource. There are no known microbes equivalent to methanogens that give off H_2 as a byproduct, although this has been discussed as a potential goal of genetic engineering. At present, however, hydrogen must be manufactured by electrolytic or chemical processes that may add a substantial cost.

The most common commercial process for hydrogen production currently in use is to manufacture it from the methane (CH_4) that makes up the bulk of natural gas. It can also be separated from water (H_2O). Methane has four hydrogen atoms per molecule compared to two for water and is the more favored feedstock. Depending on how it is produced, hydrogen comes in a number of "colors." For example, "gray" hydrogen is produced from natural gas where the CO_2 created as part of the process is simply released into the atmosphere. When this CO_2 is captured and sequestered instead, the hydrogen produced from natural gas is said to be "blue." More environmentally-benign "green" hydrogen is produced electrolytically from water using renewable electricity like solar or wind with no GHG emissions. Sometimes solar-produced hydrogen is called "yellow," and that produced using nuclear electricity is called "pink." However, in all cases the hydrogen itself is a clear, colorless gas.

In the United States, most industrial hydrogen is produced from natural gas through a process called steam reforming. This uses high-temperature steam (700 °C–1000 °C) under high pressure in the presence of a catalyst to react with methane and produce hydrogen (H_2) and carbon monoxide (CO). The CO and steam are then reacted using a catalyst in what is called the "water-gas shift reaction" to produce CO_2 and more H_2.

The steam reforming reaction is:

$$CH_4 + H_2O(+\text{added heat}) \rightarrow CO + 3H_2$$

Water-gas shift reaction:

$$CO + H_2O \rightarrow CO_2 + H_2 (+\text{small amount of heat given off})$$

A final process step called pressure-swing adsorption is used to remove the CO_2 from the produced gas stream and leave essentially pure hydrogen behind. What is done with this CO_2 to dispose of it then determines if the hydrogen is gray (vented) or blue (sequestered).

Another process called partial oxidation can be used to produce hydrogen from methane. This is essentially the same chemistry described earlier to make town gas out of water and coal before natural gas became widely available. Instead of using steam to break apart the methane molecules, the necessary heat is obtained by burning the methane with limited oxygen. The partial oxidation process gives off heat, rather than requiring heat like steam reforming, so it is cheaper to operate. The combustion products are H_2 and CO, which are then run through a water-gas shift reaction to create CO_2 and more hydrogen. Steam reforming produces more hydrogen than the partial oxidation process per unit of natural gas, but the economics may not be as favorable.

Finally, hydrogen can be made by electrolysis from water. If the electricity used for this comes from renewables or carbon-neutral sources, the hydrogen is "green." The discovery that passing an electrical current through water breaks it down into hydrogen and oxygen is credited to English scientists William Nicholson and Anthony Carlisle. They successfully split water into its component parts in 1800 using a battery-like device invented by Alessandro Volta. (The electrical volt is named for Volta.) In fact, by 1869 an inventor named Zénobe Gramme had developed a machine that produced industrial quantities of hydrogen through the electrolysis of water. Hydrogen gas accumulates at the negatively charged cathode, while oxygen accumulates at the positively charged anode. Small amounts of acid are usually added to pure water to improve conductivity.

Svante Arrhenius, mentioned back in Chap. 5 as the scientist who predicted in 1896 that carbon dioxide from the combustion of fossil fuel would affect the climate, spent much of his career as a chemist investigating the galvanic conductivity of electrolytes dissolved in water. He discovered that to varying degrees they split or dissociate into electrically opposite positive and negative ions, which are then attracted to oppositely charged electrodes. This was not only true of dissolved salts, but also of the water molecules themselves. Arrhenius won the Nobel Prize[6] in 1903 for explaining how electrolysis worked, rather than for discovering it.

Hydrogen can be used to power vehicles in two different ways. First of all, it is combustible so like gasoline, ethanol, or methane it can be used in internal combustion engines to provide power. As mentioned above, internal combustion engines running on hydrogen must be specially designed and constructed for this purpose. A second and perhaps far more useful application of hydrogen would be to provide portable, direct electric power in place of a battery with a device known as a "fuel cell." These use a relatively slow chemical reaction to produce electricity. The U.S. Department of Energy has been supporting research on fuel cells and fuel cell-powered electric vehicles (FCEVs) for many years.[7]

Fuel cells consist of negative and positive electrodes in an electrolyte. Hydrogen is fed to the negative cathode, and oxygen (or air) is fed to the positive anode. The hydrogen molecules are ionized into protons and electrons by a catalyst at the cathode. The protons migrate through the electrolyte to the anode, while the electrons flow through an external circuit, creating an electric current. The protons and electrons unite with oxygen at the positive electrode to produce water and heat.

---

[6] Svante Arrhenius – Biographical. NobelPrize.org. Nobel Prize Outreach AB 2021. Sat. 18 Dec 2021. <https://www.nobelprize.org/prizes/chemistry/1903/arrhenius/biographical/>
[7] https://www.energy.gov/eere/fuelcells/fuel-cells

There are many variations in the type of electrolyte, the operating temperature of the fuel cell, and the exact chemistry of the input gases. However, these devices do work, and in fact they have been used successfully since the 1960s to supply electric power on a variety of spacecraft. The rupture of an oxygen tank on the outbound Apollo 13 moon mission (made famous as "Houston, we have a problem…") forced the spacecraft's three fuel cells to shut down because they require both oxygen and hydrogen to work. The lack of electrical power nearly killed the three astronauts on board the spacecraft before they could return to Earth.

Fuel cell-powered vehicles are under development in China. A company called Sinosynergy was established in 2015 to commercialize fuel cell vehicles in Guangdong Province.[8] Currently they are producing buses but plan soon to expand to trucks, trams, passenger cars, ships, and even provide backup power supplies for 5G telecommunication base stations. The company has a global market share of 10% of the hydrogen fuel cell vehicles in the world, and half of the fuel cell vehicles in China. They have produced nearly 5000 vehicles so far and expect to have ten times that many on the road by 2025. Sinosynergy claims to be able to produce 20,000 hydrogen fuel-cell stacks per year and is one of the world's largest producers. Clean energy policies, lower costs for hydrogen technology, and strategic planning may grow the global market for hydrogen fuel-cell vehicles to US$20 billion by 2030, according to some assessments.

In 2022 the Chinese government released the country's first-ever long-term plan for hydrogen through 2035. Guangdong's hydrogen industry has attracted the attention of the Chinese Communist Party (CCP) leaders, who see it as a technology of the future. Sinosynergy has said that the company benefited tremendously from state support and subsidies. The Chinese government's endorsement of hydrogen has prompted at least 23 provincial governments to issue similar development plans for hydrogen energy and fuel-cell vehicles.

Some hydrogen fuel cell vehicles are available in the U.S., mostly in California where they are used to meet clean air standards. According to the EPA website, the only manufacturers of these vehicles currently are Hyundai and Toyota. There are about 13,000 hydrogen fuel-cell vehicles on the road in the United States at present, or less than 0.005% of the total. The U.S. is far behind the Chinese on fuel cell vehicle development and the hydrogen infrastructure needed for refueling.

Hydrogen fuel-cell vehicles have a range on a full tank of fuel similar to internal combustion engine vehicles, around 300–400 miles (500–600 km), which is two to three times greater than the driving range of battery-powered EVs on a full charge. Fuel cells also operate efficiently at temperatures well below freezing, another advantage over EVs, which often perform poorly in the cold. Drawbacks of fuel cell vehicles include difficulties in obtaining hydrogen and the high cost of the fuel when it can be found. Hydrogen in China is some 16 times more expensive than gasoline on a per kilogram basis.

---

[8] https://www.sinosynergypower.com/eindex.html

As mentioned earlier, the transport of hydrogen gas through existing steel pipelines is problematic because of its tendency to embrittle steel and cause fractures. One option being considered for hydrogen transport is to combine it with nitrogen and create ammonia (NH_3). There are a number of advantages to this. First, ammonia contains a significant amount of hydrogen, about 17.6% by weight. Secondly, ammonia is a common compound in nature produced as a waste product by animals and it serves as a precursor to nitrogen fertilizers used by plants. Ammonia makes up a significant part of the atmospheres of the gas giant planets in the solar system. It liquifies at the cold but not cryogenic temperature of -33.34 °C (-28.012 °F) under atmospheric pressure and can be transported as a liquid under refrigeration. It can also be dissolved in water and transported that way.

However, ammonia has some downsides. In concentrated liquid form, which is the most economical way to transport hydrogen, ammonia is both caustic and hazardous. The National Institute for Occupational Safety and Health (NIOSH) in the United States classifies it as an extremely hazardous substance.[9] Facilities that produce, store, or use ammonia in significant quantities are subject to strict reporting requirements. Ammonia was briefly considered by DOE as a potential storage medium for hydrogen on fuel cell vehicles, but even trace levels of the substance can degrade the electrolyte membranes on fuel cells and "poison" the cell. The idea was abandoned after it was determined that it would be nearly impossible to remove every last trace of ammonia from the stored hydrogen gas in an onboard fuel cell system.

Ammonia is normally produced industrially by the reaction of nitrogen and hydrogen in the presence of iron catalysts at temperatures around 400–600 °C and under pressures of 200 to 400 atmospheres. Decomposition or cracking of ammonia back into hydrogen and nitrogen is the reverse of the synthesis reaction but requires more heat. The temperature necessary for efficient cracking depends on the catalyst, although some effective materials such as nickel require temperatures above 1000 °C. Others have good conversion efficiency at temperatures in the range of 650–700 °C. The embodied emissions for these various heat-intensive processes could potentially be substantial.

A novel concept for transporting hydrogen comes from the Fraunhofer Institute for Manufacturing Technology and Advanced Materials IFAM in Dresden, Germany. They have developed a paste of magnesium hydride stabilized with an ester and metallic salt. The paste reacts with water to release pure hydrogen at a controlled rate that can then be fed into a fuel cell or used for combustion. The paste is stable to temperatures as high as 500 °F (260 °C) and can carry a significant volume of hydrogen.

The oil and gas industry is looking for new natural gas markets in a low carbon economy. Blue hydrogen is one product that is certainly under consideration. With blue hydrogen made from natural gas, a large part of the transport problem can be

---

[9] BPI Information Services, 1994, <u>NIOSH Pocket Guide to Chemical Hazards</u>, DHHS publication: DIANE Publishing Company, 400 p.; ISBN 1579790593, 9781579790592

avoided by setting up the steam reforming plants near the intended markets. The "hydrogen" can be transported in the form of natural gas through existing pipelines and turned into blue hydrogen where needed for local sales. Green hydrogen made by electrolysis from water using wind or solar electricity can also be manufactured near markets, avoiding similar transport issues.

However, the oil business thinks globally, and the production and distribution of hydrogen as a non-carbon fuel will eventually see international distribution. There are risks to transporting H_2 via ammonia, especially if the petroleum industry approaches it in their usual supersized manner and fills up gigantic tanker ships with the liquefied gas. An accident, collision, or terrorist incident near a populous port city that releases millions of gallons of highly concentrated ammonia could kill or injure thousands. Even if it's not near a port, the solubility of ammonia in water could put tons of it into the ocean after an accident and devastate marine life for miles around. Similar concerns about hazards were raised when large tanker ships began carrying liquefied natural gas (LNG) internationally. Most LNG terminals are located a good distance away from populated areas in case of accidents. The same will have to be required for ammonia terminals.

Shell Oil and the Norwegian renewable energy company Norsk Hydro have an agreement to use Norsk's existing renewable energy hubs to generate green hydrogen. The initial goal is to provide green hydrogen as a clean-burning fuel for their own operational needs. However, plans are to eventually scale-up and produce green hydrogen for the European market. Drilling company Baker Hughes has invested in the Canadian firm Ekona Power to tap into the hydrogen market. Ekona claims to have developed a natural gas reforming process that uses a thermal treatment called pyrolysis to break apart methane molecules inside a reactor vessel, producing hydrogen gas and solid black carbon. The production of solid carbon instead of the CO_2 created during the traditional steam reforming process has allowed Ekona to drastically reduce CO_2 emissions.

I think these are early stages for hydrogen, dipping in a toe if you will. We are likely to see much more interest from the fossil fuel sector as alternative energy for decarbonized vehicles becomes more prominent on world markets. After all, energy is energy, and if these companies want to survive the transition, they will need to be nimble enough to jump onto new energy technologies. Oil and gas, in particular, have always been pretty good at that.

For example, the adoption of lateral drilling and staged hydraulic fracturing for shale gas production took about 5 years from Mitchell Energy's early successes until it started being picked up by the rest of the industry on a large scale. Five years after that, shale gas was a full-fledged boom. Although there are climate skeptics and deniers who want to hold onto the status quo, by and large the petroleum industry is fairly adept at reading the tea leaves. You don't need to be much of a mystic with a crystal ball to see which way the energy industry is going. One thing I found telling was a technical conference in Houston in 2022 hosted by the American Association of Petroleum Geologists. The subject was carbon dioxide capture and subsurface storage, and the conference was quite well attended by many currently-employed petroleum geologists thinking about their future careers.

The whole idea of running vehicles and appliances on hydrogen bemuses me, because this kind of talk has been around for some time. Back in the day when nuclear power plants were first being developed, advocates were saying that nuclear electricity would be so cheap that it wouldn't even be metered. You would just pay a low, flat monthly fee and use all you wanted. This of course terrified the natural gas industry because what if electricity became so cheap and abundant that nobody wanted to use gas anymore?

In response, the natural gas industry came up with something they called the "hydrogen economy." The idea would be to use waste heat from the nuclear reactors that were soon to be on every city block to thermally dissociate water into hydrogen and oxygen. The hydrogen gas would be piped through the existing natural gas distribution infrastructure to customers, allowing the gas industry to survive. Of course, there were the ongoing issues of hydrogen embrittlement of steel pipes and the higher hydrogen flame temperature that would have required all burners to be changed out, but business models for the hydrogen economy were in development.

No one back then was worried about the climate, and I don't think anyone had considered using hydrogen as a vehicle fuel, but they probably would have gone for that as well if they had thought of it. As we all know, nuclear power stations are not located on every city block, and nuclear electricity turned out to be anything but unmetered and cheap. So at least that iteration of the hydrogen economy never came to pass. Still, I always wonder who is reading the old reports every time someone comes up with a "new" idea like hydrogen.

I've provided a number of suggestions for technology to replace fossil-fueled vehicles and decarbonize the transportation sector of the economy. We can do this technically, but I don't know how easily we can do this politically. There are a lot of people with a substantial level of vested interest in continuing to manufacture gasoline-powered automobiles and supply them with fossil fuel. Money to repair and maintain U.S. highways comes from a highway trust fund, supported by federal taxes on gasoline. Switching over to EVs, CMG vehicles, and/or hydrogen will bypass this gasoline tax, leaving the highway fund high and dry. States also tax gasoline, and in some, like California and New York, it is a significant source of revenue. How will that income be replaced, and what will be the political resistance to decarbonization until it is?

As mentioned earlier, a number of states also receive substantial income from crude oil production, such as Alaska, North Dakota, Wyoming, Texas and others. Weaning state governments off oil income may be challenging and yet another roadblock to decarbonizing transportation.

The federal government could add a substantial carbon tax on fossil fuel, making it more expensive and forcing the transportation sector to switch to lower-cost alternatives. The politics of this get murky and special interest groups abound. Raising the cost of fossil fuel would cause significant pain to individual pocketbooks and to the economy in general, at least in the short term. No one should be under any illusions that decarbonizing transportation will be easy. However, I and many others who have looked at this think that some sort of a tax on fossil carbon is the incentive needed for getting vehicles to transition away from fossil fuels.

So far, attempts at raising the cost of fossil fuel with a carbon tax have not gone well, prompting strikes in France and riots in Kazakhstan. Price increases in both places were soon rescinded. Farmers in India went on strike for months to protest an increase in the cost of diesel fuel for their tractors. When the U.S. faced fuel price increases in 2021 after the COVID lockdown ended, President Biden found it necessary to withdraw oil stocks from the Strategic Petroleum Reserve to increase supplies and reduce prices. None of this bodes well for the success of a carbon tax on fossil fuel.

I suggest as an alternative that perhaps the "carbon tax" should be built into the price of new gasoline-powered and diesel vehicles, rather than into the cost of gasoline. This will serve several functions. First, it incentivizes the purchase of electric vehicles or those powered by hydrogen or biogas, making them cheaper than fossil fuel vehicles or at least far more cost-competitive. It also encourages the engineering and development of these vehicles because they would no longer have to compete against cheap gasoline-powered cars.

Secondly, adding a carbon tax to the price of new vehicles instead of increasing the cost of gasoline allows existing vehicles to still be operated cheaply, avoiding riots in the streets over a fuel tax. Paying a carbon tax can be postponed by keeping the old gasoline-powered clunker going for as long as it will run. At some point, though, it will have to be replaced, and a high carbon tax on the cost of new gasoline and diesel vehicles will incentivize replacing it with an electric or other non-GHG vehicle. Such a tactic will ensure that the percentage of carbon neutral and zero emissions vehicles on the highway gradually increases over time, eventually reaching a majority. If people still want gigantic, diesel burning pickup trucks to "roll coal" and automakers still manufacture these, they can be obtained but at a much higher cost than at present. I think that will be enough to cause most people to seek alternatives.

Funding from a fossil fuel vehicle tax can be used to replenish the highway fund, and a modest annual highway use tax levied on non-gasoline vehicles, possibly as part of the license plate renewal process, can bring in money to maintain existing roads or construct new ones. Alternatively, all public roads can be converted to toll roads, and you would pay as you go. Given the new electronic toll reading devices on modern turnpikes, this would not be very difficult to implement.

We must act quickly, and we must act at a global scale. We can do this, and I hope that for the sake of the future of humanity that we WILL do this. The climate has a lot of inertia, and it will be challenging to get it to change course. Society also has a lot of inertia, but people are beginning to realize that we are in trouble, and something must be done. Political leaders need to hear this and be spurred into taking action. Dithering about whether or not climate change is "real" is no longer an option. I will discuss some other things that we can do in the next chapter. Through all of it keep in mind that we must not give up hope. This can be fixed.

Chapter 9
Mitigation by Geoengineering

Keywords Solar radiation management · Carbon dioxide removal

Geoengineering seeks to alter the Earth itself to respond to the climate crisis. Two of the main ideas are to reduce incoming solar radiation by releasing aerosols high in the atmosphere to cool the planet, or to remove the excess greenhouse gas, mainly carbon dioxide, that has built up in the atmosphere over the past two centuries. Aerosols injected into the stratosphere by nature during volcanic eruptions can sometimes produce dramatic cooling effects, such as the "year without a summer" in 1816 following the eruption of Mt. Tambora in Indonesia. Solar radiation management seeks to add anthropogenic aerosols to the stratosphere to overcome the most severe effects of global warming such as massive heat waves. Carbon dioxide removal from the atmosphere can be done using photosynthetic plants, although the amount of available land for planting trees is limited and up to a trillion new trees would be needed to mitigate climate change. A second option is to engineer devices to capture it using various chemical processes and specialized machinery. The captured carbon must be stored or sequestered away from the atmosphere for periods of at least a century and the longer the better. All these options are under consideration by governments, research institutions, and venture capital investors. When combined with an energy switch away from fossil fuels, geoengineering techniques promise a way to mitigate the worst aspects of climate change.

****

There is good news and bad news about the climate crisis.

The bad news is that eliminating our use of fossil energy won't be enough to avoid the climate crisis. The sooner we stop using fossil fuel, the less bad it will be, but we are well past the point of avoiding climate change altogether. Most climate scientists think the planet is definitely due for at least 2 °C of global warming, and possibly as much as 3 °C before things can be brought under control.

The good news is that there are some other actions we can take in addition to eliminating fossil fuels that will help to mitigate the climate crisis. Certain engineering measures can reduce greenhouse gas levels in the atmosphere and slow the rate of global temperature rise. If we decarbonize energy and stop adding to the problem, there is a good chance that these engineering actions implemented on a large scale can help humanity and the rest of the planet's ecosystem avoid some of the worst consequences of changing climates.

The modification of the Earth to adjust climate is called geoengineering. There are many different ideas for how to do this, some good and others not so good. Engineering on a planetary scale has been contemplated for places like Mars and Venus to make them habitable for humans. If we make a miscalculation trying to engineer habitat on Mars or Venus, as far as we know nobody is living on either of these planets at present, so no harm done and back to the drawing board. However, many people are concerned about geoengineering on the Earth, fearing that scientists are playing God by experimenting with the only home humanity has at the moment. If we make a mistake here, we could find ourselves in serious trouble.

I understand these concerns and offer two thoughts in response: first, the geoengineering techniques under consideration require active monitoring and maintenance and can be stopped at any time if necessary if things start to go awry. Second, some geoengineering techniques are much safer and more benign than others, and these are the ones we should use. Carrying out geoengineering that seeks to undo some of the direct damage caused by climate change is more helpful for addressing the climate crisis than some of the more exotic, off-the-wall ideas that have been proposed for dealing with climate. It is clear that decarbonizing large sectors of the economy will take some time. Starting geoengineering sooner rather than later can help mitigate the severity of climate change while we are decarbonizing. Climate change is a massive problem and carrying out multiple solutions simultaneously is not a bad idea.

Geoengineering for climate mitigation basically falls into two categories: controlling the temperature or controlling the amount of greenhouse gas. The temperature control option is called "Solar Radiation Management" (SRM) and the greenhouse gas management option is called "Carbon Dioxide Removal" (CDR).

I want to make it absolutely clear that geoengineering is not an excuse to continue burning fossil fuels. All the geoengineering in the world will not be able to remediate the climate crisis if fossil fuel combustion keeps adding to the problem. We must stop using fossil fuels, period. Some environmentalists object to CDR, confusing it with "carbon capture and storage" or CCS, which was indeed developed to enable the electric power industry to continue burning coal. The term CDR primarily refers to the process of direct air capture (DAC) of carbon dioxide from the atmosphere, not from a power plant smokestack. In standard pollution terminology, CCS is focused on point sources of CO_2, whereas CDR focuses on non-point sources. Other forms of CDR besides DAC include direct ocean capture, where dissolved CO_2 is removed from seawater, and carbonation, where the CO_2 reacts with soil minerals to form solid carbonates.

CCS is designed to capture carbon dioxide from point sources like power plants and prevent coal combustion gases from entering the atmosphere. Even in the absence of fossil fuels, it still has potential uses for capturing emissions from industrial operations like cement plants or steel mills. The coal program in the DOE

Office of Fossil Energy and Carbon Management (FECM) has been funding CCS research for years, including about half of my research budget when I worked at the DOE National Energy Technology Laboratory (NETL). One of my duties there was to investigate geologic formations and structures that could sequester captured CO_2 away from the atmosphere.

The stated DOE goal for CCS is 95% carbon capture with only a 15% increase in the cost of electricity (COE). When I expressed this to people who actually worked in the electric power industry, they would often tell me, "It's good to have goals." The reality is that 95% carbon capture on a coal power plant raises the COE to levels more expensive than nuclear power (refer back to the COE chart in Fig. 9.9). The economics of CCS have always been challenging in part because coal is so artificially cheap to begin with. A former director at NETL used to point out that it cost more to have a truckload of topsoil delivered than a truckload of coal. Coal is literally cheaper than dirt.

Because of these costs, the CCS program evolved into Carbon Capture, Utilization, and Storage (CCUS) in the hope that finding a profitable use for all this captured carbon dioxide will help with the economics. Unfortunately, there is little unfulfilled industrial demand for CO_2 at the moment, and this is a problem for DAC as well. The Warrior Run coal power plant in Maryland mentioned earlier has the carbonated beverage market on the East Coast pretty well locked up. Most industrial operations that use CO_2 obtain it from ethanol fermentation or ammonia plants, where it occurs as a byproduct, or from wells in Mississippi that produce natural carbon dioxide at a lower cost than CCUS.

* * * * * * * * * * * *

Solar Radiation Management (SRM)

The "Year Without a Summer" was an unusual cooling event that occurred in 1816. The weather remained gloomy and cold throughout the entire season. New England saw snow in June, freezing temperatures in July, and a killing frost in August. There were massive crop failures in North America and Europe, leading to food shortages for both people and livestock. The price of food that was successfully grown rose sky-high.

In Europe, the leaden skies, cold weather, and frequent rain trapped the writer Mary Shelley, her poet husband, Percy Bysshe Shelley, and their friend, the poet Lord Byron in their chalet during a vacation trip to Lake Geneva in Switzerland. With little else to do, they held a writing contest. Whatever Percy Bysshe Shelley and Lord Byron might have written for this competition is not known, but for her part, Mary Shelley produced a book-length manuscript titled <u>Frankenstein; or, The Modern Prometheus</u>, which was published 2 years later in 1818. Most readers are probably familiar with this Gothic horror story of science gone wrong. It was set in a bleak, dark, and stormy environment that Mary Shelley presumably described by merely glancing out the window of her Swiss villa.

The odd weather in the summer of 1816 was caused by volcanic ash blocking sunlight. On April 10, 1815, Mount Tambora in the Dutch East Indies (now Indonesia) produced the largest volcanic eruption in recorded history, ejecting over 200 cubic km of material into the atmosphere. (In comparison, the 1980 Mount St. Helens eruption in Washington ejected about one cubic km of material into the air.) There had been a number of other major eruptions over the 3 years prior to 1815 in the Philippines, Japan, and the Caribbean that also ejected large volumes of ash into the atmosphere. Some of this residual material was still present when Tambora erupted and added significantly more of it. The total amount of volcanic dust, ash, and aerosols in the stratosphere was unprecedented, creating what was described as a "dry fog" high in the sky that was observed in the summer of 1816. This fog reddened and dimmed the sunlight so much it was said that sunspots could be observed on the disk of the sun with no need for a filter to block the light.

The tiny particles of rock material and sulfates that were shot high into the atmosphere by the Tambora eruption persisted for many months and had a global effect on climate. Ash blocked sunlight from reaching the ground resulting in an average worldwide temperature drop of as much as 3 °C, although some scientists think it may have been less, perhaps around 1 °C. In either case only a few degrees of lower temperature were required to produce the radical events that typified the Year Without a Summer. Thus, when scientists warn that greenhouse gases may increase global average temperatures by 2 or 3 °C, it is important for readers to understand that even though this sounds like a small amount, it can have profound effects.

The predominant geoengineering strategy for SRM seeks to cool the Earth by mimicking Mount Tambora and emplacing fine particulate matter into the upper atmosphere, reducing the amount of solar radiation reaching the ground. This process, known as stratospheric aerosol injection, is being taken seriously enough by the U.S. National Academy of Science that they requested the U.S. government to allocate up to $200 million to investigate how SRM might offset the effects of climate change, releasing a report on the subject in 2021.[1] Most SRM research in the United States is currently funded by philanthropic foundations and individuals.

SRM studies that are focused on stratospheric aerosol injection are being carried out in the United States, China, Sweden, Norway, India, and Germany. Humans could reproduce a volcanic eruption by releasing aerosol particles approximately one μm (1/1000th of millimeter) in size into the stratosphere, where they would remain suspended for about a year. For comparison, "dust" particles are defined as particulate matter 10 μm in size, and "smoke" particles are defined as 2.5 μm in size or smaller. The aerosol particles in the stratosphere would be 40% smaller than the upper size limit for smoke.

The aerosols might consist of sulfates, which occur naturally in the atmosphere, and one idea is to simply add more sulfate aerosols to the natural background concentration to reflect additional sunlight. Other candidates for aerosols include

---

[1] National Academies of Sciences, Engineering, and Medicine, 2021, Reflecting Sunlight: Recommendations for Solar Geoengineering Research and Research Governance: Washington, DC; The National Academies Press. https://doi.org/10.17226/25762

carbonates or silicates, which are essentially finely powdered rock. Still another idea is to use tiny particles of black carbon, i.e., soot. Scientists involved in the research claim that about 50 aircraft flying at an altitude of 20 km (66,000 feet) near the equator would be sufficient to disperse enough aerosols to offset one or two degrees of global warming. Conceivably, a single nation like the U.S., China, or Russia would have the capability of carrying this out on their own without consulting the rest of the world. If this sounds like the premise for a techno-thriller novel, it very well could be.

Despite the numerous scientific papers from these various studies proposing the benefits of uniform and moderate levels of stratospheric aerosol injection, some questions remain about possible negative effects. These include the potential for uneven dispersion of aerosols, which would create greater cooling effects in some locations compared to others. There are also possibilities that aerosols could contribute to air pollution, alter rainfall patterns, or affect the ozone layer.

As far as I'm concerned, the biggest worry about stratospheric aerosol injection is the least predictable. What happens if there is a massive, Tambora-sized volcanic eruption a few weeks or a month after we've seeded the upper atmosphere with aerosols and other particulates to block solar radiation? The unexpected addition of hundreds of cubic km of volcanic ash and aerosols on top of the human aerosols already up there may block far more sunlight than anyone ever intended, sending the Earth into a deadly freeze that devastates crops, destroys ecosystems, and kills people. The 1816 Year Without a Summer may look like a picnic in the park by comparison and our techno-thriller will have become a full-blown disaster novel.

We can't stop a volcanic eruption. We can't easily clear aerosols out of the upper stratosphere once they have been injected. (Have you ever tried clearing smoke from a room?) We are not very good at predicting volcanic eruptions, so we can't even schedule a stratospheric aerosol injection program around them. Proponents will argue that the probability is low for a major eruption to take place during the year or so when human stratospheric aerosols will be floating around in the atmosphere. This is true, but low is not zero. If we continue to inject aerosols year after year to combat the climate crisis through SRM, the probability of an eruption increases over time, eventually becoming a sure thing. Remember, risk is the probability of an event times the consequences. The consequences of such an eruption would be dire, making stratospheric aerosol injection an unacceptably high risk in my opinion. Other, much safer geoengineering options are available.

Nevertheless, researchers at Harvard University School of Engineering and Applied Sciences are planning to use balloons to loft small-scale, controlled experiments of stratospheric aerosol injection to provide data for computer models. The experiments were planned to be carried out in Sweden with the collaboration of the Swedish Space Corporation but are temporarily on hold until input is obtained from more stakeholders. I don't think it is wise for this to ever proceed beyond the computer modeling stage.

Other ideas for SRM include increasing cloud cover in the lower atmosphere to cool the planet by reflecting more solar radiation back into space. This is called marine cloud brightening and thickens clouds in the lower atmosphere over the ocean to increase their reflectivity. If you've flown above clouds in daylight, you know they can be quite bright. Sunlight including short-wave IR that is being reflected back into space from cloud tops is not warming the ground.

Australian scientists are investigating the potential for marine cloud brightening to reduce the warming that has damaged and bleached the Great Barrier Reef. Marine cloud brightening targets low ocean clouds at altitudes of about 800 meters (2500 ft). Tiny droplets of sea water sprayed high into the air evaporate and leave behind nanometer-sized salt crystals. These are caught in the updrafts feeding moisture to the clouds and act as seeds on which cloud droplets can condense. This brightens and thickens existing clouds, reflecting more solar energy back into space. Some indirect evidence that marine cloud brightening works is based on observations that ships commonly inject particulate matter into clouds from their smokestacks, making them thicker and brighter.

Cloud brightening can be problematic because clouds can also trap and reflect heat. Water vapor absorbs the infrared wavelengths radiated from the warm Earth into space almost as effectively as other GHGs like CO_2 and methane. I learned this the hard way when living in Las Vegas back when I worked on the Yucca Mountain Project. I have been an avid bicyclist for decades, but during the summer it is simply too hot in Las Vegas to ride during the daytime. Fortunately, the streets are well lit, and the retirement community near my home became very quiet after 9 PM. I would bicycle through these streets at night, rarely encountering any cars, motorcycles, dogs, pedestrians, or anyone else. One of my older friends who lived in the community called it "the graveyard."

After one especially hot July day when the temperature peaked around 117 °F (47 °C), it clouded over right about sundown, and I decided to go for a bike ride that night. It was still incredibly hot, even though it was quite dark. I sweated and sweltered through the ride, finally cutting it short and heading back home when I ran out of water (my bike carries two bottles). I turned on the news, and in the weather segment the air temperature was reported to be 109 °F (43 °C). This was at 11:15 PM, and I have to admit I was impressed. During my ride an hour or two earlier it must have been between 110 and 115 °F. The clouds acted like a blanket over the Las Vegas valley, trapping and reflecting the heat from a very hot day well into the night.

High, thin clouds tend to trap heat because they let sunlight through to warm up the Earth and then reflect back the longwave IR radiated from the warm Earth. Thick, low clouds just make for a gloomy day underneath, and don't let enough sunlight through to warm up anything. So, a third idea for SRM consists of consolidating the water ice crystals in high, thin clouds to make them fall to lower levels of the atmosphere and dissipate. This is a natural process that can be observed in the "virga" falling from mares tail cirrus clouds, and there may be ways like cloud seeding to cause this artificially.

One method for SRM that I think has an acceptable level of risk is to place some large sunshades in low Earth orbit. These can be similar to the thin, flexible sun shields deployed to shade the Webb Space Telescope. They can be made of thin sheets of plastic film coated with reflective aluminum and launched in a rolled up or folded manner that deploys in orbit. Such a project would be hugely expensive because many square kilometers of shade material are needed to create an appreciable effect on global temperatures. An additional concern is that low Earth orbit is already getting crowded with a myriad of other uses, and there simply may not be room for numerous, large sunshades. However, if a volcanic eruption or another event does darken the upper atmosphere, the orbiting sun shades can be removed much more easily than aerosols injected into the stratosphere.

* * * * * * * * * * * *

Carbon Dioxide Removal (CDR)

Even if we stopped all fossil fuel combustion tomorrow, we would still be required to remove an estimated 730 billion metric tons (804 billion short tons) of carbon dioxide from the atmosphere by the end of this century to limit global warming to the 2 °C goal of IPCC. A billion metric tons is known as a gigaton (GT), and this nomenclature (often with the British spelling "tonne" to denote that it is metric) is typically used when discussing massive amounts of carbon dioxide removal. Assuming that we start doing large-scale CDR by 2030, about 10 gigatons of CO_2 will need to be removed from the atmosphere annually until the year 2100.

Under the more realistic scenario that fossil fuels will continue to be burned until full decarbonization is achieved by mid-century, CDR at about ten GT per year before 2050 can help stabilize GHG levels and keep actual warming within the 3 °C range. After 2050, CDR would be required to double to 20 GT per year to meet the end-of-century goal. This is actually a good thing, because it will allow time for carbon capture technology to be fully developed and deployed. The captured carbon must be removed from the carbon cycle by permanent isolation or sequestration from the atmosphere for at least a century, and the longer the better. Options for sequestration are geological, terrestrial, and oceanic. These are addressed below.

CCS to capture and store carbon dioxide from point sources is generally an engineered technology. In contrast, CDR can be biological, geochemical, or engineered, or a combination of these processes. Biological DAC can be as simple as planting trees that will take CO_2 out of the air by photosynthesis and incorporate it as carbon in the wood. Geochemical DAC includes things like the enhanced weathering of volcanic soils to react with atmospheric CO_2 and lock it down as new minerals. Engineered DAC uses constructed devices, sometimes called "artificial trees" (a name that I detest and shall not repeat after this single usage) that capture carbon dioxide from the air using chemical or cryogenic methods and sequester it away from the atmosphere. Dissolved CO_2 can also be removed directly from the oceans by various means to allow seawater to uptake more of the gas from the air.

Biological DAC is the option most favored by environmentalists, but it has a number of drawbacks that must be overcome before it is viable. The first is that nearly a trillion trees (that's a 1000 billion, for those who need a conversion) must be planted to remove the many gigatons of CO_2 from the atmosphere that are necessary to mitigate climate change. Simply finding enough suitable land area to plant that many trees is a significant challenge. Trees are durable, adaptable, and have a very wide growing range from the tropics to nearly the poles, but they don't grow everywhere. The ocean, for example, covers 70% of the Earth's surface but trees only grow on land. So right out of the gate we only have 30% of the globe to work with. But even on land, trees don't grow in deserts, nor on high mountains above the "treeline" where it gets too cold. The same is true in polar regions – there are impressive boreal forests in Canada and Siberia, but as one approaches the Arctic these give way to tundra and eventually ice. There are no trees at all on the continent of Antarctica, and hardly any plants. Is there even enough forest land on Earth to accommodate a trillion trees? How will it affect other land uses like agriculture, grazing, and urban development? Saying that we can grow our way out of the climate crisis by planting trees may be comforting to some, but the numbers just don't add up. Where will we put them?

In fact, we have the opposite problem in that substantial areas of land are actually being cleared of trees in places like South America and southeast Asia to harvest the timber and to open up forest land for agriculture. Stopping this land clearing would be helpful to the climate, but it is not likely to happen given politics and economics.

The second problem with biological DAC is ensuring that the captured carbon remains sequestered from the atmosphere for at least a century and hopefully for much longer. The carbon will remain sequestered in the wood for the life of the tree, which can extend several centuries. Even if the tree is harvested and the wood is used, for example, to make furniture, the carbon will stay sequestered as long as the wood remains intact. However, the more frequent and extensive wildfires that are expected under future climates will release the CO_2 as wood and brush burn. The trees could be planted in geographical areas that are less prone to wildfires, but these have been shifting with the climate and wildfires are now occurring in unusual places like Greece. So designating an area for tree planting where they are likely to survive and prosper is largely guesswork.

Unburned vegetation like dead leaves, fallen tree trunks, sticks, and grasses that die and rot in the air will oxidize and release captured CO_2 back into the atmosphere. To effectively sequester the carbon, dead vegetable material must be stored in anoxic environments, such as deep ocean sediments or swamp muds where it eventually becomes coal. Some ideas have suggested injecting liquefied vegetable matter into old oil wells or storing woody material in deep, abandoned mines. I have concerns about both the effectiveness and the scalability of these processes to sequester hundreds of GT of carbon.

An idea for biological DAC that has been around for a while suggests fertilizing the oceans with iron and nitrogen to create an algal bloom. Algae are marine plants that photosynthesize, so they will remove CO_2 from the air. In most parts of the deep ocean far from land, algal growth is limited by the absence of nutrients. Adding these can cause rapid growth. It is necessary for the algae to then be consumed by animals ranging from krill to whales, and the carbon processed into fecal pellets that fall to the deep ocean floor and remain sequestered. The entire process seems to be a bit overly complicated, and to rely on specific animal behavior. As far as I know it has not been attempted beyond a few experiments.[2]

Geochemical DAC envisions capturing CO_2 out of the atmosphere and sequestering it as carbon in soils. One way to do this is to mix small particles of mafic volcanic rock such as basalt into the soil as an amendment. Basalt is composed of calcium and iron-rich minerals that are unstable at the Earth's surface and weather in the presence of water and oxygen. The weathering process produces ions that will react with atmospheric CO_2 to form carbonate minerals such as calcite or siderite in the soil. Basalt weathering also produces a class of hydrated sheet silicate minerals known as clay minerals that can retain organic carbon in soils from decayed plants. Some scientists have talked about pulverizing the basalt surfaces of entire volcanic islands to speed up the weathering process and capture carbon. I think this might be more harmful to the local environment than climate change itself and is perhaps a case where the cure is worse than the disease. Geochemical DAC, at least as applied by humans appears to be a generally small-scale process that as presently designed will be unable to contribute much to the many gigatons-per-year CDR requirement needed to mitigate climate change.

Engineered DAC removes CO_2 from the atmosphere by directly capturing it using cryogenic methods, chemical reactions, or gas separation membranes, which are still in the experimental stage. The cryogenic method works by freezing the CO_2 out of the air. Carbon dioxide forms "dry ice" at temperatures that are cold (-109.3 °F or -78.5 °C), but still much higher than temperatures required to liquefy other gases like nitrogen or oxygen. The method requires very dry air because water vapor in humid air will freeze out with the dry ice and cover everything with regular ice. The energy requirements for freezing carbon dioxide out of the air are relatively high compared to other options.

Although many people think of CO_2 as a chemically-inert gas (it is after all used in fire extinguishers to put out electrical fires), it does react with various substances. Chemically-engineered carbon dioxide removal from the air using solid sorbents has been around for decades but was generally limited to applications where breathing air is restricted, such as on submarines and spacecraft. The process used is called "carbonation," and is carried out with chemicals that react with the CO_2 and bind to it chemically. The two most common processes use either calcium hydroxide or sodium hydroxide as the reactants. The spent chemicals can simply be replaced with a fresh batch, or the reaction can be reversed using heat or other means to "recharge" the reactant for re-use.

The carbonation reaction with calcium hydroxide is as follows:

$$Ca(OH)_2 + CO_2 \rightarrow CaCO_3 + H_2O$$

---

[2] Coale, K.H., Johnson, K.S., Fitzwater, S.E., et al., 1996, A massive phytoplankton bloom induced by an ecosystem-scale iron fertilization experiment in the equatorial Pacific Ocean: *Nature*, v. 383, issue 6600 (Oct 10), p. 495-501; doi: 10.1038/383495a0.

The end products of the reaction are water and calcium carbonate that immobilizes the carbon. $CaCO_3$ is the formula for the mineral calcite, which is the main component of limestone. Limestones are extremely stable and have been around for hundreds of millions of years. A similar carbon capture process uses sodium hydroxide as the reactant:

$$NaOH + CO_2 \rightarrow NaHCO_3$$

The end product of this reaction is sodium bicarbonate, more commonly known as baking soda. It is more soluble that calcite and less suitable for long-term sequestration, but as a process for CDR it is quite effective. These simple chemical reactions could be scaled up for DAC to reach the gigaton-per-year levels needed to mitigate the effects of climate change. The carbon dioxide can be released and sequestered separately by recharging the reactant, or the stable mineral end products can be left as they are to keep the CO_2 out of the atmosphere.

Another common carbon capture process uses a class of chemicals called amines derived from ammonia. The chemistry between CO_2 and amines is an acid–base equilibrium reaction where the amine reacts with CO_2 to form what is called a carbamic salt.[3] The carbamate bond in the salt can be broken by heating or by exposure to low pressures, releasing the CO_2 for storage and recharging the amine for re-use. These processes are known respectively as temperature-swing and pressure-swing adsorption. A physical separation technology is under development to filter CO_2 directly out of a gas stream using a selectively permeable membrane.

All the carbon capture technologies work better if the CO_2 is more concentrated first. Thus, CCS on fossil fuel combustion has a somewhat easier time of it than DAC because flue gases contain much higher concentrations of CO_2 than the atmosphere. A process called oxy-combustion that burns coal or natural gas in nearly pure oxygen rather than air produces even higher concentrations of CO_2 in the flue gases of up to 70%. It also increases the cost of combustion, because of course obtaining pure oxygen is not free.

Although CO_2 in the atmosphere is at record high values of 420 ppm, this is only 0.042% of the air by volume. From a chemical engineering standpoint, the capture of such low concentrations of any gas is a formidable technical and economic challenge. Engineered DAC processes must contact large volumes of air to capture reasonable amounts of CO_2. This is done typically by using large banks of fans to move air across the collecting surface. Some designs use already-installed air handling equipment, like the heating-ventilation-air conditioning (HVAC) systems on commercial buildings to move air past the collectors, and other designs use the wind. Air handling is a major energy expense on engineered DAC systems, and significant amounts of carbon zero electric power are required to operate a large bank of fans.

---

[3] Yamada, H., 2021, Amine-based capture of CO2 for utilization and storage: *Polymer Journal*, v. 53, p. 93–102.

A company called Climeworks has been operating an engineered DAC plant outside of Zurich, Switzerland since 2017 that removes about 900 tons of carbon dioxide per year. Canada-based Carbon Engineering has been working on DAC since 2015, developing a 1 ton/day pilot unit, and a third company, Global Thermostat based in New York, is removing about 1000 tons (1 kiloton) of CO_2 per year. Climeworks recently began operations in Iceland at a large, centralized DAC complex called Orca.[4]

The Orca facility consists of eight collector units that use a temperature-swing adsorption method; each unit has a rated CO_2 capture capacity of 500 tons per year, or a total of four kilotons annually. The collectors are arranged around a centralized processing area that provides electric power and handles the captured carbon dioxide. The collectors draw large volumes of air into the capture devices with banks of electrically-operated fans. The heat necessary to remove the CO_2 from the collectors and recharge them is also electrical. The power to run this DAC facility is supplied by the nearby Hellisheidi Geothermal Power Plant. Abundant geothermal power is one of the advantages of operating in Iceland. Geothermal power is carbon-free and renewable, so it does not contribute any additional GHG into the atmosphere.

All these operations are currently in the "pilot" phase in my opinion, because a kiloton here and there is really small potatoes. To put this in perspective, around 31 gigatons of anthropogenic CO_2 were released to the atmosphere worldwide in 2020. The amount of CO_2 captured and stored by all of humanity that year using DAC and CCS was only around 5 million tons (5 megatons), or approximately 0.016% of emissions. Clearly, we must do better. If CDR is to make a difference in the climate crisis, it must be scaled-up massively and dramatically.

DOE has recently launched a program called the Carbon Negative Earthshot under President Biden's Executive Order 14008. This is the U.S. government's first major effort to take on direct air capture CDR and is reportedly funded at $10 billion. Hopefully this funding will advance the technology in a meaningful way to make a significant difference. (DOE has been funding CCUS under the coal program for point source capture for many years.) The goal of the Carbon Negative Earthshot is to begin removing several gigatons of CO_2 from the atmosphere by 2030 at a cost under $100 per metric ton.

I am associated with the development of a distributed carbon capture system that we have named Carbon Blade.[5] I can describe the workings of this system in more detail because of my involvement, but this is not a pitch for one system over another. There are many other designs and a number of other DAC facilities are focused on reducing CO_2 concentrations in the atmosphere. CDR is a huge undertaking that requires many hundreds of gigatons of greenhouse gas to be removed and sequestered. From the moment that I first understood the magnitude of this task, I have maintained that gigaton levels of capture will only be achieved by employing multiple technologies, and all the various DAC projects, proposals, and ideas out there

---

[4] https://climeworks.com/roadmap/orca

[5] https://www.carbon-blade.com/

should be cooperating, supporting and learning from one another, and bringing breakthrough technologies online as quickly as possible. The climate crisis needs all hands on deck and collaboration is more effective than competition. CDR is a very large and very deep pool, and there is plenty of room in it for everyone to swim.

Rather than a large, centralized facility like Orca, distributed DAC systems like Carbon Blade use equipment that is portable and mobile to capture carbon. There are a number of advantages to this. First, the Carbon Blade units are about the size of a marine cargo container, and they can be placed directly on the site where the carbon is to be sequestered. This avoids the need for costly pipelines to transport carbon dioxide sometimes hundreds of km from central DAC facilities to sequestration sites. Second, the power demands of Carbon Blade are much lower than a centralized DAC system, and instead of requiring line power from a nearby geothermal power plant, the Carbon Blade units are self-powered with small wind turbines and solar panels. This means they are location agnostic and can be set up essentially anywhere. The capture device for carbon dioxide, known as the air contactor, consists of a series of membranes mounted on vertical wind turbine blades on top of the unit (hence the "blade" reference) that spin under a light breeze. Rather than the multiple, powered fans used by Orca and other centralized DAC facilities to bring air into their contactor units, Carbon Blade uses the wind.

Each of the Carbon Blade cargo container-sized units are designed to capture about a metric ton of CO_2 per day. Scaling up to a gigaton per year capacity will be achieved by deploying multiple units; our calculations indicate that about 4 million will be required. This may sound like a lot until you realize there are nearly 300 million cars on American roads alone. Building 4 million devices will not be a big strain on global industrial capacity.

The capture technology is relatively simple. Sodium hydroxide solution (NaOH) is used to pull carbon dioxide out of the air:

$$NaOH + CO_2 \rightarrow NaHCO_3$$

The resulting sodium bicarbonate solution ($NaHCO_3$) is then reacted with sulfuric acid (H_2SO_4). If you've ever done the vinegar on baking soda experiment as a kid to make a "volcano," you know that sodium bicarbonate reacts with acid (in the case of vinegar, acetic acid) to release bubbles of gas. That gas is CO_2, which is then sequestered:

$$2NaHCO_3 + H_2SO_4 \rightarrow NaSO_4 + 2H_2O + 2CO_2 \left(gas\right)$$

The sodium sulfate ($NaSO_4$) left in solution after the gas is removed is a non-hazardous salt that can be regenerated back into sodium hydroxide and sulfuric acid through the use of a device called an electrodialysis bipolar membrane, or EDBM. These employ an electrical field to separate the positively charged cations and negatively charged anions in a solution, pulling them through oppositely charged membranes and concentrating them on the other side. The EDBM is a proven technology used commercially for chemical processing and seawater desalination.[6]

$$NaSO_4 \text{ in EDBM} + \text{electric current} \rightarrow NaOH + H_2SO_4 \text{ (separated by the membranes)}$$

A schematic of the Carbon Blade capture system is shown in Fig. 9.1. We initially thought the spinning capture blades could power an electrical generator until we talked to an actual wind turbine engineer and learned that the wind speeds and forces required are quite different. The present design uses separate small wind turbines along with solar panels for power generation.

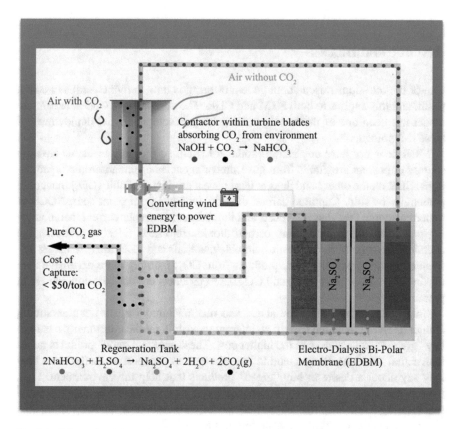

Fig. 9.1 Schematic of the carbon capture system design for Carbon Blade
Source: Dr. Hunaid Nulwala/Carbon Blade Corporation; used with permission

---

[6] Valluri, S., and Kawatra, S.K., 2021, Reduced reagent regeneration energy for CO2 capture with bipolar membrane electrodialysis: *Fuel Processing Technology*, v. 213, p. 106691.

Unlike the proposed SRM techniques of cloud modification or stratospheric aerosol injection, CDR is a much safer method of geoengineering that can be closely controlled and monitored. It will take a long time to reduce carbon dioxide levels in the atmosphere and progress will be incremental. We will have plenty of opportunities to observe how this is going and make appropriate adjustments as needed.

Perhaps an even more important point is that CDR seeks to remediate the cause of the problem, whereas SRM is focused on merely addressing the symptoms. It's the difference between fighting a fever with an aspirin versus taking an antibiotic. The climate crisis and global warming are the symptoms of excess GHG in the atmosphere from fossil fuels. Instead of trying to cool the planet down with potentially risky techniques, isn't it a better idea to just reduce the GHG that is causing the problem?

****

Revenue and BECCS

One of the questions raised about geoengineering is how to monetize it to make a profit, and this applies to both SRM and CDR. The absence of a workable revenue model has been one of the factors holding back development and deployment of these technologies.

SRM does not have any actual products to sell, so the only available revenue sources at present are grants from government agencies or philanthropic organizations. CDR on the other hand does at least have a product available (CO_2) that could potentially be sold. Captured carbon dioxide, whether from point source CCS or nonpoint source DAC has high costs and limited markets in our current manufacturing processes. Ideas for turning carbon dioxide into new, saleable products like chemical feedstocks, fuels, plastics, durable materials and other uses are facing difficult economics. Making these products from CO_2 is simply more expensive than making them from petroleum, and CO_2-based products are not cost-competitive at present.

Some companies have gone ahead and manufactured various items including diamonds from converted CO_2 in an attempt to cash in on the consumer's desire to buy "green" products despite the higher cost. These are "feel good" products rather than actual solutions. People tend to buy whatever is cheapest no matter what they may say about a desire to buy "green" products that help the environment. Plus, there is the issue of scale. By mid-century, the world needs to be capturing and sequestering something like 10 billion tons of carbon dioxide annually to make a difference in climate. That's a mountain of diamonds.

One way to monetize CO_2 is to find a cheap process to convert the gas into a hydrocarbon so it can be used as fuel. Recycling the combustion product of a hydrocarbon fuel back into a new fuel is a true example of what is called the "circular economy," where the goal is to convert CO_2 from a waste product into something useful. If it gets turned into methane and burned as a fuel, the combustion products would be net zero and putting the CO_2 back into the atmosphere will allow it to be re-captured and recirculated. Thus, the circular economy. I don't know if existing methanogens would accept carbon dioxide as an organic feedstock, but microbiologists are always telling me that genetic engineering can do almost anything. The genetic engineering community hasn't shown a lot of interest in working with methanogens so far, but perhaps DOE and other funding organizations could put some money behind this.

Another potential source of CDR income would be government subsidies in the form of direct grants and tax credits, and from the sale of carbon offsets to organizations interested in buying these. As mentioned previously, the U.S. tax credit for each metric ton of carbon dioxide captured and sequestered as specified in section 45Q of the Internal Revenue Code is at best $50. With the costs of carbon capture using current DAC technology ranging between $100 and $200 per ton, $50 per ton isn't likely to draw many takers. Congress has discussed raising the 45Q tax credit as high as $175 per ton, but at this writing it has not yet done so.

Potential CDR revenue from carbon offsets is uncertain. Pricing varies between buyers, and the market for these is totally unregulated at present. Some DAC companies are planning to use carbon offsets as their primary revenue stream, and more than a few appear to be optimistic that the price for captured carbon will rise steeply in the near future. I would hedge my bets on this at least a bit. However, it is telling that at least a few venture capital firms are willing to take a risk by funding DAC startups. Some are even buying carbon offsets in advance to give the industry a boost and get in on the ground floor of what they expect will soon become a lucrative market. I concede that I might be wrong here about uncertainty over the future of carbon offsets, and frankly, I couldn't be happier if I was.

Unfortunately for the climate, the major profitable market for captured CO_2 at present is enhanced oil recovery (EOR). As described earlier in relation to induced seismicity, this process attempts to re-mobilize residual oil in depleted petroleum reservoirs with a pressure surge or flush through the reservoir to encourage additional oil flow to production wells. The pressure surge can be achieved by injecting water into the field in an operation called a waterflood, or by injecting a gas like CO_2. Carbon dioxide injected into depleted oil reservoirs will tend to remain in the subsurface and not be produced with the oil. Revenue is supplied by both the recovered oil and the 45Q tax credits for sequestering the CO_2 underground. Neither is all that great fiscally on its own, but the economics are favorable with the two combined. Thus, oil companies receive an IRS 45Q tax credit for sequestering CO_2 even though the process of doing so releases additional fossil fuel destined to make more greenhouse gas. I know. Go figure.

EOR using captured carbon dioxide is being done on old oilfields in Texas, Louisiana, and Wyoming, including several projects partially financed by the Department of Energy. The State of Wyoming, which contains both abundant coal reserves and many old oil fields has in fact constructed a CO_2 pipeline network from several coal-fired power plants equipped with carbon capture to supply the gas for EOR on depleted oil fields at Salt Creek (refer back to Fig. 3.1) and other areas in the eastern part of the state. The economics of enhanced oil recovery do require relatively high oil prices, so the use of CO_2 for EOR comes and goes as prices change globally.

The search for a carbon neutral or carbon negative emissions method to achieve CDR with a revenue stream that would make it profitable leads us to BECCS, or bio-energy with carbon capture and storage. As explained earlier, plants use photosynthesis to pull carbon dioxide out of the air and make sugars and starches. If these plants are then turned into a biofuel like methane gas, it can replace natural gas to generate electricity and provide a source of revenue. As discussed previously, the process is carbon neutral if the combustion products are emitted into the air because the biofuel is merely returning the CO_2 back to the carbon cycle. However, if the power plant is equipped with CCS to capture and sequester the emissions, then the process becomes carbon negative.

A negative emission means the net removal of carbon dioxide from the atmosphere. Adding CCS to a biogas power plant for the BECCS process will remove carbon dioxide from the atmosphere while producing electricity as a saleable product. The added cost of CCS can be covered with the 45Q tax credit, sales of carbon offsets, and if necessary, a slight increase in the cost of electricity (COE). If all current U.S. natural gas CC power plants were converted to biogas and equipped for BECCS, this would get CDR into the vicinity of the 10 gigatons per year recommended by the IPCC at a modest capital cost, and with little or perhaps even no increase in the COE.

I spent several decades working with the natural gas industry to develop shale gas. At the time, it was the right thing to do because we were in an energy crisis. However, now we are in a climate crisis, and it is time to replace natural gas with biogas and use BECCS to capture the carbon dioxide as negative emissions.

* * * * * * * * * * * *

Carbon Sequestration

Once captured from a smokestack or directly from the atmosphere, carbon dioxide must be kept isolated or sequestered from the atmosphere for long time periods to have a positive effect on climate. The minimum isolation period is a century, but the longer the better. This is known as durability, and it is an important consideration for government and philanthropic funding groups supporting CDR, as well as venture capitalists pre-buying carbon offsets.

The term "sequestered" in American English usage means separated or isolated. It is most commonly heard when a jury is said to be sequestered while hearing evidence and deliberating a verdict in a criminal trial. It means that they must be kept isolated from outside news sources and opinions that could potentially influence their decisions.

In the CCS program, the Department of Energy switched from the use of the term sequestered to calling the carbon dioxide isolation process "storage" instead. The reason for this is because "sequester" has a different context in French, where rather than isolation, it implies being held against one's will, similar to kidnapping. Given the international nature of the CCS program, DOE did not want to offend French-speaking partners with what was thought to be a rather harsh term for isolating CO_2.

I've always had an issue with the term carbon "storage" because it implies putting something in a place where you can get it back later. When you put furniture in a rented storage unit, you are eventually going to want to retrieve it. However, if you take an old couch to the dump, that pretty much means you don't want it back. Carbon dioxide is a lot more like the old couch at the dump than furniture in a storage unit, and storage is just not the right term.

When I was at DOE, we floated the idea of calling it "carbon disposal," but that was quickly shot down. The agency was trying to kick off the CCUS program at the time to find uses for captured carbon that would improve the economics of capture. "Disposing" of it made it sound like a waste product with no further use. The DOE managers felt that this could reduce industry enthusiasm for their promotion of carbon dioxide utilization, which was already perilously low because as mentioned previously, there just aren't a lot of markets.

Thus, have I learned to accept the logic of government bureaucrats. For this book I prefer the term "sequester" because it precisely describes what must be done with captured CO_2: isolate it from the atmosphere. Even if it has to be kidnapped.

The IPCC says we need to sequester gigatons of CO_2, so it would be nice to know how much space we are going to need. Keep in mind that CO_2 is a gas, so depending on how much it is compressed, it will occupy different volumes. We can start by exploring the volume of a metric ton (1000 kg) of CO_2. Some of the visualization conversions below are a bit imprecise, but still in the ballpark and will provide an idea of the volumes.

CO_2 gas at the standard temperature of 25 °C or 77 °F and a standard pressure of 1 atmosphere has a density of 1.836 kg/cubic meter. Thus, one metric ton of the gas under standard temperature and pressure conditions occupies a volume of 545 cubic meters (19,235 cubic feet). This can be visualized as a cube about 8 meters (26 feet) on a side. It can also be represented by a sphere some 10 meters (33 feet) in diameter.

If compressed to 4 atmospheres (60 psi) a metric ton of CO_2 will occupy a volume of 136 cubic meters (4808 cubic feet). This can be visualized as a cube about 5 meters (16 feet) on a side. If compressed to a supercritical fluid (explained in more detail below), one metric ton of CO_2 occupies 2.65 cubic meters (93.6 cubic feet). This can be visualized as a cube about 1.4 meters (4.6 feet) on a side.

If stored as dry ice at a density of around 1.6 kg/liter, a metric ton of CO_2 would occupy about 0.625 cubic meter (22 cubic feet). This can be visualized as a cube about 0.85 meters (2.8 feet) on a side. A generously-sized kitchen refrigerator has a capacity of 22 cubic feet.

There are three main options for sequestering captured CO_2: (1) geologic isolation deep underground, (2) terrestrial sequestration in soils, vegetation, or manufactured materials on the surface, and (3) ocean sequestration in deep seawater. These are discussed individually below.

Geologic Sequestration

Injecting captured CO_2 deep into the subsurface will keep it isolated from the air. The gas can be placed in depleted conventional oil and gas fields, unmineable coal seams, depleted shale gas wells, deep saltwater aquifers, or in volcanic basalts. Each of these have advantages and disadvantages that are discussed separately below. Successful field experiments have been carried out in all of them except shale gas wells. The performance of shale for CO_2 sequestration has been tested in the laboratory.[7]

Storing carbon dioxide deep underground requires dealing with pressure effects. Pressures increase with depth in both rocks and in the water filling the pores of the rocks. Essentially, the weight of the material above presses down with greater force at greater depths, increasing the pressure. Rock pressure, called the lithostatic pressure gradient, increases at a rate of about 1 pound per square inch (psi) per foot of depth. The metric equivalent is 22.6 KPa per meter. Fluid pressure, called the hydrostatic pressure gradient, increases by about half a psi per foot. This requires compressing carbon dioxide gas to inject it into deep rocks.

At elevated pressures, carbon dioxide becomes a "supercritical fluid." Starting at atmospheric pressure, the gas gradually compresses as pressure is increased until it reaches the so-called "critical point" defined by a combination of temperature and pressure. At this point, the gas suddenly condenses into dense liquid in equilibrium with the gas phase. A slight change in pressure or temperature from the critical point will make it go all liquid or all gas depending on the direction. The critical point for CO_2 is 88 °F at 1070 psi (31 °C at 7.38 MPa). In rock formations under normal hydrostatic pressure gradients, this pressure and temperature combination occurs at depths of about 2500 feet (800 m). Substantially more CO_2 can be stored in a rock as a supercritical fluid than as a gas, providing a significant advantage for carbon sequestration in subsurface pore space depending on the geologic storage options available. It is important to note that supercritical CO_2 is also a powerful solvent and can affect downhole seals and barriers.

Depleted Oil and Gas Fields Before they were drilled and produced, these held significant volumes of oil and natural gas in the subsurface over geologic time periods. The storage medium consists of porous rocks such as sandstone or limestone sealed within a geologic trap like a fold or an offset along a fault. As long as the

---

[7] Hong, L., Jain, J., Romanov, V., Lopano, C., Disenhof, C., Goodman, A., Hedges, S., Soeder, D., Sanguinito, S., and Dilmore, R., 2016, Factors Affecting the Interaction of CO_2 and CH_4 in Marcellus Shale from the Appalachian Basin: *Journal of Unconventional Oil and Gas Resources*, v. 14, p. 99–112.

integrity of the trap and seal were not compromised by pressure cycling during oil-field operations, the pore space in the depleted field could potentially store CO_2.

Most conventional oil and gas reservoirs have a layered density structure with a gas cap at the top overlying petroleum-saturated rock, which in turn overlies porous rock containing saltwater brine. Basically, oil floats on water and gas floats on oil. As the gas and oil are produced from the reservoir, the brine moves upward under a "water drive" to fill the pore space. Some of the brine is produced with the oil and gas during the life of the field. This is known as the "water cut," and the gas and oil can be readily separated from the brine by gravity in a surface tank. Disposal of the saltwater adds a cost to oil production, and if the water cut gets too high and oil prices too low, the well may be shut in. Eventually, the invading water will split apart the remaining subsurface oil into individual, non-mobile droplets that remain trapped in the pores of the reservoir rock, and only saltwater is produced from the well. At that point, the oil well is said to have watered-out. Significant amounts of residual oil often remain behind, and EOR operations can be used to produce additional oil if the price is high enough.

Depleted wells will often change hands many times as old owners divest themselves of liability and new owners try to extract whatever minimal amount of oil is left. A well may produce a few barrels a week over extended time periods in what are called "stripper" operations. These typically consist of one person with a pickup truck making the rounds to visit their wells. When depleted, state regulations require that specific actions be taken to legally plug and abandon (P&A) the well. However, the cost of having a workover rig and crew come onsite to properly P&A the well is far greater than the annual fee paid to the state for renewing a lease on a stripper well. A non-productive oil well may have the lease renewed annually for decades by a stripper well operator after a well has been shut in. When owners retire or die, if no one takes over responsibility for the wells they are often abandoned and forgotten.

Abandoned wells that have no known owner are said to be "orphaned." There are an estimated three million abandoned and orphaned wells in the United States dating back to the first oil boom started by Colonel Edwin Drake in 1859. Pennsylvania alone has more than 100,000 old wells. So does Tennessee. The 1920s oil drilling boom in California saw intensive drilling operations in the Los Angeles basin, and today there are hundreds of abandoned wells under the streets, in back yards, and next to houses in the older sections of Los Angeles.

These wells are not a benign presence; many emit methane, a fire hazard and a more powerful GHG than CO_2. Some wells may also emit VOCs and toxic gases like hydrogen sulfide that can cause health problems to nearby residents or even kill at certain concentrations. An isolated, abandoned well in a Pennsylvania farm field is one thing, but when these wells are located in densely populated urban areas like LA, many more people are put at risk. Properly plugging and sealing abandoned wells is an important public health and climate issue in the U.S.

An old oilfield with abandoned wells is not producing any income, and carbon sequestration can provide a source of revenue to the land owner. Even an oilfield with active stripper wells can use sequestration as a source of income. If a stripper well operator can make more money from carbon offsets and 45Q tax credits than by producing a few barrels of oil per week, they might be interested in injecting CO_2 into their wells.

I have been collaborating with professors and graduate students at the South Dakota School of Mines and Technology to identify microbes that can survive at the elevated pressures and temperatures deep in the Earth. We have found some promising candidates from a deep mine and from geothermal vents on midocean ridges. The goal of the research is to genetically engineer these microbes to convert carbon dioxide into solid carbonate minerals like calcite using a natural enzyme called carbonic anhydrase. Storing captured carbon as a solid, stable mineral phase is the ultimate durable sequestration.

Carbon dioxide dissolved in water creates carbonic acid:

$$H_2O + CO_2 \rightarrow H_2CO_3$$

Replacing the hydrogen ions in carbonic acid with calcium or another cation present in oilfield brines converts it to solid minerals.

$$H_2CO_3 + Ca \rightarrow CaCO_3 (\text{calcite})$$

$$H_2CO_3 + Mg \rightarrow MgCO_3 (\text{magnesite})$$

$$H_2CO_3 + Na \rightarrow NaHCO_3 (\text{nahcolite})$$

$$H_2CO_3 + Fe \rightarrow FeCO_3 (\text{siderite})$$

This mineralization can happen inorganically over time through aqueous geochemical reactions. It is much quicker when performed by organisms, however.[8] Creatures from clams to corals extract calcium carbonate from seawater to build their shells. There are several classes of algae and bacteria that can do this as well. The SD Mines researchers are seeking first to find microbes that can survive under the high temperatures and pressures at the depths of conventional oil and gas reservoirs, and then genetically modify these organisms to precipitate carbonate minerals. The process is known as biomineralization.

Microbes that can survive in extreme conditions are called extremophiles. Temperature tolerant species known as thermophiles can be found naturally in hot springs like those in Yellowstone National Park. The vivid colors in Yellowstone's Grand Prismatic Hot Spring, for example, are caused by extremophiles living in the hot volcanic waters (Fig. 9.2). Other microbes can tolerate high pressures, high salinity, and harsh pH conditions, and a few can tolerate nearly all of these extremes.

---

[8] Bhagat, C., Dudhagara, P., and Tank, S., 2017, Trends, application and future prospectives of microbial carbonic anhydrase mediated carbonation process for CCUS: *Journal of Applied Microbiology*, v. 124, p. 316–335.

If we can get microbes to biomineralize carbon dioxide into calcite or other carbonates within the pore space of a depleted oil reservoir, the carbon will be sequestered and immobilized deep underground as solid carbonate minerals for geologic time periods. This is an extremely durable form of carbon sequestration. In addition, filling up this pore space with minerals will seal it and block the migration of methane, VOCs, hydrogen sulfide and other gases through the rock. This will stop fugitive emissions of methane into the atmosphere from depleted and abandoned oil wells, elegantly solving two problems simultaneously.

Depleted, conventional oil and gas fields exist across the U.S. and around the world and represent a very large volume of subsurface pore space that can contain substantial amounts of DAC carbon dioxide. Worldwide capacity estimates of depleted oil and gas reservoirs for carbon sequestration may be as high as 14,000 GT according to the Global CCS Institute.

An additional advantage of depleted oilfields is that they possess existing wells that can be used for CO_2 injection. The cost of drilling new wells for the geological sequestration of carbon dioxide is not a trivial expense. In 2016, the EIA estimated that the cost of drilling oil and gas wells ranged from $100 to $200 per foot for vertical drilling, and $400 to $800 per foot for laterals.[9] Costs are presumably higher these days but using the $200 per foot number as an example, drilling a new vertical well to the 2500-foot minimum depth for carbon dioxide to become a supercritical fluid would cost half a million dollars. Even if an existing well needed some repairs to the casing or cement before being used for CO_2 injection, this would still be cheaper than drilling an entirely new well.

Fig. 9.2 Vivid colors from extremophiles living in Grand Prismatic Hot Spring, Yellowstone National Park
Source: Photographed in 2019 by Dan Soeder

[9] Energy Information Agency (EIA), 2016, Trends in U.S. Oil and Natural Gas Upstream Costs: U.S. Department of Energy, Washington, DC 20585, March 2016, 141 p.

One of the downsides to using depleted oilfields is that they do contain a lot of old wells. Each one of these could provide a potential leak and pathway to the surface for injected CO_2 if the old wells have deteriorated cement or corroded casing. Obviously, assessing the physical condition of wells in a depleted oilfield is important before injecting CO_2. Pressure cycling during the production of the oilfield also can sometimes fracture the natural seal of the geologic reservoir, providing an additional pathway for CO_2 to leak from the geologic trap.

DOE calls the process for keeping track of injected subsurface CO_2 monitoring, verification, and accounting (MVA). Monitoring techniques like gravity measurements, seismic data, computer modeling, and analysis of satellite data are used to track pressure changes in and above a subsurface reservoir receiving CO_2 injections. MVA zeroes in on anthropogenic CO_2 and monitors the movement of any escaped CO_2 plumes or pressure fronts. Ultimately it seeks to verify containment effectiveness to protect human health and the environment.

Migration will not be a problem once the gas is biomineralized into a solid carbonate form in the pore space; leakage concerns occur during whatever time period the microbes might need to carry out the conversion. The SD Mines researchers estimate that this will take place over a period of several weeks to a few months, and CO_2 will be mobile during that time. MVA activities are expected to be active until the carbonate mineralization is complete.

Coal Seams The organic carbon that makes up the bulk of coal is capable of adsorbing CO_2, providing a significant amount of storage (refer back to Chap. 3 for an explanation of adsorption versus absorption). The adsorbed gas is present as a thin film held under a relatively high density and the amount of adsorbed gas in a coal can be significant.

Both carbon dioxide and methane adsorb onto organic matter in coal. Because of the different shapes of the molecules (CO_2 has a more linear shape than CH_4), about three times more carbon dioxide than methane can adsorb onto coal. One of the CO_2 utilization technologies we were investigating for CCUS when I was at DOE was to employ carbon dioxide to displace the adsorbed methane in coal and produce it as natural gas while sequestering the CO_2. This runs into the same problems of using CO_2 for EOR in that you are still producing fossil GHG while you are trying to sequester it.

Methane in coal seams has long been a nightmare for underground miners because of its explosion hazard. This was a potential new source of natural gas investigated by DOE after the energy crisis. Known as coalbed methane, or CBM, it was first produced as a gas resource in the 1980s in Alabama. The story I was told is that a mining company in the Black Warrior basin was drilling wells into an underground coal seam far out in front of the mining face to try to purge the methane from the coal so they wouldn't have gas problems when they mined it. Someone noticed a natural gas transmission pipeline nearby and hit on the idea of selling the waste mine gas to the gas company instead of flaring it off and CBM was born. CBM has been produced up and down the Appalachian basin, in a big way on coal seams in eastern Wyoming, and other places as well.

To get the methane gas to desorb from the coal so it can be produced, the pore pressure has to be lowered. In coal seams, this is achieved by de-watering the coal. A coalbed methane production field looks a lot like an oil field, with reciprocating pump jacks pulling sucker rods up and down. They are not pumping oil, however, but water. The gas comes up with the water and is separated from it by gravity in a separator tank. The produced water must then be disposed of.

This can be a problem because some of the produced waters are pretty nasty. Coal seams contain toxic metals like selenium and mercury, sulfide compounds that oxidize and form sulfuric acid, and a number of hazardous organic substances. All of these come up with the produced water and must be handled. The saving grace is that the coal seams with the most toxic fluids are also those that produce the least amount of water, and coal seams that produce substantial water usually produce cleaner water. In the Powder River basin in Wyoming, for example, the copious produced water associated with CBM operations was clean enough to use for crop irrigation.

There have been a lot of environmentalist objections to CBM, mainly related to the way companies handle and dispose of produced water. I was invited a few years ago to give a lunchtime talk on shale gas and fracking at a Coalbed Methane Forum hosted in Morgantown, WV by a colleague of mine at West Virginia University. We were in the buffet line when a group of people behind us (who had no idea that I was the speaker) began discussing the topic. One woman said she was really interested to hear about fracking because she thought it was an amazing technology. Her companions were appalled, pointing out that the success of fracking on Marcellus Shale gas production was overwhelming the CBM industry in northern West Virginia. She agreed but said that fracking was also monopolizing the interest of the environmentalists and they were leaving the CBM producers alone.

To sequester CO_2 in coal, a seam where CBM has been produced would provide adsorption sites for the CO_2. There is no actual empty space underground, and in order to put CO_2 down there, something has to be removed to make room. DOE has suggested that so-called "unmineable" coal seams that are too deep or too thin for commercial coal mining are potential candidates for sequestration. If these seams already contain methane, and most probably do, then it will be displaced by carbon dioxide and migrate off somewhere. Exactly where is the question, but if it ends up in somebody's basement and blows up their house (and this has happened), then the injection company could face major liabilities. Likewise, even if the displaced methane harmlessly makes its way to the surface and then into the atmosphere, it is a GHG with 86 times the global warming potential of CO_2 over 20 years, and 28 times the potential of CO_2 over a century.

Another problem is that the adsorption process can cause the coal to swell, cutting off flow paths and making it difficult for large amounts of CO_2 to enter the seam. Sequestration of carbon dioxide in coal seams has been investigated in the laboratory, and there have been some field tests to inject CO_2 into coal. These were not sequestration investigations as such but were carried out to determine if CBM production could be enhanced by CO_2 injection.

Depleted Gas Shales Organic-rich shales that have produced gas through the process of hydraulic fracturing may provide another option for CO_2 storage once they are depleted of natural gas.[10] Shale is noted for having a dual-porosity structure, with the bulk of the pore volume consisting of very small void spaces between mineral grains, within mineral structures, and within organic components. The second part of the pore system are the natural cracks and manufactured hydraulic fractures that contain less than 1% of the pore volume but provide high permeability flow paths for the gas to move to the well bore.

This dual structure of pores that store versus fractures that flow is why shale is noted for producing a lot of gas early on, but production falls off sharply after a short time period. The formation produces gas at much slower rates over very long times. The early stages of production drain gas from the fractures, but the bulk of long-term production occurs as gas migrates from the tiny shale pores into the fractures and then to the well. For example, some old, vertical, unfracked shale wells in New York state along the Lake Erie shoreline have been producing small but steady amounts of gas for more than a century. Shales typically produce some water cut with the gas, but they are not prone to watering-out like conventional wells.

An advantage that depleted gas shales share with depleted conventional oil and gas reservoirs is that existing production wells can be used for CO_2 injection. Storing carbon dioxide in depleted gas shales is the opposite of production. Filling the fracture flow paths is fast and easy, but this is followed by a tedious wait as the gas slowly makes its way into the tiny pores. Shale also has an adsorbed component of gas; like adsorption in coal it is stronger for carbon dioxide than for methane. The adsorbed gas would give shale a storage advantage over conventional, depleted oil and gas reservoirs because it would be able to hold more gas in the same volume of pore space. However, filling that pore space would take time. Working with shales demands patience.

Although I have asked this question of the shale gas industry many times, it remains unclear exactly when a gas production company would declare a shale gas well to be "depleted" and available for carbon dioxide injection. Since these wells don't water out and have been known to produce steady amounts of natural gas for decades, there is no hard and firm cutoff point. The best answer I've ever been able to get from industry is that "it depends on the price of gas." As some point the amount of gas produced is too small and the price is too low to justify the continued operation of the well. But nobody could tell me exactly when that would be.

The possible effects of hydraulic fracturing on the integrity of shale for CO_2 storage are largely unknown and mostly unstudied. There have been a few laboratory experiments on the interaction of carbon dioxide with shale, but as far as I know no field measurements. The shale gas industry has been strongly opposed to any

---

[10] Levine, J.S., Fukai, I., Soeder, D.J., Bromhal G., Dilmore, R., Guthrie, G., Rodosta, T., Sanguinito, S., Frailey, S., Gorecki, C., Peck, W., and Goodman, A., 2016, U.S. DOE NETL methodology for estimating the prospective CO_2 storage resource of shales at the national and regional scale: *International Journal of Greenhouse Gas Control*, v. 51, no. 8, p. 81–94.

injection of CO_2 into a producing gas shale. They have refused to agree to any field experiments on their shale wells.

An old shale gas well from the 1980s Eastern Gas Shales Project called MERC-1 that was located on the grounds of the DOE National Energy Technology Laboratory in Morgantown, WV would have been perfect for a CO_2 field injection test. This well was drilled nearly 8000 feet (2.4 km) deep and produced small amounts of gas from the Marcellus Shale for 40 years. Unfortunately, the lab director ordered it to be plugged and abandoned before we could obtain funding to do the investigation, and it was an opportunity lost. Until there is access to a depleted shale well, understanding how carbon dioxide might be sequestered in shale remains limited to lab experiments and computer models.

Deep Saline Aquifers Porous sedimentary rocks at great depths contain saltwater rather than the fresh groundwater found in shallower aquifers near the surface. Fresh water from rain and melted snow on the surface infiltrates into the ground and percolates downward to recharge shallow aquifers. Fresh water only gets down to moderate depths, however, and porous rocks at depths below about 500 meters (1640 ft) typically do not receive the freshwater recharge. These deep rocks contain saltwater, known as connate water, which is often the residual seawater that was trapped in the sediment when it was deposited in an ocean or inland sea.

The saltwater in deep sedimentary rocks was produced to supply salt back in pioneer days for settlers who needed to preserve food but were too far from the ocean to easily obtain sea salt. By the mid-nineteenth century there was plenty of saltwater drilling going on in the Appalachian basin. As described back in Chap. 3, Samuel Kier was producing saltwater in Pennsylvania when he discovered that the petroleum coming up with his brines could actually be sold as a patent medicine instead of being discarded as an undesirable waste product. He then came up with the idea and the design of a kerosene lantern that used refined petroleum for illumination.

The oldest sedimentary units at the bottom of many basins often consist of coarse sandstones deposited on top of the basement rocks. These are generally described as arkose, which are sandstones composed of quartz and feldspar. Feldspar is an unusual mineral in a sandstone because it is much more prone to weathering than quartz. Most feldspar that erodes out of igneous rocks turns into clay minerals upon exposure to air and water, and eventually ends up being deposited as shale.

The geological explanation that I learned as an undergraduate for the basal sediment in basins being coarse arkose is that when the granitic continental crust began down warping into a sedimentary basin, the granites themselves eroded as water flowed in. The crumbling rocks sent quartz and feldspar grains tumbling a short distance into the bottom of the basin to be deposited as sand-size sediment. The short transport distance and brief travel times preserved the feldspar and quartz as coarse, angular grains. Later sediments deposited on top of the arkose came from many different sources outside the basin as streams were established. The much longer travel times of these stream sediments rounded off quartz grains, weathered feldspars into clay, and deposited a variety of different sedimentary rocks into the basin.

In the Appalachian basin, the basal arkose is the Late Cambrian age Potsdam Sandstone, described by legendary New York geologist H.P. Cushing as "coarse grit, white or buff, in massive beds, much of it cross-bedded and ripple marked, with very coarse conglomerates at the base.[11]" In the Illinois basin and Michigan basin, the basal sedimentary unit is the Late Cambrian Mount Simon Sandstone, similar in age to the Potsdam and also described as an arkose. In the Wind River basin in western Wyoming where I took my geology field camp course, the basal sandstone is the Middle Cambrian age Flathead Formation, which I remember as an arkose.

The deep basin-center rock units can be used to sequester substantial amounts of CO_2 as a supercritical fluid in solution under high pressure. The biggest obstacle is the cost of drilling deep wells, which is not a trivial expense especially at great depths. Using the EIA drilling cost estimate presented earlier of $200 per foot of depth, drilling a well to the roughly 15,000-foot depth required to reach the basal Mount Simon Sandstone in the center of the Illinois basin would easily cost three million dollars. Although this is the normal cost of doing business for the oil and gas industry, it is beginning to get a bit spendy for a government carbon sequestration project.

Tracking the carbon dioxide-saturated plume for MVA as it migrates at great depths is also extremely challenging. Many of the monitoring and measurement techniques used on shallower formations are not as accurate or don't work very well at extreme depths. In addition there is a concern that highly pressurized CO_2 in deep brines will produce a strong carbonic acid solution, with potential effects on well casings, cement, seals, and the target formation itself that are poorly understood.

DOE got involved in a CCS project a few years ago called FutureGen, which was supposed to generate electricity in central Illinois from a "clean coal" plant equipped with CCS, and then sequester the captured carbon in deep saline aquifers. This was the first attempt that I know of to use deep aquifers for carbon sequestration. As an interesting aside, these deep saline aquifers have been used for decades to dispose of dangerous chemicals and toxic waste.

FutureGen was a public-private partnership that was intended to demonstrate the engineering for adding CCS to a coal-fired power plant. The original concept announced by the Bush Administration in 2003 was to build a near zero-emission, 275-MW coal-fired power plant near Mattoon, Illinois to prove the feasibility of CCS while producing both electricity and hydrogen. The CO_2 was supposed to be sequestered in the St. Peter and Mount Simon sandstones, both of which are deep saline aquifers in the Illinois basin. The St. Peter is an Ordovician-age sandstone that is younger and shallower than the Mount Simon but still quite deep.

Things didn't go as planned. The project was repeatedly delayed by permit problems, landowner protests and lawsuits from environmental groups, and it was still in

---

[11] Cushing, H.P., 1901, Report of the State Geologist for the year 1899: New York State Museum Annual Report, no. 53, p. 37-82., Ann. Rpt. Board of Regents, Univ. New York.

the development stage when funding was pulled in January 2008. However, the 2008 housing crisis and stock market collapse later that year led the Obama Administration to establish an economic stimulus program in 2009 focused on what was called "shovel-ready infrastructure."

Even though it wasn't exactly shovel-ready, the FutureGen concept was revived, restructured, relocated, and restarted under the name FutureGen 2.0. A different, smaller coal-fired power plant in Meredosia, Illinois was retrofitted for oxy-combustion to concentrate the CO_2 in flue gases and make capture easier. Construction was scheduled to begin in 2009, with full-scale plant operations to begin in 2012. The goal was to capture and sequester 1 million metric tons of CO_2 per year for 4 years to demonstrate the feasibility of CCS.

DOE was on the hook for a billion dollars, with the remainder of the estimated $1.65 billion cost being covered by private funding. The private part of the FutureGen public-private partnership was the FutureGen Industrial Alliance, a non-profit consortium of coal mining and electric utility companies. For various reasons, the Alliance was unable to come up with their required share of funding, and construction was delayed time and again.

This project was now under a tight schedule deadline because it was funded with federal stimulus money. The Obama economic stimulus program was intended to kick-start the economy by providing a short-term boost through national infrastructure projects that ran for 2–5 years, not decades. As the delays piled up on FutureGen, a few exploratory boreholes were drilled at the sequestration site and cores were obtained that eventually ended up at the DOE National Energy Technology Lab. Finally in February 2015, the Department of Energy withdrew funds and suspended FutureGen 2.0 because the project was unable to commit to spending the stimulus funds by the required 2015 deadline.

We got some nice deep saline aquifer core samples at NETL out of all this, but to date, no one has demonstrated the feasibility of sequestering CO_2 in these rock units. Deep saline aquifers could potentially be used in some situations where there are no other choices for sequestration, but shallower options are often better and cheaper.

Basaltic Lava Rocks   These rocks are created from hardened lava flows at the surface of the Earth. Molten rock is known as magma when it is underground and lava when it erupts. Rocks made of lava are called extrusive, and rocks made of magma that cooled underground are called intrusive.

Igneous rocks are defined by composition and texture. Extrusive rocks such as rhyolite cooled quickly on the surface of the Earth and have a fine texture of small crystals. Intrusive rocks like granite cooled more slowly underground and have a coarse texture of large crystals. Granite and rhyolite can have the same composition, and possibly even be derived from the same magma body, but because granite cooled underground and rhyolite at the surface, they have different textures and thus are different rocks.

The composition of igneous rocks has two basic end member types – mafic and felsic – with a variety of intermediate chemistries. The most common mafic rock is basalt, solidified from a metal-rich, silicate melt that originated as magma either deep in the crust or in the mantle. The most common felsic rock is granite, composed predominantly of silica-rich quartz, potassium feldspar and biotite derived from shallower depths.

Mafic magmas tend to be thin, free flowing, and erupt readily, and most of the rocks with this composition are extrusive. The Hawaiian Islands and Iceland are both made largely of basalt. In contrast, the silica-rich felsic magmas are thick, viscous, pasty, and do not erupt very easily, tending to become emplaced deep underground where they cool slowly. These underground masses of solidified magma are called plutons when smaller and batholiths when large. The slow cooling allows large crystals to form, creating coarsely-crystalline rocks like granite. The granite that makes up the Sierra Nevada Mountains in California and is famously exposed at El Capitan and Half Dome in Yosemite National Park is the top of a batholith.

When lava is extruded at the surface and exposed to the air and cold ground, the melt will often congeal rather quickly into amorphous volcanic glass. This then crystallizes some years later into a fine-grained rock. Gases trapped in the hardening lava create frozen bubbles called vesicles. As the magma rises through the Earth, it begins to cool, and different minerals will often crystallize from the melt at different temperatures. An erupting lava may sometimes contain a slush of mineral crystals known as phenocrysts embedded in the molten rock. Such rocks are said to have a porphyritic texture. Figure 9.3 shows a porphyritic Hawaiian basalt with gas vesicles and green phenocrysts of the mineral olivine, known as peridot in gemstone form. The presence of olivine suggests that this high temperature magma originated deep in the mantle and the olivine began to crystallize as it rose through the crust and cooled a bit before erupting. The famous "green sand" beach on the south end of the big island of Hawai'i gets its distinctive color from sand containing weathered olivine phenocrysts.

Basalt rocks store carbon dioxide by carbonation. Because the minerals that make up basaltic rocks originate so deep within the Earth, they are not stable at the surface and weather easily in the world of wind and water. Basalts are typically composed of plagioclase feldspar and pyroxene minerals. Plagioclase is a mixture of the calcium end member anorthite, $Ca(Al_2Si_2O_8)$ and the sodium end member albite, $Na(AlSi_3O_8)$. The pyroxene mineral group consists of a suite of complex Ca, Na, Mg, and Fe aluminosilicates. Olivine is also common in basalts, consisting of the magnesium end member forsterite (Mg_2SiO_4) and the iron end member fayalite (Fe_2SiO_4). These minerals release Ca, Mg, Fe, and Na cations into solution as they weather. The cations replace the hydrogen in carbonic acid (H_2CO_3) to form carbonate minerals. This is the inorganic version of the biomineralization process described previously for sequestration in depleted oil fields.

Although basalts contain the correct minerals for creating carbonates, they are often highly fractured, and it was not clear early on how well the CO_2 would be contained within the rock while undergoing conversion to mineral form. The Icelandic company Carbfix ran some field experiments over the past decade to inject CO_2 into Iceland basalts with the support of the European Union and the

Fig. 9.3 Porphyritic Hawaiian basalt with green olivine phenocrysts and vesicles formed by trapped gas bubbles. Coin for scale is a U.S. quarter dollar, 23 mm in diameter
Source: Photographed on Hawai'i in 2018 by Dan Soeder

U.S. Department of Energy. The carbon dioxide was injected into a naturally fractured, vesicular basalt (Fig. 9.4). Surprisingly, instead of requiring decades as expected to obtain results, the researchers discovered that substantial amounts of carbonate minerals formed in as little as 2 years.[12]

Carbfix concluded that basalt rocks could contain the CO_2 long enough for it to react and mineralize, and the company partnered with Climeworks in 2017 to sequester captured carbon dioxide. Climeworks constructed the Orca centralized DAC facility in Iceland about 30 km east of Reykjavik for access to the Hellisheidi Geothermal Power Plant operated by Reykjavik Energy and to use the Carbfix injection site a few kilometers away to sequester the carbon dioxide. The geothermal plant has the capacity to generate 303 MW of electricity and also supplies hot water for Reykjavik's district heating. The captured carbon dioxide is dissolved in water to form a carbonic acid solution, which is then injected into the basalt. The Carbfix injection wells and associated monitoring wells range in depth from 50 m to 2000 m.[13] Converting CO_2 into carbonate minerals is the most secure and durable method for sequestering it away from the atmosphere.

As mentioned previously, the Orca plant is designed to remove about four kilotons of CO_2 from the atmosphere every year. This is of course a drop in the bucket compared to 31 gigatons of annual emissions, but it is a start. More importantly, it is a learning experience. Climeworks is looking to scale-up, and the next plant will be ten times bigger with an annual CDR capacity of 40 kilotons. All of the engineered CDR technologies were developed with planned scale-ups in mind, but

[12]Matter, J.M., Stute, M., Snæbjörnsdottir, S.Ó., et al., 2016, Rapid carbon mineralization for permanent disposal of anthropogenic carbon dioxide emissions: *Science*, v. 352, no. 6291, p. 1312–1314.

[13]Pogge von Strandmann, P.A.E., Burton, K.W., Snæbjörnsdóttir, S.O. et al., 2019, Rapid CO_2 mineralisation into calcite at the CarbFix storage site quantified using calcium isotopes: *Nature Communications*, v. 10, no. 1983; https://doi.org/10.1038/s41467-019-10003-8

Fig. 9.4 Iceland basalt
core on the left shows
vesicles filled with
precipitated white
carbonate minerals; core
on the right shows porous
vesicular basalt before
carbon dioxide was
introduced to the system
**Image source: Carbfix,
U.S. Department of
Energy (public domain)**

before one can scale-up, it is important to build the system, test that all the components work together, and optimize performance. This takes time, and I don't expect
to see a substantial deployment of engineered CDR until about 2030.

Basalt is one of the most common rocks on Earth. It makes up the crust under the
oceans, and forms the bulk of oceanic islands, like Iceland, Hawaii, the Azores, the
Aleutians, Japan, and many others. The mid-ocean ridges are enormous underwater
mountain chains that wind around the globe at the boundaries of tectonic plates and
are composed almost totally of basalt. There are also large basalt deposits on continents, including the Columbia River Basalt in eastern Washington state, the Deccan
Traps in India, and the enormous Siberian Traps, the eruption of which is thought to
be a primary cause of the Permian-Triassic extinction event. (The name "traps" for
these rock formations describes the step-like structure of the landscape from the
layered basalt flows. It is derived from the Swedish word "trappa" for stairs.)
Sequestering carbon dioxide as carbonate in basaltic rocks is not likely to run short
of material.

Other sequestration options are discussed briefly below:

Terrestrial Sequestration

CO_2 can be isolated in soils, vegetation or utilized in durable construction materials and other products that will keep it out of the atmosphere for an extended period of time. Most terrestrial sequestration ideas revolve around reforestation and using soils and biomass in forests to sequester CO_2 as organic carbon. The numbers on this are challenging on two fronts.

First, in order to reach the 10-gigaton per year carbon capture and sequestration goal that the IPCC says is necessary to hold warming below 3 °C, it is possible that a trillion trees will need to be planted. Assuming we can even do this, how long will it take to reach this level of sequestration? Trees grow slowly and a forest can take decades to mature.

The second problem with sequestration in trees is deforestation. Instead of planting trees, people are actually cutting them down at an increasingly rapid pace. Deforestation in the Amazon, in Africa, and in southeast Asia is clearing vast swaths of land for agriculture and to harvest valuable tropical hardwoods. Re-planting these areas with commercial timber may restore the "forest," but it is not as efficient at sequestering carbon as the soil and biomass in the old-growth forest ecosystem.

Sequestration in soils generally seeks to store the carbon dioxide as organic carbon, which often happens naturally and gives enriched soils their dark brown to black color. One enhanced method for doing this is to add a substance called biochar to the soil. Biochar is defined as "the solid black carbon material obtained from the thermochemical conversion of biomass in an oxygen-limited environment." This is exactly how charcoal is made for your barbecue grill, and if you mix it into the soil instead of burning it to cook bratwurst, it will sequester the carbon for many years.

I have reservations about the effectiveness of biomass and soils alone for terrestrial carbon sequestration. I just don't think forests can be scaled up big enough on our land-limited and resource-limited planet to offset 3 °C of warming. The amount of wood and other biomass waste that can be converted into biochar is also limited. However, the volume of CO_2 that must be removed from the atmosphere to mitigate the climate crisis is so large that any and all methods will help. Planting trees may not be THE single solution to the climate crisis, but it certainly can contribute to a solution.

As discussed earlier, concrete emits major amounts of CO_2 during the manufacturing process as limestone is converted into calcium oxide. If this GHG is captured at the point of origin at the cement plant and sequestered, then the concrete will be carbon negative, taking up CO_2 from the atmosphere as it cures. Concrete made with CCS can definitely help remove carbon dioxide from the atmosphere, and humanity uses a lot of concrete. However, we don't use enough to capture gigatons of carbon per year, so although CCS concrete can be part of the solution, again, it won't be the whole solution.

The incorporation of CO_2 into other durable materials besides concrete is often discussed as part of the circular economy, where atmospheric GHG is turned into useful products and chemicals. One idea is to use genetically engineered microbes to create graphene, carbon fiber, carbon nanotubes, or other desirable materials out of captured CO_2. Products could also be made using chemical engineering technology. For example, CO_2 can be used as a feedstock to make ethylene, which can then be made into a host of materials ranging from gasoline to plastic bags. The carbon in these materials came from the atmosphere, so any emissions would be carbon-neutral and could simply be re-captured and made into new products. None of this is quite ready for prime time at the moment because the economics of these processes remain wildly uncertain. At present, the cheapest source of ethylene is condensate produced as a byproduct of fossil natural gas production.

Ocean Sequestration

Carbon dioxide will remain a supercritical fluid in the ocean under the pressures at depths greater than 1 kilometer. Supercritical CO_2 has a density greater than the surrounding seawater and will dissolve and disperse without coming to the surface. There has been talk about sequestering CO_2 in the deep oceans by injecting it from pipelines on deep water oil drilling platforms or from ships.

Carbon dioxide already dissolves into seawater from the air as part of the carbon cycle, and because it is stored in water as carbonic acid, higher inputs since the Industrial Revolution have increased the acidity of the oceans. Ocean acidification is a significant concern because of the damage it does to shelled creatures in particular and ocean biota in general, most of whom evolved to tolerate a fairly narrow range of neutral to slightly alkaline pH conditions. The risks of major environmental impacts on sea creatures from injecting large quantities of CO_2 into the deep ocean are high. It could be especially dangerous to any organisms that found themselves uncomfortably close to a concentrated CO_2 injection point.

Adding CO_2 to the deep ocean water also runs the risk of what is called "turnover." The most disastrous example of this to date occurred in 1986 at Lake Nyos in Cameroon, Africa. The lake sits in a volcanic crater, which emitted gases into the lake bottom and supersaturated the bottom waters with CO_2. On August 21 of that year, something disturbed the equilibrium of the lake and the top and bottom waters flipped, bringing the supersaturated waters to the surface and releasing over a megaton of CO_2. Being heavier than air, the gas flowed down a valley, displaced oxygen, and suffocated 1746 villagers and thousands of livestock within 25 km (16 miles) of the lake.

The most frightening part of this is that no one knows what triggered the turnover. Speculation ranges from a landslide to excessive rainfall on one side of the lake. We know there are climate mechanisms like "El Nino" that bring deep ocean waters to the surface off the Chilean coast in winter that enhance fishing (and cause weird weather in North America). What if an El Nino event brought up gigatons of carbon dioxide that we thought were sequestered? Loading up the deep ocean with CO_2 when we don't know if it will stay there is a bad idea.

Ocean sequestration is in the modeling and research stage at present with no actual field tests. In my opinion, it should remain there.

* * * * * * * * * * *

Net-zero pledges from industry appear to be counting on vast expanses of imaginary forest to sequester carbon dioxide. Environmental groups like ActionAid and Oxfam have reviewed the net zero pledges from the oil and gas sector that invoke nature-based DAC solutions like planting trees and concluded that if the entire industry adopted the standards proposed by industry leaders Shell, BP, Total, and ENI, "it could end up requiring land that is nearly half the size of the United States, or one-third of the world's farmland."

Many companies have discovered that saying you are going to achieve net zero (at some poorly-defined time in the future) helps to calm down a nervous public; meanwhile you can continue business as usual and emit GHG. Especially egregious are the claims by oil and gas companies that they are going to achieve net zero on scope 1 and scope 2 emissions but say nothing about scope 3. As defined in the earlier discussion on Life Cycle Analysis, scope 1 emissions are those produced by the company itself in daily operations, scope 2 are emissions from outside utilities that supply energy to the company, and scope 3 are upstream emissions from the suppliers of materials procured by the company and downstream emissions from consumers who use the company's products. For an oil company, the scope 3 down-stream emissions as consumers burn gasoline and diesel fuel far overshadow any scope 1 or scope 2 emissions. In my opinion, pledging to go net zero on scope 1 and 2 while ignoring scope 3 is greenwashing, plain and simple. Net zero means all emissions, not just a select few.

Global net-zero is the point at which all the carbon dioxide emitted by coal-fired power plants, diesel trucks, gas furnaces, and all other human fossil fuel use is bal-anced by removing carbon dioxide from the atmosphere. At present, we are nowhere near balanced.

First off, we don't have the technology available at the scales needed to offset these GHG emissions. For example, the eight collector units at the Climeworks Orca plant in Iceland have an annual capture capacity of 500 tons each, with a total CDR capacity of four kilotons per year. The original Climeworks DAC plant outside of Zurich, Switzerland removes about 900 tons of carbon dioxide per year. Carbon Engineering reports their DAC pilot facility captures 1 ton/day or 365 tons annually. Global Thermostat claims to be removing about 1000 tons of CO_2 per year. Adding all these DAC efforts together produces a total annual carbon dioxide removal value of 6265 tons, which is 0.00002% of emissions.

Secondly, trying to remove GHG from the atmosphere while merrily continuing to burn fossil fuel is futile, ineffective, and foolish. Carbon dioxide emissions in 2020, which was actually a low year because of the COVID pandemic, were still 31.5 billion tons. To achieve net zero against this much carbon with Orca plants would require over 7 million more of them. Stopping deforestation and planting billions of trees might capture and sequester 10 gigatons per year, which is less than a third of the annual emissions. Clearly, doing something on the emissions side of the equation is required if anything on the capture side is going to be of help.

****

Many of the numerous young and start-up companies presently developing technologies for CDR are planning to use the sale of carbon offsets as a significant portion of income in their business models. Carbon offsets are generally parsed into quantities of one metric ton. They can be purchased by individuals or by companies to balance carbon dioxide emissions from business operations. There are two types of carbon offset markets: the voluntary market and the compliance or "involuntary" market. The voluntary market at present largely creates carbon offsets through agriculture or forestry practices, such as planting trees and caters to people who are concerned about climate change or feeling guilty about taking a long aircraft flight. As mentioned previously, the market for voluntary offsets is completely unregulated. It is definitely *caveat emptor*.

The most prominent company that has attempted to set standards for the voluntary carbon market and validate carbon offsets is Verra, based in Washington, D.C. They use what is called a Verified Carbon Standard for quality assurance based on specific accounting methodologies tailored to the type of CDR project, independent auditing of their accounts, and a registry system. While admirable, it doesn't have the force of a law.

The involuntary or compliance market for carbon offsets is regulated by a government cap on the amount of annual emissions that various industrial sectors can release. Companies with GHG emissions below the limit can either save their carbon credits for future use or sell them to others. If a company exceeds their allowable emissions, it must buy carbon credits from another company to stay under the limit. This process is called cap-and-trade, and so far the closest the United States has come to implementing it was in 2009 under the Obama Administration.

The bill was known as the American Clean Energy and Security Act of 2009 (ACES), sponsored by Representatives Henry A. Waxman (D-CA) and Edward J. Markey (D-MA). The proposed cap-and-trade plan was similar to one implemented by the European Union called the European Union Emission Trading Scheme. The bill passed in the House of Representatives on June 26, 2009, by a vote of 219–212. It died in the Senate however, having never been brought to the floor for discussion or a vote.

The bill would have required a 17% reduction in GHG emissions by 2020, and an 83% reduction by 2050, in line with former President Obama's stated goals. It would have made electric utilities meet 20% of their demand through renewables by 2020. It would have funded research and development on new, clean energy technologies and energy efficiency, including $90 billion for renewable energy, $60 billion for carbon capture and sequestration, $20 billion for electric and other advanced technology vehicles, and $20 billion for basic scientific research and development. I think we would be a lot farther ahead than where we are on the climate crisis if these research initiatives had been funded a decade ago.

The bill died because the White House insisted that the message be crafted around green jobs rather than climate change, judging that Americans would respond better to a positive economic message instead of dire warnings about a climate crisis. However, the prospect of green jobs did not necessarily appeal to voters who already had jobs, which was most of them, and people tended to focus more on the impacts to the cost of electricity instead. The EPA had published an evaluation that the legislation would cost American families about $13.20 a month, or $160.60 a year to reduce carbon dioxide emissions. That doesn't seem like a huge sum of money, but apparently it was a bridge too far.

Cap-and-trade is not impossible. The Environmental Protection Agency has run an acid rain mitigation program for many years that cuts emissions of sulfur dioxide from coal-fired power plants through a cap-and-trade program. Sulfur dioxide emitters can sell or save excess sulfur dioxide credits if they are below a permitted level of emissions or buy credits if they exceed emission levels. The only U.S. state with a cap-and-trade law for carbon dioxide is California, where the goal is to reduce GHG emissions 40% below 1990 levels by 2030. Carbon credits can be purchased within limits to stay under the emissions cap.

The ACES bill was supposed to be as simple as these other laws, but industry input and increasing demands made it ever more complicated. Some compromises were necessary because the transition to a cleaner economy was going to cost far more in midwestern and southern states dependent on coal for electricity supplies. Allowances for the coal industry and other GHG emitters were required to get enough congressional votes behind cap-and-trade to pass it. However, this backfired with environmentalists, who viewed it as a corporate welfare program for coal companies and other polluters. By the time it came up for a vote in the House, the Waxman-Markey bill was more than 1200 pages long, and loaded with provisions for certain favored corporations, complicated schemes for carbon trading, and numerous loopholes for existing coal power plants.

Senate Majority Leader Harry Reid knew the ACES bill would never get the 60 votes needed to pass a divided Senate. This was a few years before the glut of shale gas overwhelmed the coal electricity market, and at the time a substantial amount of American electricity was made by burning coal. The economics simply did not appeal to a lot of senators from industrial states. Senator Reid spiked the bill and it never even made it to the floor for debate.

However, we can't dodge this forever. National leaders are tasked by Article 6 of the 2015 Paris Agreement with developing methods for reducing GHG emissions. The World Bank reports that 64 carbon compliance markets are currently operating around the world, including in the European Union, China, Australia, and Canada. The United States is conspicuously absent. We could have been a leader with the ACES bill back in 2009. Instead, we are not even in the game.

Chapter 10
The Energy Future

Keywords Despair · Presumption · Hope

****

A better energy future needs clean energy. This means energy that is sustainable and does not add to the growing burden of greenhouse gas in the atmosphere. There are multiple options for clean energy including wind, solar photovoltaic, solar thermal, geothermal and solar assisted geothermal, nuclear technology both new and old, biofuels especially bio-methane, blue and green hydrogen, and even limited fossil fuels if equipped with CCS technology. New technologies like nuclear fusion, ocean current turbines, tidal power, and ocean thermal energy conversion are on the horizon. There is no real debate about what constitutes "clean" energy, although a few people stubbornly insist there is.

I am mystified at how some people can pine away for the "good old days." As Billy Joel famously sang, "The good ole days weren't always good, and tomorrow ain't as bad as it seems." The point of the song is that it's good to remember the past, but you can't re-live it. Steven Pinker, a psychology professor at Harvard University has said, "The best explanation for the good old days is a bad memory."

All of us have seen a lot of things during our time on this planet, both good and bad. I was born in Cleveland in 1954; the same city and the same year that disk jockey Alan Freed invented and popularized the term "rock and roll" on radio station WJW for a new type of popular music. I like to think that this was not a coincidence(!).

The 1950s were the days of atmospheric nuclear testing, and in fact my birth year was marked by the Castle-Bravo test of the first operational U.S. hydrogen bomb at Bikini Atoll in the Pacific. This event was notable not only as the first thermonuclear detonation by humans but also because the actual yield of the device, a thousand times larger than the Hiroshima bomb at 15 megatons, was twice the calculated yield of 6–8 megatons. A lot of surprised nuclear weapons designers went back through their equations to look for flaws in the math. The Castle-Bravo test created a huge fireball that vaporized a lot of seawater and irradiated the chlorine from sea salt, spreading the radiation around the world. Radioactive chlorine-36 isotopes can

D. Soeder, *Energy Futures*, https://doi.org/10.1007/978-3-031-15381-5_10

still be detected in sediments, and in fact these have been proposed as a time marker bed for the start of the Anthropocene.

The atmospheric nuclear tests kept escalating, culminating in 1961 with the detonation of the Soviet RDS-220, a device commonly called "Tsar Bomba" (Russian for the King of Bombs). This weapon had a yield of 50 megatons when exploded over the Soviet test site on the Arctic Ocean island of Novaya Zemlya, and it remains to this day the largest explosion ever set off by humans. It blew out windows in Finland, hundreds of miles away. Despite being treated with heat-reflective paint, the aircraft that dropped the bomb was damaged by the intense heat but managed to survive. The most frightening thing about the Tsar Bomba was that the original design capability was 100 megatons. It was dialed back to 50 megatons for the test because the Soviets didn't think they could contain a test of the full-yield weapon.

People started talking about banning atmospheric nuclear testing soon after this explosion. A treaty that prohibited all test detonations of nuclear weapons except for those conducted underground was signed in 1963 by the Soviet Union, the United States, and the United Kingdom. The two other nuclear powers on Earth at the time, France and China, were invited to join but refused. However, virtually all nuclear testing has been done underground since then, and in 1992, the United States stopped nuclear testing underground as well. The tests were still going on during my first 2 years on the Nevada Test Site at Yucca Mountain, and I felt a few of them. The shaking of the building from the effects of a human-made device detonated some 30 miles (50 km) away truly brought home the power of nuclear weapons.

The Cold War was in full swing in the 1950s and 1960s, and I remember the tension. Even as a kid, the hair trigger nuclear alerts between the U.S. and the Soviet Union and the potential for imminent thermonuclear destruction came through into our daily lives. Some generals back then believed that a nuclear war was winnable if only the first strike logistics could be properly worked out somehow. Fortunately, this dangerous fantasy fizzled out a decade later when it was replaced by the doctrine of "mutually assured destruction" where both sides recognized that a nuclear war could not be won and must never be fought.

I was in third grade when JFK was assassinated (which I clearly remember) and in middle school during the summer of love, LSD, hippies, and flower power. I find it amusing to see how hippy culture has been adopted by some of today's youth, who almost don't seem to realize that their grandparents were living it 50 years earlier. I was in high school for the civil rights marches, the RFK and King assassinations, anti-war rallies, the Kent State shootings (my sister had a friend who was enrolled there but was unhurt), and the moon landing, and in college for Watergate and the energy crisis. After college came the Iran hostage crisis, the Challenger disaster, the collapse of the Soviet Union, the Oklahoma City bombing, the 9–11 attacks, the wars in the Middle East, and finally the climate crisis. Most readers have similar stories; some longer and others shorter. But like it or not, these things have shaped and changed all of us as we live through the interesting times of the ancient Chinese curse.

When I visit the neighborhoods in Cleveland these days where I grew up and went to school, I don't recognize them anymore. It's a whole new town, and a different town. The skeleton of the old town is still there, but much that is new has been overlaid onto it with the passage of time. As just one example, the small maple saplings that I remember city workers carefully planting and staking along my street to beautify the neighborhood when I was a child are now massive, full-size trees (and serve to remind me that I might be getting just a bit old). No one can go back to the past, or even stand still in the present. Fossil fuel is the energy past. The only option for the energy future is to move forward.

Dealing with the climate crisis means clean energy and geoengineering to repair the atmosphere. Individuals and companies that fail to move forward with the program and continue living in the past risk becoming irrelevant and being left behind. There is an object lesson here from a century ago, when the passenger railroads and the telegraph companies were the "big oil" behemoths of their day. Both thought they would remain essential forever. I used to know an elderly gentleman who had been a conservative accountant raised in New England. When I knew him as a retiree in the early 1970s, he was feeling fiscally secure because he had invested heavily in the railroads and telegraph companies. After all, what could be safer?

It was true, then. A hundred years ago, the idea that someone might want to fly coast-to-coast instead of taking the train was absurd. Although it was possible to fly that far, early aircraft were small, dangerous, slow, and uncomfortable. Coast-to-coast flights and flights across the ocean were publicity stunts, performed by the likes of Charles Lindbergh, Amelia Earhart, and Wiley Post. For practical travel, trains were far more civilized, with Pullman sleepers and dining cars. The railroads considered themselves to be a critically important industry that the American public could not live without.

Telegraph companies likewise considered themselves essential because telephone exchanges were only local and paper letters took too long. If you needed to communicate quickly with someone across the country, telegrams were the only option. Well, here we are a century later. You can still ride a train coast-to-coast with some difficulty, and believe it or not, you can still send a telegram (they remain important in some parts of the world). But when was the last time you actually did either of these things? Exactly. Energy companies that insist on staying strongly committed to fossil fuels will soon become equivalent to someone trying to sell telegraph services in a room full of people holding cell phones.

Caring about the future can be difficult. Psychologist Edward Wasserman says that humans have evolved like other animals to deal with common everyday problems in the here and now.[1] This means that our evolutionary background pushes us to address urgent issues like food and shelter, and our emotions reward us for solving an immediate problem like finding the ripest berries. The human ability to contemplate the distant future and say, open a retirement savings account that provides no immediate benefit is a much more recently evolved trait. We can think about the distant future, but it is not emotionally rewarding.

---

[1] Wasserman, E.A., 2021, As If By Design: How Creative Behaviors Really Evolve: Cambridge, U.K.: Cambridge University Press, 250 p.

Animal intelligence researcher Justin Gregg uses the term "prognostic myopia" to describe this disconnect between the human ability to think about the distant future and our inability to actually feel strongly about that future. The term literally means nearsightedness about the future. As an example, Gregg cites the importance society places on easing a gasoline shortage by increasing the number of oil and gas leases, even though climate science shows that we may very much regret doing this several decades from now. He concludes that this impulse is both unforgivable and completely understandable.

Aristotle defined virtue as "a mean between two vices, that which depends on excess and that which depends on deficit." For example, an excess of the virtue courage is recklessness, and a deficit of courage is cowardice. True courage lies in understanding the danger but doing what needs to be done anyway, and is found between these extremes. Thomas Aquinas in the thirteenth century argued that hope also lies between two vices: presumption is the excess of hope, while despair is the deficit.

Presumption is the unrealistic confidence that everything is going to be just fine. In the climate crisis, presumption is a delusional optimism that we can just keep doing things the way we have been, mine coal and burn diesel, and never worry about any consequences. This is very strongly related to the human impulse to preserve the status quo. People resist change and the presumptive person finds no reason to address climate because they don't want to see it as a problem. A common presumptive response to climate disasters is that these are just "weather."

Despair is the sinking feeling that the climate problem can't be fixed no matter what we do, and we're all doomed. The person who despairs doesn't want to address the climate crisis either, because they see no point in even trying. Despair is the Borg message to the Enterprise that "resistance is futile." It is as dangerous as presumption. At both extremes, taking action to deal with the crisis is seen as either unnecessary or useless. So nothing gets done.

Aquinas classified hope as the mean between presumption and despair, and also as the more realistic response. By Aquinas' definition, hope is grounded in some desired future that is possible to achieve but also challenging. Hope requires us to be aware of what we are up against, and to also have a clear goal about what we want to achieve. Looking at it this way shows us the end point, the pitfalls along the way, and allows us to develop a strategy for avoiding those pitfalls. Once we understand the actual difficulties we need to overcome to achieve a desired goal, the path to that goal becomes much more apparent. In my opinion, this exactly describes the situation with the climate crisis, and gives us hope.

The prevalence of anxiety and despair among younger people that governments are not doing enough to address the climate crisis was described back in Chap. 5. Fortunately for the future, there are a lot of young people who have not given up hope. A film called "To the End" shown at Sundance in 2022 documents the efforts of a group of young people who work on climate change initiatives.[2] Included are Congresswoman Alexandria Ocasio-Cortez (D-NY), Varshini Prakash, the execu-

[2] https://www.sundance.org/blogs/festival-blog/check-out-these-sundance-supported-films-at-the-2022-festival/

tive director of the Sunrise Movement, Alexandra Rojas, the executive director of the Justice Democrats, and Rhiana Gunn-Wright, Climate Policy Director at the Roosevelt Institute.

On the subject of engaging with political leaders on climate change Prakash said, "It was an opportunity. We knew we weren't going to end up with Bernie's (Sanders) climate platform, but we were able to improve on and add a whole host of different climate policies...in those moments, you remember when you are in that room, you have an opportunity as young people to bring our values and vision and policies that matter to us to the table." The other side of this coin, of course, is that if you are not engaged and absent from the table, then none of your visions or values are going to be heard. She added, "We have to hold on to hope even in the moments that feel the darkest. Because otherwise, how is anything going to change?" Another ray of hope is former presidential candidate Andrew Yang, who is famous for his proposal to give every American "universal basic income." Fed up with both Democrats and Republicans, Yang has formed a new political party called "Forward" that pledges tolerance, technology, and fact-based governance, all of which are necessary for dealing with the climate crisis. The name is derived from his motto: "Not left, not right, but forward."

As mentioned previously, scientists also have anxieties about the climate crisis. Many of us are close to the issues with a good understanding of just how bad things are and how much worse they could actually get, and in response we developed a sense of agency to work on solutions. In addition to writing this book, my sense of agency was the co-founding of Carbon Blade. Our team believes that we possess a tangible, workable method for removing excess carbon dioxide from the atmosphere. More importantly, we believe that building and deploying millions of these units will make a real difference for future climates. You could say we have hope. This hope has given me a mission, a focus, and a purpose. We have defined a path to achieve that goal – not easy and strewn with technical and financial pitfalls, but nonetheless a path. Some might say that my motivation is little more than an optimistic delusion that I can help fix the planet. My response is that I would rather try something and fail than not try anything at all. We cannot continue with business as usual. Failure is always an option, but quitting is not. I dedicated this book to my grandson for a reason – because of our failure to act, the climate crisis will be a problem for his generation and succeeding generations in the decades ahead. We must do what we can to address it today.

Like many people, I am personally disappointed at the lack of progress dealing with climate change. We should have started 40 years ago when James Hansen and Michael Mann first sounded the alarm bells. Sometimes the only vindication a scientist gets is the right to say, "I told you so." While Dr. Hansen and Dr. Mann might smile at that, it doesn't do much to help in our present situation. I decided to write this book because I have hope that we can solve this problem and that educating people about what is at stake will mobilize some folks to act. We need people to get angry and concerned, and to let political leaders know that we are angry and concerned.

Martin Luther King and John Lewis had no idea if the civil rights movement would succeed when they started it. After Lewis got beaten to a pulp for leading a march across the Edmund Pettus Bridge in Selma, Alabama, it certainly looked like they were headed for failure. But he hung in there, along with Dr. King and dozens of others, and they persisted. Sadly, Dr. King didn't live to see the success of voting rights, anti-discrimination laws and other civil rights. But U.S. Congressman John Lewis did, and his advice to get into "good trouble" for what you believe resonates through the climate crisis. Dr. King's message of non-violence and non-silence is how we achieve success.

I live in West Virginia and the railroad track behind my office has kilometer-long trains passing through regularly that are piled high with West Virginia coal to be burned for electricity or steel making. I counted 125 loaded coal cars on one recent train (Fig. 10.1). Seeing all this coal go past my window and knowing that it is destined to become carbon dioxide in our atmosphere is admittedly depressing.

It is impossible to work on climate or energy and not have some feelings about climate change. Every scientist I know who works in this area has been forced to come to terms with it. My own feelings were ambiguous for many years. I was concerned about the climate of course, but I still had the "energy crisis" mentality that it was more important for us to continue to increase domestic sources of energy, and that meant oil and gas. The solutions I had heard for addressing climate change were simplistic, impractical, or both. An earnest young man at an environmental conference once assured me that the path forward was to enforce energy conservation and only generate electricity with wind and solar. How then do you manufacture cement and purify pig iron? How do you power a vehicle? How do you even make baseload electricity?

Some solutions essentially advocate adopting the technology of the sixteenth century, trading cars for ox carts, airplanes for sailing ships, and gas furnaces for

Fig. 10.1 A long coal train passes through Rowlesburg, West Virginia on a sunny winter morning. (*Source: Photographed in 2022 by Dan Soeder*)

firewood. There are far too many people on the planet these days for us to live a sixteenth century lifestyle, which could still be resource-intensive (remember King James' edict about saving trees for ships?), inefficient, and not very productive. Without our current level of technology, billions would starve, freeze, dehydrate, or die from illness. Some people answer that the Earth would be better off without humans and we should just stop breeding and go extinct. I disagree. A dirty house should be cleaned, not burned down. I do advocate for humanity to leave Earth, but via spacecraft to become a multi-planet species, not by dying off. The next chapter looks at how humans can expand throughout the solar system.

I crossed the Rubicon on climate in 2017 (better late than never, right?). I had done a bit of research at NETL a few years earlier on the potential for local geothermal energy to supply secure energy for U.S. military facilities. I was thinking about it again as a potential project on a remote tribal reservation and out of curiosity I attended a DOE Geothermal Technology Office (GTO) program review meeting in Denver to learn what people were doing. For the first time I saw an energy solution - a real solution - in engineered geothermal, which can be deployed just about anywhere to provide large amounts of energy while adding zero GHG into the atmosphere.

Whoever came up with the idea of adapting lateral drilling and staged hydraulic fracturing from the shale gas industry to use on geothermal resources deserves some major accolades. This was a genius move that made the whole operation to tap geothermal heat from "hot, dry rock" at least an order of magnitude more feasible than using vertical wells. It could definitely work, and EGS program manager Lauren Boyd and her colleagues at GTO deserve a huge amount of credit for supporting it.

I remember coming home from the Denver GTO meeting with a new understanding that while yes, we needed more energy, what we really needed to protect the climate was sustainable, carbon-neutral energy. Engineered geothermal was maybe not THE answer, but it was AN answer, and it was the first one I became aware of that was likely to work at the scales needed.

Other options also became apparent. Adding a solar heat loop to EGS would boost the temperature in shallower wells, suggesting that a SAGE system could increase the versatility of EGS and allow even wider applications. Because EGS can be located anywhere on Earth as long as the drillholes are deep enough to reach hot rocks, the possibility occurred to me that the coal burners in existing thermoelectric power plants could be replaced with a geothermal hot loop. If we simply change out the heat source that is used to make steam, we don't need to build new power plants to decarbonize electricity. This is huge.

Small nuclear reactors could also replace coal burners. We fit them into submarines, why not power plants? New nuclear technology is small, efficient, and can be mass produced in a factory. Surely it can be made small enough to fit under a boiler and produce as much heat as a coal fire. Likewise, replacing natural gas with biogas will allow efficient CC powerplants to keep operating. If an existing thermoelectric power plant could retain all of its expensive generating and power distribution infrastructure and use a geothermal or nuclear heat source, or run on biogas, we could decarbonize electricity quickly and cheaply.

Would any of this actually work? I don't know – it needs to be tested and evaluated. However, we do know that it doesn't defy the laws of physics and obtaining commercial levels of electricity from these resources is simply a matter of scaling-up. The economics of retrofitting an existing power plant with a carbon-free heat source has got to be cheaper than abandoning the entire facility and building completely new wind and solar infrastructure. In fact, replacing natural gas with biogas in a CC power plant doesn't require any modifications at all.

In my opinion, GTO ought to be given billions and told to get after EGS in a crash program. What they actually have are millions for one geothermal field test site and things are moving slowly. Nuclear technology needs to be freed from stalled permit applications and unresolved waste disposal issues, including an empty, five-mile long tunnel in Nevada. Projects like the Gates-Buffet joint venture between TerraPower and PacifiCorp to build the Natrium liquid sodium-cooled reactor should be supported. Biogas could use some funding for genetic engineering of methanogens and facility design. These technologies could decarbonize electricity within a decade, but we have to jump start them with some technical and regulatory advancements. At the moment those are not happening.

No one should give up hope. As of this writing, Congress has passed and the President has signed a bill called the Inflation Reduction Act that contains $369 billion over ten years to combat climate change. The bill supports clean energy projects, climate-related research, and incentives for consumers to adopt clean energy technology. Some environmentalists have complained that it doesn't do enough, but everyone agrees that it's a start and a huge improvement over what we had before, which was nothing. My recent experience working with renewable energy and carbon dioxide removal has made me aware that a lot of really smart and very dedicated people are working hard to solve both the energy and climate crises. It will take some time, but not a huge amount of time. Everyone working on these things knows that there is a looming deadline, and that we can't be lackadaisical about this. Many people are pushing hard. The oil and coal industries are pushing back, but it is an act of desperation on their part. Denying climate change exists and that fossil fuel combustion is responsible for it is becoming an increasingly untenable position. I think we will see significant progress in the next 10 years.

It is important to remember that previous generations faced existential threats almost as bad as climate change. The first half of the twentieth century featured the Great Depression sandwiched in between two very brutal world wars. These events alone or in combination could have destroyed civilization, but they did not. People adjusted, adapted, and displayed resilience.

The last world war was followed by a hair-trigger Cold War nuclear standoff that lasted for decades and could have destroyed all life on Earth, but it did not. However, it did come damned close at least once. I lived through the Cuban missile crisis as a child, and the one thing I remember about it is that my parents were scared. When you're a kid, your parents seem fearless. To see them really frightened by television newscasts made more of an impression on me than anything else. As an industrial city, Cleveland was certainly a prime Soviet target. Watching my father try to figure out how to seal off our basement against fallout was something I'll never forget.

Fortunately for us all, Khrushchev and Kennedy worked things out and humanity survived. Looking back on it now through the lens of history, it is apparent that yes, we should have been frightened. In fact, we should have been terrified. This is probably the closest the world ever came to a full-scale nuclear war, and if not for a couple of fateful decisions made by a handful of critical people on both sides that could easily have gone the other way humans would have loosed hellfire upon the world.

It's okay to be concerned and maybe a bit scared about the future. Both climate change and nuclear war could make large parts of the planet uninhabitable. These are serious threats. But it is important for the people who are frightened about the climate crisis to understand that their fears affect those around them, and especially those who look up to them. Turning to agency instead of just fretting about climate can convert the energy of fear into the energy of action, and it can also reassure others. Writing to your political leaders to deal with the climate crisis may not be as dramatic as my father trying to seal the basement, but it will probably be a lot more effective.

The young people of today who are giving up on the future need to know that climate CAN be fixed. Technology got us into this crisis, and technology will get us out. I firmly believe that the future will be more utopian than dystopian, but we need to push it in that direction. Clever thinking and innovation have saved humanity from great dangers in the past, and there is a lot of clever thinking and innovation being applied to the climate crisis. Unlike some of the other crises, we know how to fix this one. The key is changing "it can be fixed" into "it will be fixed." We need a strong dose of determination and stubbornness, and perhaps some "good trouble." Instead of sitting around wringing our hands in despair, let's get to work.

****

So what needs to be done to get to some kind of reasonable energy future? The best way to start is to take an inventory of the situation and try to figure out a logical and responsible path for dealing with it. We can divide nations into two broad categories: the developed world and the underdeveloped world. Two problems immediately become apparent: (1) nearly all of the GHG emissions in the atmosphere to date have come from the developed world but the underdeveloped world is suffering the brunt of the effects; (2) the underdeveloped world, by and large, wants to become more like the developed world.

The simple truth is that the wealthiest 10% of world population uses half of the Earth's energy and resources. The other half is spread out among the remaining 90%. The underdeveloped world doesn't have the money or the means to develop itself with fancy new energy technology. Two centuries of ruthless fossil fuel exploitation by the developed world has left the atmosphere with no more capacity to take additional GHG emissions and options for the underdeveloped world are limited. They can't afford to develop nuclear or EGS, but if they expand the use of cheap and easy fossil fuels, the climate goes over the edge. In that case, they are basically shooting themselves in the foot because they will suffer the worst impacts of the climate crisis, such as heat waves, sea level rise, fierce storms, and severe droughts. Not surprisingly, this set of circumstances has created a huge amount of resentment.

The underdeveloped nations also have the least ability to cope. Sub-Saharan Africa is on shaky ground for food, energy, and economics even on a good day. Climate-exacerbated droughts and heat waves are wilting crops, killing livestock, and driving human populations into refugee status. After a killer heat wave, some of the worst monsoon floods in history have devastated Pakistan. In the South Pacific, entire island nations such as Tuvalu may go underwater with sea level rise. The small country of 12,000 occupies nine coral islands where the highest elevation is 15 feet (4.6 m) above sea level. Tuvalu foreign minister Simon Kofe dramatically addressed the COP26 summit by standing knee-deep in ocean water to emphasize the concern over the future of his citizens.

In my opinion, it is the responsibility and duty of the developed world to create and commercialize the sustainable energy technologies that will help the underdeveloped world modernize. Why? Because it is our fault that they can't do this the easy way. Fossil fuels provided the energy for Samuel Kier, John D. Rockefeller, Henry Ford, Thomas Edison, Admiral John Arbuthnot Fisher, Winston Churchill, and many others to improve technology. Some of them also became quite wealthy. And the future was left to shift for itself.

As of this writing there are still nearly a billion people in the world with no access to electricity, along with an additional three billion people (some 40% of the world population) that continue to rely on unhealthy fuels like wood and coal stoves for heating and cooking. A stated goal of the United Nations is for reliable electricity to be made available to everyone in the entire world by 2030. Is this realistic? It is if we want it to be.

There are three approaches we can take to address sustainable development for all of humanity under the shadow of climate change: first, the wealthiest 10% can conserve and use less of the Earth's current resources, leaving more for the remaining 90%. Efforts in this direction so far have not been promising as people refuse to give up their luxuries and conveniences. We are probably not going to conserve our way out of the climate crisis.

Second, the wealthiest 10% can expand the development of new, sustainable, carbon-neutral resources, and design these specifically for use in the developing world. One example I gave earlier for communications was the cell phone system. As they get telephones, remote villages in Africa are bypassing the landline system altogether and just going straight to cellular, avoiding utility poles, miles of copper wire, and inefficient central exchanges. With no legacy systems to accommodate or grandfather in, they can adopt the newest technology.

There is no reason underdeveloped nations can't do the same thing with energy. For example, instead of copying the nationwide electric grid system in developed countries that requires miles of expensive copper wire, it might make more sense in less developed countries to set up independent, local wind-solar or geothermal power systems on a village-by-village basis. One possible design could be a centralized battery system that is recharged by wind and solar and provides continuous DC power at a standardized voltage to individual homes for lighting, heating, cooking, and other uses. Other designs are of course possible depending on the local resources.

Having a village-by-village electrical system allows the most efficient designs to be used locally without the need to tie it all together on a national grid. Powering up one village or city at a time is also more approachable than trying to electrify the whole country all at once and allows for a learning process. Generating power from solar, wind, or geothermal sources within each village eliminates the need for long-distance electrical lines that are highly vulnerable to storms or wildfires. Standardizing voltages and other power parameters allows people to move between towns and still use their electrical appliances. The development such "indigenous" power sources in each village will greatly improve climate resilience by making the electrical system local instead of national and with no reliance on long-distance transmission lines the power will stay on after a disaster even in villages that are remote and isolated.[3] Individual farms in the countryside can be supplied with their own stand-alone electrical systems. Clever people in the developed world need to be working in partnership with the U.N. and the developing world to think up tailored, carbon-neutral energy solutions.

The third thing we can do is to expand the total inventory of energy and material resources available for humanity beyond what is available on Earth by going into space and gathering space resources. I've devoted the last chapter of the book to this issue because I think it might be the ultimate answer.

The underdeveloped world has two major issues upon which they expect the developed world to act. The first is to provide a transition to a global clean energy system that will lead to the growth of material wealth and jobs for all. The second is for the developed world to aid underdeveloped nations in responding to "loss and damage" - the devastation caused by climate change. Addressing loss and damage recognizes that certain countries, such as small island states like Tuvalu mentioned earlier have contributed the least to the climate crisis yet are bearing the greatest burdens from it in the form of extreme storms and sea level rise. The COP27 summit in Egypt in November 2022 is largely focused on loss and damage.

Underdeveloped countries have a valid point about loss and damage and have been asking for help for a decade. A few high profile celebrities have taken up the cause and shown up to visit, but almost nothing tangible has been done. Until the wealthy nations of the world are prepared to do their part and lead the way on dealing with the climate crisis, the poorer parts of the world can be expected to dig in their heels when it comes to cutting emissions. How is it fair to tell a tiny island nation like Tuvalu that they can't develop their economy using fossil fuel when China emits over 12 gigatons of CO_2 per year and India looks ready to follow suit? (Refer back to Fig. 6.4). Wealthy and fashionable celebrities jetting around the globe may provide entertaining reading on the tabloid web pages, but the reality is that a private jet only gets about 4 miles to a gallon of fossil fuel. This is hardly a

---

[3] Soeder, Daniel J., Sawyer, J. Foster, and Benning, Jennifer, "Off the Grid: Developing resilient energy infrastructure from indigenous resources:" Hawaii University International Conferences on STEM/STEAM and Education, Honolulu, HI, June 5–7, 2019.

shining example to hold up to underdeveloped nations while telling them they must use only renewable energy sources.

One solution for Tuvalu and other island nations is to use DAC and biomineralization to convert atmospheric CO_2 into solid carbonate minerals to build up land surfaces and increase the elevation of low-lying atolls. This would be cheaper than dredging and much cheaper than bringing in fill materials from elsewhere. The atoll bedrock is already made of reef carbonate so plating more carbonate on top of it won't change anything but the height of the island. There is also a certain poetic justice in using CO_2 emitted by the industrialized world to save Pacific islands while reducing atmospheric GHG levels at the same time. This could be done with self-powered units like Carbon Blade and the cations available in seawater.

The United Nations Development Program has a plan for fair energy pricing reform that includes phasing out fossil fuel subsidies and putting a price on carbon. This aligns with the other stated U.N. sustainable development goals: eliminate poverty and hunger, improve health, education, and gender equality, provide clean water, energy, and economic growth to all, support industry, innovation, and infrastructure, including sustainable cities and communities, and responsible consumption and production, take action on the climate, and protect the Earth's ecosystems. These are great goals but achieving them all will be a challenge.

The author Kim Stanley Robinson published a novel in 2020 called "The Ministry for the Future" in which a fictional United Nations agency charged with solving climate change struggles with the science, politics, and economics that swirl around the issue. The story postulates that carbon dioxide levels in the atmosphere begin to decline by 2053 but getting there is grueling. Sea level rise, wildfires, and fierce storms have killed or displaced hundreds of millions of people, climate refugees have overwhelmed other nations, economies have collapsed, and most frightening of all, terrorists from stricken countries with nothing to lose launch revenge attacks against those they hold responsible for the climate crisis. Robinson invokes the decarbonization of energy and use of geoengineering to stabilize the climate, but in his story everything is done in a rushed, haphazard manner. Humanity grudgingly makes adjustments and barely avoids the worst climate catastrophes, but it is far from a well-considered, strategic plan. I'm afraid Kim Stanley Robinson's novel hits a little too close to the truth and is probably less fiction than it is prophecy. Former President Barack Obama included it on his list of the best books for 2020.

The problem we are facing is that much of the underdeveloped world aspires to live like Americans. China and India are building robust economies and large middle classes that want consumer goods and technology, including motor vehicles and electronics. The cold reality is that the United States has the largest economy but only 5% of the world's population. Yet the U.S. consumes 25% of the world's energy resources. China has the second-largest economy with a quarter of the world's population and could easily use up the rest. Then what happens? The U.S. economy runs on fossil fuel, but the developing world cannot copy our economic model because the math just doesn't work. There is not enough fossil fuel remaining on Earth for everyone to live like Americans. There isn't even enough GHG capacity remaining in the climate for everyone to live like the Chinese. We must find energy alternatives and time is not on our side.

Sustainable energy technology must be implemented in the next two to three decades, or we risk technological collapse. This will happen if we continue business as usual and pretend that climate change is nothing but a Chinese hoax. The laws of physics don't care about anyone's politics. The climate crisis will deepen. Droughts, storms, heat waves, wildfires, and sea level rise will get worse. More climate refugees will wander the world looking for safe havens. Fossil fuel will be in short supply after we pass Hubbert's peak oil, especially if more people than ever are consuming it. There will be resource wars, ironically using up whatever scarce resources are left. As people fight over the remaining oil, the remnants of civilization will center around the last few energy resources and fiercely protect them. Mad Max, we are here.

To avoid such a bleak and dystopian future we have to move forward with sustainable, zero carbon or carbon neutral energy. Existing energy companies that continue resisting this change are headed for what Leon Trotsky famously called the "dustbin of history."

So what is the energy future? A reasonable path in my opinion is to power the world with 100% zero carbon or carbon neutral clean electricity. This would include wind, solar, engineered geothermal, new technology nuclear, and methane biogas, plus maybe some other technologies that have yet to be invented. We should try to electrify everything as much as possible, including home heating and cooking that currently uses wood or coal.

After clean electrification, existing gasoline and diesel vehicles should be modified to run on biogas or bioliquids, and new vehicles should be limited to biofuels, hydrogen, or batteries. Coal mining should end. Petroleum and natural gas production should be limited to non-combustion uses, such as feedstock for chemicals or fertilizers. Any fossil fuel that is still burned must have CCS with 100% capture. Energy jobs should transition into geologic storage of CO_2, the proper P&A of old oil and gas wells, restoring the land at surface coal mines, sealing underground mine workings to stop acid mine drainage, and reclaiming drilling sites. Finally, we should work toward climate justice, building urban infrastructure for clean and affordable public transit, biking, and walking options.

Despite the best of intentions, it is still possible to do this badly. One of my friends in California has been lamenting a state mandate that new homes be "all-electric" including so-called "affordable" housing. The all electric idea is becoming popular among local governments, especially in California because it makes city councils look like they are gallantly taking action against the climate crisis. All they are actually doing is banning new natural gas hookups.

It is true that natural gas emits carbon dioxide when burned and methane when it leaks, and domestic gas lines and appliances are notoriously prone to small leaks. Stanford University researchers even found methane emissions from gas appliances that were turned off.[4] Still, many people like to cook with gas and the gas utilities have decreed that banning hookups is unfair. Local governments are getting sued as a

---

[4] Lebel, E.D., Finnegan, C.J., Ouyang, Z., and Jackson, R.B., 2022, Methane and NOx Emissions from Natural Gas Stoves, Cooktops, and Ovens in Residential Homes: *Environmental Science & Technology*, vol. 56, no. 4, p. 2529–2539.

result. The truth is that most natural gas in the U.S. is used for generating electricity, not for baking birthday cakes. Investing in nuclear power or engineered geothermal electricity is a much more effective way to cut natural gas usage. The relatively small volumes of gas used by domestic utilities can be made carbon neutral by switching to methane biogas. No one's furnace or hot water heater will notice the difference.

Although banning natural gas sounds good for the climate, the reality is that it is virtually impossible to go all-electric on any normal-sized home in California and not have some months where the electric bill tops $500. The state has set residential electricity prices at 33–35 cents per KWH to encourage conservation. This is expensive electricity. (For comparison, my residential electric rate in West Virginia is about 9 cents per KWH.) Building affordable housing for people with low incomes and then forcing them to use costly electricity just doesn't make sense. It will only result in sending some people to the poorhouse.

Another consequence of the sky high electric prices in California is that virtually everyone who could afford it has installed residential solar panels and is selling power back into the grid at 35 cents per KWH. Wealthy people with grid-tied solar panels on their roofs are able to offset a major percentage of their monthly electric bill this way. Low income people who don't have solar are forced to pay the full amount. Recognizing that this looks bad from all angles, the California Public Utility Commission is considering making solar less economically attractive by instituting expensive fees for connecting to the power grid. If this happens, one result is that homeowners who can do so are likely to go completely off the grid and set up independent home power stations using solar, wind, and batteries while thumbing their noses at both the utility company and the state regulators.

Distributed, grid-tied residential solar power is an efficient way to supply zero carbon electricity into the grid. If California manages to tie this up in bureaucratic knots, it will be bad for the climate and set a dismal example for other places. People who disconnect from the grid create an underfunded grid for those who remain, and California's electric grid is already overloaded and unreliable. If new homes are forced to go all-electric without the presence of distributed power generation from home solar systems, the full load will have to be taken up by utility generating plants and the current grid. This may produce even more unreliable electricity if the grid can't handle the load. Hours of downtime will drive additional people to cut the cord and underfund the grid even further, making everything worse. It becomes a death spiral.

My friend, who is both a brilliant analyst and an ardent environmentalist suggests that tariffs and incentives are needed to bring the California electric system into some semblance of order. He would like to see the state go all-electric using a significant amount of distributed zero-carbon generating capacity (like residential solar and wind) along with battery storage for baseload power. Focusing tariffs on the wealthy along with incentives to tie distributed power sources into the grid will reduce the burden on the poor and move the state toward all-electric homes equipped with solar panels that are mostly self-sufficient. This scheme can use the electrical distribution system that is presently in place with only minor improvements and capacity additions.

****

Ending government subsidies for fossil fuel ought to be a no-brainer. These were implemented during the OPEC oil embargo to protect U.S. companies from financial losses and to encourage them to replace foreign oil imports with additional domestic production. It was a good idea at the time, but the necessity for it is long past. Still, oil industry lobbyists push for this every year with their favorite congressperson at budget time, and it has been hanging around for almost 50 years. Because hey, who doesn't like free money? As far as I'm concerned, this "free" money is taxpayer dollars that could be better spent elsewhere on a thousand different critical issues that take priority over the balance sheet of an oil company.

A carbon tax of some sort is in the future but remains unpopular. It is literally a political hot potato that no one wants to touch. Politicians have discovered that proposing to raise the cost of fueling cars and heating homes is not a good way to win elections. In a 2019 poll, the Washington Post found that 51% of respondents opposed a $2 monthly tax on residential electricity to fund climate programs, and 71% opposed a $10 monthly tax.

Instead of a direct tax, perhaps there is an indirect solution to this conundrum. One idea would be to make industrial GHG emissions illegal and require coal and gas-fired utilities and industries to capture and sequester all of their CO_2 combustion products instead of releasing them into the atmosphere. Such a mandate could be imposed by the EPA under some new, revised "clean air law" passed by Congress but worded to provide political cover for timid congresspersons. The rule would force utilities to implement CCS, cause steel mills to convert from coke to hydrogen, and require cement factories to capture emissions. These actions will almost certainly result in an increase in the cost of fossil electricity, steel, and concrete. Such price increases are likely to be short-term, however, as industry adapts to the new technologies and improves efficiencies, and electric utilities continue to seek and develop lower cost, carbon neutral power sources.

Simply prohibiting GHG emissions in the electric power sector will make sustainable and carbon neutral/carbon zero solutions like biogas, geothermal, or nuclear more economically attractive as replacement heat sources for existing thermoelectric generating plants. CCS is expensive and not likely to get much cheaper and utilities will look at options beyond fossil fuels if forced by the economics. Most people pay little attention to the source or the rates of their electricity, but they would certainly notice a line item like carbon taxes added to their electric bill. Imposing an emissions ban on the utilities rather than trying to tax consumers directly will avoid the ugly political battles that come with a "carbon tax." It would also give the utilities an incentive to get serious about seeking carbon neutral or carbon negative generating technologies with robust GHG reductions.

For automobiles, as suggested in Chap. 8 the "carbon tax" could be built into the price of a new, gasoline or diesel powered vehicle to incentivize the purchase of zero carbon vehicles. A high price for carbon emissions on new gasoline powered vehicles will make them less attractive (although still available if you are willing to pay for it) and ensure that the overall fleet of vehicles on the road will gradually become zero carbon. Not adding a carbon tax to gasoline will allow those who are holding on to older gasoline powered vehicles to continue driving them without taking a serious hit to the pocketbook. Attempts at raising the cost of fossil fuel in other countries have resulted in strikes and riots. Revenues from a carbon tax could be used to convert existing gasoline-powered vehicles over to compressed or liquid methane biogas, also described in Chap. 8. This conversion will not only reduce carbon dioxide emissions but other emissions as well, leading to much cleaner air in cities. A more forceful approach could simply impose a ban on GHG emissions from vehicles, similar to that suggested above for utilities. The California Air Resources Board recently mandated that new vehicles sold in the state must be emission-free by 2035. This ruling used the Clean Air Act Waiver, a policy that allows states to impose vehicle emission standards that are stricter than the federal standards.

****

Why have we been so slow to deal with the climate crisis? Along with the political obstruction and manufactured uncertainty, the climate crisis doesn't really have a "do or die" deadline. Unlike some other existential threats such as an asteroid on a collision course with Earth, climate change does not provide a deadline date where we must act or be doomed, nor offer us a clear boundary between success and failure. Like frogs put into a pot of water where the temperature is raised slowly, we don't realize the pot is boiling until it's too late. Because we have already put an excess of GHG into the atmosphere, we are going to see some degree of climate change in the future no matter what we do. Acting on it now will certainly make it less bad but won't allow us to avoid it altogether. This lack of a hard deadline has made it easier for political leaders to avoid tough and potentially unpopular actions and kick the can down the road. But the longer we wait the worse it will get.

Some people claim that the technology needed to address climate change is not fully developed, including long-range electric vehicles, direct air capture of CO_2, and zero carbon electric power generation, and we need to wait until these are mature before tackling climate change. Some of this is coming from the fossil fuel industry, who would of course like to delay the introduction of carbon-neutral technologies for as long as possible. While there certainly are some aspects of these various technologies that could use further work, such as improved energy storage, better vehicle batteries, and hydrogen-powered steel plants, holding out for the perfect technology just adds delay while the climate crisis grows worse. We can't allow the perfect to be the enemy of the good and wait for nuclear fusion or zero point energy to save the day. *Tempus fugit*. We need to use what we have and get started, and if improvements come along as we go, they can be added as necessary.

Possibly the most serious cause underlying the delay in decarbonizing the economy is the perception that doing so will impose significant changes on the American way of life and many people are resistant to those changes. The public is far more climate-aware now than a decade ago, and although quite a large number of people still think climate change is a hoax, the majority of Americans take it seriously and agree that something needs to be done.

When it gets down to brass tacks, however, transitioning to a green energy economy will require some real sacrifices by some sectors. In some parts of the U.S., the fossil fuel industry is the only source of well-paying jobs available to people without a college degree. Coal jobs in West Virginia and Wyoming, and oil and gas jobs in Texas, North Dakota, and Pennsylvania pay better than just about any other private industry. This isn't just workers toiling in mines or on drill rigs – it also applies to support workers.

For example, during the Marcellus Shale gas drilling boom in Pennsylvania and West Virginia between 2008 and 2018, people with a commercial driver's license (CDL) could make a very good salary hauling gravel and equipment out to shale drill pads and hauling wastewater back in for disposal. As a result, local communities faced a sudden shortage of school bus drivers, and the state highway departments couldn't find anyone to drive snow plows, both of which also require a CDL.

The economic gains are not comparable in green energy jobs. The median annual wage for a solar-photovoltaic installer in the U.S. was $44,890 in 2019 and the salary of a wind-turbine technician was $52,910 compared to median wages between $70,310 and $81,460 in the fossil fuel sector. Male workers with high school as their most advanced degree have faced a shrinking job market for the past 40 years as America de-industrialized and sent manufacturing jobs overseas. Fossil-fuel workers and others in the skilled labor pool have little reason to believe that a green energy economy will be good for them. They are probably correct; evidence suggests that a post-carbon economy will not provide the highly paid, unionized jobs of fossil energy and a basic social welfare safety net.

In principle, the salaries and benefits from green energy jobs can be just as good as those in fossil energy jobs, but in reality, this has not been the case. Green energy companies need to step it up here if they want to attract both workers and enthusiasm from the fossil energy industry. Competitive salaries, the presence of unions, and less greed on everyone's part will help the green energy transition move forward. Provisions are also needed for workers in the fossil fuel industry to apply their skills to other work, which could include reclamation of surface coal mines, plugging abandoned oil wells, and drilling new wells for the sequestration of CO_2.

The climate crisis is upon us, and natural disasters are going to get worse and more frequent. If we continue to delay decarbonization, we will soon reach a point where our children, grandchildren, and great-grandchildren will be condemned to lower standards of living or worse. Each year that we delay results in more deaths in the regions bearing the brunt of climate change and increases the risks of global catastrophe. Responsible decarbonization is rapid decarbonization.

Decarbonizing quickly requires the acceptance of some disruptions to the status quo. People are going to have to get over the NIMBY syndrome and allow for the construction of solar farms, wind turbines, biogas plants, electric transmission lines, and other green infrastructure. Despite people wanting new facilities to be "elsewhere," they all have to go somewhere, and in doing so it will turn up in someone's backyard. Who's backyard matters. NIMBY attitudes in the past have been responsible for many environmental justice issues, where factories and plants were pushed onto marginalized and minority communities who suffered from environmental degradation without receiving many benefits like jobs or tax breaks. New green infrastructure must make an effort to avoid this and not take advantage of the disadvantaged. Replacing coal and gas burners at fossil fuel power plants with geothermal, nuclear, and biogas will minimize some of this because that infrastructure is already in place.

All of this has a cost. The upper middle class and the wealthy must be willing to pay higher taxes, and everyone must be willing to pay higher electricity rates and higher transportation costs. Even geothermal retrofits on existing power plants and modifications to existing vehicles for bi-fuel CMG will cost something. The repeated delays in responding to climate change have made the actions we take now more expensive than if we had started all this 40 years ago. Despite our best efforts, it is still possible that catastrophic climate change will affect our grandchildren anyway. However, the benefits of decarbonization to the economy, public health, and ecology are likely to outweigh the costs, suggesting that this is the most responsible and rational course of action.

That being said, this is where we run into some serious headwinds. Many American citizens who claim to be well-informed about climate change do not necessarily support the measures required to minimize it. A Reuters survey in 2019 found that 69% of Americans believed the U.S. should take "aggressive" action to combat climate change, but only 34% were willing to pay $100 more per year in taxes to finance it. A Washington Post survey obtained similar results.

Even some environmentalists who claim to be committed to addressing climate change steadfastly refuse to prioritize it over other concerns. A recent referendum in Maine to approve a power line that would bring zero carbon hydropower into the state from Canada was rejected by voters. The hydropower would have offset some three million metric tons of CO_2 annually, but rejection of the measure was supported by three of Maine's top environmentalist organizations: Environment Maine, the Natural Resources Council of Maine, and the Sierra Club. They decided it was more important to preserve Maine's "natural beauty" from an ugly power line.

There are a million and one excuses for not acting on climate change. One is that we can't act unless the science is 100% certain. When politicians suggest that they need more evidence before taking action, it's often a form of science denial. Climate science is about 99% certain, and for all practical purposes that is close enough to definite. As mentioned earlier, there is now far more evidence in the scientific literature for climate change than there was for CFCs damaging the ozone layer, yet the world managed to pull together on that crisis and develop the Montreal Protocol. The mathematical probabilities and consequences of a dire future if the climate crisis is not addressed are more than enough incentive to get it done.

Another excuse is that technology will save us, so we don't really need to do anything. A technological solution will help, but any solution will require changes in policies, lifestyles, and everyday practices. Technology does no good unless it is implemented. New nuclear technology and engineered geothermal could make major inroads into decarbonizing electricity, but political support and research funding for them has been weak. So far, politicians and citizens have taken the easy path of talking about new climate technologies, while refusing to make the hard choices needed to actually develop and implement those technologies.

A third common excuse from skeptics is that climate actions will ruin the global economy. Dealing with the climate will certainly cost money but it won't be as expensive as dealing with coastal flooding, displaced populations, intense drought, wildfires, and severe storms. As Stanford University public policy expert Robert Reich has famously stated, the cost of climate action may be high, but the cost of inaction will be far higher.

****

The 2021 COP26 climate summit in Glasgow, Scotland offered insights into some possible energy futures and how various nations intend to achieve these. It was promoted as the last chance to limit global warming to 1.5 °C, which in my opinion is already in the rearview mirror. We will be doing well to keep it below 2 °C, but I think it is going to climb a bit higher than that before things are finally brought under control. The problem is that 1.5 °C is considered the threshold for significant climate change impacts and disruptions. As such, we are probably on the hook for some interesting times ahead no matter what we do, but we can at least try to make it less bad.

The recommendations that came out of COP26 will be familiar to readers who have gotten this far in the book. They are all things I've already mentioned, but it's always good to be vindicated. The following list was compiled by the BBC:

1. Keep fossil fuels in the ground
2. Cut methane emissions
3. Switch to renewable energy
4. Abandon petrol and diesel as vehicle fuels
5. Plant more trees
6. Remove greenhouse gases from the air
7. Give financial aid to help poorer countries

So the question now is not so much what to do, given the shopping list above, but how to do it? How do we motivate people and governments to take actions, some of which might be rather painful, to deal with the climate crisis? As I have noted several times in this narrative, no matter how painful it may be to deal with the climate crisis, not dealing with it will be far worse. Convincing people of that is a good first step.

This presents another conundrum to politicians. If you cause pain to your constituents now to avoid much greater pain at some unspecified time in the future, how is that going to help you at election time? Unless you have an extremely enlightened electorate, it won't. People will blame you for the immediate pain, such as high gasoline prices, expensive electric bills, etc., without giving you any credit for the future good. They will very likely turn you out of office at election time, especially if you face a clever opponent who can exploit this. Political leaders agree that action must be taken on the climate, but political survival demands that it be done by some future prime minister, president, or chancellor, not the current one. It's a lot like the sign in a saloon that perpetually says, "Free beer tomorrow." This is how we ended up having 26 COP climate summits, with more probably on the way. It should have been solved after COP #1.

Because the climate affects everyone on Earth, there is a perverse sociological incentive for people in large groups (in this case, all of humanity) to not pitch in to address shared problems but leave it to others. The concept is called "collective action theory" and was developed by an economist named Mancur Olson in the late 1960s.[5] The problem with leaving the work for others is that if everyone thinks this way, nothing gets done.

For example, imagine a small boat with a leak that is slowly sinking. If there are only a few passengers it is difficult for any of them to avoid doing their fair share of bailing. Every full bucket counts toward survival, and everyone can see who is taking their turn. Everybody has an incentive to pitch in and help, and social pressure from others in the group keeps them engaged. The boat stays afloat until it can get back to a dock.

Now imagine instead a gigantic cruise ship with thousands of passengers and a hole in the hull. The efforts of one individual to help bail out the boat are not going to make much difference and may not even be noticed. A few leaders will set up bucket brigades in a heroic effort to save the ship. Some folks will pitch in, but many other passengers will stand back and remain aloof, unsure of how or where to help, or even if their help is necessary. The subset of heroes will work to save the ship, and perhaps they will be successful. If this is the case, the aloof passengers have benefitted handsomely from the work of others at the lowest possible cost to themselves.

This is how most human beings behave according to Olsen and other supporters of collective action theory. Olson has some critics, but we have probably all seen something like this at least once in our lives. People in large groups will typically stand back and let someone else address the problem. If it works, they enjoy the benefits. However, it doesn't always work.

There is a danger that this strategy will backfire if there are just not enough people engaged in helping with the bucket brigade and the ship ends up sinking with all aboard. The heroes and the aloof passengers will both suffer equally, and those who did not participate will discover too late that their aloofness was counterproductive.

The climate crisis at the moment suffers from a collective action problem. There appears to be a degree of hesitation among world leaders about stepping forward to deal with climate change. Everyone seems to be waiting for somebody else to take the lead. Even in cases where there is little political risk, such as a president with

---

[5] Olson, M. 1971, The Logic of Collective Action, Cambridge, MA: Harvard University Press, 208 p.

term limits who will not be seeking re-election or a dictator appointed for life, leaders have been reluctant to institute the actions that are needed to address climate.

Centralized coercion is one way to overcome obstacles to cooperation among individuals. Pay taxes or the IRS will prosecute you. Register for the draft or go to jail. Unfortunately, there does not seem to be a centralized coercion mechanism to prod people into action on climate. Without being unnecessarily alarmist, both world leaders and ordinary citizens must understand that we are in real danger here and everyone needs to work together to solve the crisis. As Marshall McLuhan stated, "There are no passengers on Spaceship Earth. We are all crew."

The consensus among world leaders, all of whom signed the 2015 Paris Accords, is that each nation must make an effort to limit the rise in global temperatures. However, there is a fear that by going first to end the domestic use of fossil fuels when other nations do not will place one's country at an economic disadvantage. This was the reasoning used by President Trump to pull the United States out of the agreement. Each leader is looking out for their own country and tries to meet the minimum requirements of the agreement while imposing the smallest possible costs on their economies and political systems. It's the classic collective action problem where other nations are expected to absorb the burden. Countries end up negotiating minimal commitments that they have no real intention of upholding in the end.

This concern about other nations taking advantage of the first country to eliminate fossil fuels was made explicit in an amendment offered by Representative Mike Rogers (R-MI) for the U.S. 2009 American Clean Energy and Security Act described in the previous chapter. It was defeated in committee along with more than a dozen others that are known as "message amendments." These are added with no expectation of passing, but they provide headlines and keep the representative's name in the news. Rogers' amendment proposed canceling the Act unless China and India adopted similar cap and trade standards for carbon, essentially requiring that we all join hands and jump together. It would be nice if everyone in the world could do this simultaneously, but others are holding back with the same sense of distrust, and time is wasting. Somebody needs to make the first move.

No single country appears to be willing to undertake costly climate mitigation measures by itself. The problem is too big to be solved by unilateral action in any case. The solution must be international but there is no global agency that can enforce recalcitrant states to meet climate standards. Russia, Venezuela, and Saudi Arabia are unlikely to stop exporting petroleum and natural gas to the rest of the world. No one is going to shut down Australian mines because they continue to export coal or ban the export of products from China made with coal-fired electricity. India intends to keep industrializing, as do China and other nations.

Climate commandos forcing governments to act is nothing but a pipe dream of immature environmentalists. Actions that are outside the normal political process or outside the law tend to backfire and have the opposite effect intended. The activist group Extinction Rebellion often calls for the use of "non-violent direct action and civil disobedience" to prod governments into action. During the COP26 summit, climate activists disrupted the ceremonial induction of the lord mayor of the city of London. The action stopped traffic and forced some people to walk, reminding the very group that they were trying to persuade how much faster they could have reached their destination by burning petrol in an automobile.

Most people in industrialized societies want modern technology and the conveniences associated with it. Climate activists who expect people to give up their technology and the energy required to run it are not going to get much traction. A more productive approach is to find alternative energy technology that is not damaging to the climate.

To achieve an energy future with a stable climate, abundant, inexpensive and sustainable zero-carbon power, and a high level of technology, the leaders in government, business and industry must have an incentive to take action. It is obvious that they will maintain the status quo unless pushed. It is also obvious that even a threat to the survival of human civilization is not enough of a motivation. So once again, we circle back to the question of how do we do this? How do we get the world to stop burning fossil fuels, start taking carbon dioxide out of the air, and help vulnerable populations resist the ravages of climate change?

Here's an idea: what if energy technology made a giant leap forward and alternative energy sources were not just a replacement fossil fuels, but an actual, significant improvement? What if they were cheaper, more efficient, more reliable, more resilient, more convenient, and much cleaner? In this case the market would push hard for the adoption of non-fossil energy sources and fossil fuel would become as obsolete as the telegraph. This type of technology could be developed single-handedly in the United States without waiting for others to join us and without putting ourselves in economic jeopardy by adopting climate austerity measures.

But wait, you say. Aren't we already doing energy technology development? What has DOE been up to for the last 40 years? Well, yes, there has been technology development. But a lot of it has been marginal improvements done hand in glove with the fossil fuel industry and the electric power industry that mainly focused on modifying already-obsolete technologies to survive in a climate-constrained world. For example, DOE spent decades and billions of dollars trying to figure out economical ways to capture carbon emissions from coal-fired power plants. Wouldn't it have been a better idea to just find a carbon neutral or carbon zero heat source like compact nuclear reactors or engineered geothermal that could substitute for coal under the boiler? If these could be made to work it would solve the emissions problem in one fell swoop.

Of course if you own a coal company that supplies fuel to power plants, you would not be happy with that particular solution. This has been repeated time and again with coal, oil, and gas. Coal was seen as an abundant resource in the U.S. during the energy crisis and DOE spent millions trying to improve on a process developed by the desperate Germans during WWII for turning coal into liquid fuels. It never became cost-competitive with petroleum but was strongly supported by the coal industry. Shale gas eventually was successful, to the delight of the gas industry, who quickly ramped up natural gas production to levels exceeding those of Russia. Shale gas solved the energy crisis but left us dependent on fossil fuel for at least an additional decade. It also displaced a lot of coal, to the displeasure of the coal industry.

One of the goals of DOE is called "technology transfer," which means getting new energy technologies adopted by industry. As such, they have consistently tried to be too accommodating. Industry doesn't like risk, and the unfamiliar is risky. But to develop the energy future, we need to take some risks and focus on some totally fresh ideas.

The technology I'm talking about is new in terms of application, but maybe not so new in terms of chronology. Many of these ideas have been kicking around for years, and I think we have finally reached a technology readiness level where they can be implemented with small amounts of additional development. These are the kind of things that show up on the internet as "life hacks:" a new and better way of doing something using existing materials and processes.

I came up with a life hack a few years ago. For decades I've been annoyed that when I put my eyeglasses on after getting out of the shower, they steam up in the humid bathroom air and I can't see anything. I wanted to come up with a solution that was simple, cheap, effective, and used materials that I had on hand, so I sat down and thought about the physics. Water condenses on the lenses in the glasses because their temperature is below the dewpoint temperature for humidity in the air. Raising the temperature above the dewpoint would prevent condensation. My solution was to add a tall plastic container to the shower like a tumbler for iced tea (no glass in the shower!). When I'm adjusting the temperature, I fill the container with moderately hot water and set it aside. Then when I get into the shower, my glasses go into the container of hot water. When I dry them off and put them on afterward, they are warmed well above the dewpoint temperature and get zero condensation. It works like a charm, and you are welcome. So what kind of life hacks can we do for the energy future?

What if you had an electric car with a 400 mile (644 km) battery range? That's the same range as a full tank of gasoline. Some electric vehicles currently offer half this range, but battery tech is improving almost weekly. A large battery range would cure the "range anxiety" people have about electric cars.

What if you could flash recharge that battery in 10 minutes? People spend longer than that getting a cappuccino at the gas station/convenience store on a road trip. Quick recharging technology is developing rapidly. There are even some experiments being carried out in Detroit to build induction coils into roadways to recharge electric vehicles while driving. If these were emplaced under a stretch of road at 100-mile (160-km) intervals along interstate highways, for example, one could drive coast to coast without having to stop to recharge.

What if you made your own methane biogas at home from yard waste like grass clippings and dead leaves? You could add food waste or any other kind of organic matter you can imagine (and I'll leave that to your imagination, but maybe cleaning up after the dog could have some energy benefits). If you had a home supply of compressed biogas, not only could you fuel your CMG bi-fuel vehicle in your garage, but you could also run your beloved gas stove in the kitchen even if you were forbidden to connect to a natural gas utility. Remember, methane is methane. There is no difference in composition or heating value between biogas and natural gas.

Along similar lines, what if you could make your own green hydrogen from tap water with a solar panel on the roof, and a compressor in your garage? You could fill your fuel cell or hydrogen combustion vehicle independently at home and never go near a service station. The advances discussed in Chap. 8 from the Fraunhofer Institute in Dresden, Germany with stabilized magnesium hydride paste for storing hydrogen makes this idea possible, affordable, and safe. While you are adding solar panels on the rooftop to make hydrogen, why not add enough to simply power your house? Improvements in battery and solar technology are rapidly making it more possible than ever to go off-grid. Hybrid systems that include both small wind turbines and solar panels provide even more reliable power to charge a home battery system. A home running off-grid on a battery recharged by renewables is extremely resilient to storms, wind, blizzards, ice, fallen tree limbs, and all the other meteorological events that cause power outages.

If an EV performed better than a fossil fuel vehicle, had a greater range, could be recharged at a lower cost and in less time than it takes to fill up a gas tank, and had a lower sticker price than a gasoline-powered vehicle because of a carbon tax, EVs would make serious inroads into the market in about 5 years, especially if we could mine the necessary exotic metals and rare earths from electronic waste in landfills. CMG bi-fuel modifications on existing gasoline vehicles would make them carbon-neutral, much cheaper to operate, and clean up the air in cities. Hydrogen fuel cell vehicles with ranges and performance better than fossil fuel vehicles would also make serious inroads into the car market, especially if green hydrogen from solar power was available at home for free. These technologies are all close to being commercial and a shot of government money with a dash of favorable policy could move them forward quickly. The vehicle sector could be decarbonized within a decade.

If you live in an apartment in the city, the landlady is obviously not going to allow you to hang solar panels off the balcony and you are stuck with the city electric supply. In this case, municipal utilities can be encouraged to switch to non-fossil energy to improve resilience, efficiency and save costs. Instead of moving a pile of coal through hoppers to feed a boiler and then dealing with the ash and smoke, imagine instead if the steam powering the turbines was made with geothermal heat, possibly solar-assisted, brought into the boiler by a hot loop from an array of engineered geothermal wells drilled behind the power plant. Once installed, this system would be dependable, inexpensive to operate, clean, and simple. The coal trains could stop coming and the generators would continue to hum along.

If there is no room for an EGS facility, perhaps new technology nuclear is an option. Refrigerator-sized, standardized, efficient, factory-made reactors that use a thorium-based fuel cycle and molten salt cooling could directly replace the coal burner under an existing boiler, making steam and producing electricity with no carbon emissions. If we resolve the nuclear waste repository issue at Yucca Mountain by building a freshwater distillation plant powered by the heat from high level nuclear waste in monitored, retrievable storage, nuclear technology could really take off.

Municipal-level methane fermentation systems could make biogas that would directly replace natural gas. As described earlier, feedstocks could include municipal wastewater (refer back to Fig. 6.3) and produce a useable product that will reduce the overall volume of waste that must be disposed of. Biogas methane could be directly distributed through an existing town natural gas utility system, providing carbon-neutral fuel without costly retrofits to home appliances or furnaces. Biogas methane could also be used to run efficient, combined cycle power plants. The CO_2 emissions would be carbon-neutral, but if the plant was equipped with CCS, the emissions become carbon-negative. Generating power with a multitude of BECCS systems like this can make a significant dent in atmospheric carbon dioxide levels fairly rapidly.

Will any of these things work? Are they practical? Do they produce enough energy? What do they cost? All good questions. They work on paper and in theory. None of them violate the laws of physics. As for practicality and cost, we won't know until we have some existing prototypes to test for performance. People are working on these, and they are in various stages of development. Some need more work than others, but at least several are pretty close.

I think that instead of trying to address the climate crisis with government policies that only result in stubborn disagreements and a deadlocked Congress, perhaps we can approach this from an advanced technology perspective. The executive branch of government has considerable latitude on the specifics of where and how to spend appropriated research budgets. DOE should stop funding fossil energy research altogether, and focus their money on EGS, biogas, hydrogen, and new nuclear technology, as well as CDR and sequestration. Other federal agencies like DOD, Commerce, Interior, Agriculture, and Transportation could also support relevant projects. At present, research on some of these technologies is not funded at all, while others are funded on a shoestring.

Given the potentially dire consequences of the climate crisis, I would advocate for a so-called "moonshot" effort to push these technologies toward commercialization, which has always been a stated goal of DOE. The end result will be energy that is cheaper, more efficient, more reliable, more resilient, carbon neutral, and much cleaner than fossil fuels. When these energy technologies become better than fossil fuel, they will replace it. If they become a lot better, they will displace it, and fossil fuels will become "alternative" energy.

Economists call this a "disruptive technology." Think about the I-phone as an example. Steve Jobs at Apple didn't need a government policy. He invented it, improved it, and marketed it. The smart phone he developed was so much better and more useful than any other existing landline or cell phone out there that it quickly took over the market and became a standard. Samsung came out with the Android as competition and so did other companies. Now everybody has a smart phone. I know this because I was one of the last holdouts. Smart phones are a classic disruptive technology, changing how we communicate, take photographs, access the internet, and reference calendars, clocks, and calculators. It is time for a similar disruption in energy.

If new energy technologies are paired with policies that make it more expensive to emit GHG into the atmosphere, the economics improve. Placing a carbon tax on emissions from a coal-fired power plant makes BECCS more attractive. Boosting the cost of a diesel truck by adding a carbon tax to the sticker price makes an electric option more economic. I don't think we should count on this, however. A carbon tax would be great if it happens, but there is still a lot of resistance among fossil energy lobbyists and members of Congress with ties to the coal and oil industries. It might be better to just develop and deploy these new technologies on their own merits, and if a carbon tax does come along, it will sweeten the pot.

Once the United States does this, everybody else on the planet will copy our ideas and move them forward globally. This has happened with every other innovation we have ever come up with, from Hollywood movies to rock-and-roll music to fast food restaurants to smart phones. I am no longer surprised at seeing a Dunkin Donuts in Tokyo, hearing "Highway to Hell" in Germany (performed by an oom-pah band in lederhosen that was especially bizarre) or seeing a Marvel superhero movie playing in a London cinema. The number of people in China with smart phones is staggering. The customs officer at the airport used hers to translate instructions to me, since I only know two words of Chinese (hello and thank you – but you'd be surprised at how far you can get with just that). Maybe we can use American global popular culture in a positive way to influence climate.

This is a U.S. responsibility thanks to the legacies of Edwin Drake, Samuel Kier, Anthony Lucas, John D. Rockefeller, Henry Ford, and countless others. The United Kingdom was responsible for introducing coal to the world, which was widely adopted in the U.S. The United States was responsible for introducing petroleum and natural gas. There is no holding hands and jumping together. America must go first.

****

Finally, what about carbon dioxide levels in the air? There are a number of technologies, both biological and engineered, for removing CO_2. The gas can be sequestered underground in geological formations as discussed in Chap. 9, but it is hard to make a business case for this unless you are using it for EOR. Which, of course, defeats the entire purpose.

Revenue sources for sequestering captured carbon include the IRS 45Q tax credit. As mentioned earlier, Congress has been discussing an increase but at present, the maximum credit is $50 per ton for sequestration in deep saline aquifers, which is also one of the most expensive options due to the high cost of deep drilling. Another source of revenue is from the sale of carbon offsets. These are voluntary and unregulated, so from a revenue standpoint it is hard to predict how much money one might receive, or when. The carbon offset market is expected to expand widely in coming years, so this may turn out to be a significant source of revenue for CDR operations.

I believe that from a sequestration perspective, the absolutely best and safest way to store carbon dioxide out of the atmosphere is to convert it to solid carbonate minerals. These minerals make up the bulk of rocks like limestone, which have been a stable part of the geology for hundreds of millions of years (Fig. 10.2). Limestone can be dissolved by acidic groundwater, which gives us spectacular caverns, but in most cases it is an extremely stable material for locking down carbon dioxide. The early atmosphere of the Earth was rich in carbon dioxide, and limestones deposited in the Precambrian and early Paleozoic eras were responsible for removing most of this and sequestering it as solid rocks. This tried and true method seems to be the best option for anthropogenic CO_2 as well.

Converting CO_2 into carbonate is relatively easy and can occur both organically in the shells of marine animals like clams and corals, or inorganically through geochemical reactions (refer back to Fig. 9.4). It is inert, non-toxic, non-combustible, and does not migrate. This helps considerably with the MVA paperwork. Limestones have value as a product, and carbonate minerals precipitated from CO_2 could potentially be sold as concrete aggregates, soil amendments, building materials, or other uses. Finding sales markets for carbonates produced from CDR is very helpful to the economics.

The circular economy is of course the holy grail for CDR revenue, where captured carbon is turned into useful products with favorable economics. This significantly increases the financial incentive for removing it from the atmosphere. The carbonate minerals mentioned above are one thing that might have sales potential, but there are many other products that can be made from carbon dioxide. The idea is to build products from carbon that is part of the carbon cycle, not derived from fossil fuels. This is a carbon-neutral solution because if any of this material gets back into the atmosphere, that is where it came from in the first place.

The circular economy is an organic chemistry and chemical engineering problem. I am not a chemist, and it is unclear to me what all is involved here but

Fig. 10.2 A bluff of 340 million year old St. Louis Limestone along the Mississippi River in Missouri. (*Source: Photographed in 2021 by Dan Soeder*)

replacing petroleum as a chemical feedstock with carbon dioxide is apparently not all that farfetched. Actual chemists tell me there is no physical reason this can't be done, and the only thing holding it back is the economics. Costs are expected to improve over the next few decades.

Carbon dioxide supplies will become cheaper as the amount recovered from the atmosphere increases. Petroleum feedstocks will become more expensive due to a reduced demand for fossil fuels and a greater cost for recovery after passing peak oil. It is conceivable that by 2040 carbon dioxide from DAC will be the major feedstock in the chemical industry. People who support the idea of a circular economy claim that it could quickly become a multi-billion, and maybe even a trillion dollar market. God bless them all and I hope it works.

The bottom line for addressing the climate crisis is this: We must demand the decarbonization of electric power, the availability of affordable zero-carbon vehicles, and the removal of carbon dioxide from the atmosphere. If it can't be accomplished with policy, it should be done with research. Developing economical carbon zero or carbon neutral energy can be a disruptive technology that undercuts fossil fuels and renders them obsolete. The conversion of captured CO_2 to carbonates or other solid materials encourages the growth of the circular economy and displaces petroleum as a feedstock for plastics or chemicals. Decarbonizing energy over the next decade and developing markets for captured CO_2 by mid-century can restore the balance in our atmosphere and keep us from crossing over into the Weleftthescene.

Science communicators must explain to the public not just what scientists know but how they know it. Educators must teach students how to think critically, source scientific information, and evaluate it. Finally, policymakers must make decisions based on sound scientific evidence, not feelings, whims, or the interests of corporations. If we do these things, the energy future looks bright.

Chapter 11
Leaving the Cradle

Keywords Growth · Degrowth · Space habitats

****

Back in the day, if you were a wealthy industrialist you could escape the bad air, contaminated water, and filthy streets created by your factories and mills if you had a nice mansion in an upscale neighborhood or a well-appointed country estate. Most of the rich folks had these and thereby avoided living with the environmental consequences of their industry. The low level workers who couldn't afford a horse for transportation were forced to live within walking distance of the factory in lower income industrial neighborhoods or tenements and cope with air and water problems. This was the beginning of what we now call environmental justice or EJ: the placing of industrial facilities in poor neighborhoods where people are forced to live with them. New technology to address the climate crisis, like DAC facilities, wind or solar farms, and geothermal installations must be sensitive to EJ concerns wherever they are located.

Owning a horse back in the old days to get out of town was a luxury. I have horses on my West Virginia farm and even though they spend a lot of time grazing in the pasture, feeding them is not cheap. Horses were a mainstay for work and transportation in nineteenth century cities, but maintaining one was a major expense and beyond the means of most workers. It is interesting to see just how nearby some of the old worker housing is located to the factories and mills in grimy industrial cities like Cleveland and Pittsburgh. Many of the houses were only a few blocks away and literally got covered in soot from the blast furnaces.

Rich New Yorkers had estates in the Hudson River uplands away from the city. Wealthy Philadelphia families spent their summers in the cool air of the New Jersey shore. The DuPont chemical family in Delaware built a road from Wilmington to the beach (now designated "DuPont Highway") and put a summer estate at the end of it. Baltimore and Washington elites coped with the summer miasma of the tidal Potomac River and Chesapeake Bay by escaping into the Blue Ridge Mountains. (The presidential retreat at Camp David on Catoctin Mountain is an example.) Rich people today buy tropical islands in the Caribbean or ranches in Montana. However,

D. Soeder, *Energy Futures*, https://doi.org/10.1007/978-3-031-15381-5_11

the climate crisis is different from other environmental issues. It affects all of the Earth and every part of the Earth. These days, there is nowhere to go and no escape.

John D. Rockefeller had a large estate called Forest Hill located in Cleveland Heights, well above the refineries and steel mills of Cleveland, which were largely confined to the "flats" in the Cuyahoga River valley. The Rockefellers used a Victorian mansion on the property as a summer home, which they called the "Homestead." The family moved to New York City in 1884 but continued to return to Forest Hill each summer. After he became a widower in 1915, Rockefeller seldom returned to Cleveland and the Forest Hill "Homestead" mysteriously burned down in 1917. When Rockefeller died in 1937, his son gifted 235 acres of the Forest Hill acreage to the city the next year as a park and subdivided the rest into residential and commercial properties.[1] The Second World War put construction on hold, and in 1948 Rockefeller Jr. sold Forest Hill to a developer. John D. Rockefeller's signature estate is now a park and an entire upper middle-class neighborhood, called appropriately, Forest Hills. I had a childhood friend who moved there from my old neighborhood after his father got a new, high-paying job. Their house was quite impressive.

The Rockefellers also had a mansion along "Millionaire's Row" on Euclid Avenue in Cleveland. This was an impressive stretch of nineteenth century mansions that extended from about E. 22nd Street to University Circle. Between 1860 and 1920, this stretch of Euclid Avenue housed numerous mining, shipping, steel, oil, coal, and railroad magnates and was one of the wealthiest areas in the world. It was known internationally as the "Showplace of America." Sadly, only a few of the original mansions still exist. The 45-room Mather Mansion, built by iron ore mining and shipping magnate Samuel Livingston Mather was acquired by Cleveland State University where I went to college as an undergraduate. It was built in 1910 and at the time was the largest and most expensive home in Cleveland. The university now uses it as a conference center, but when I was there it was used for classrooms. I still remember the incredibly ornate hand-carved woodwork and the massive central staircase.

The mega-rich are still with us, and one of the remarkable things about the twenty-first century has been the development of commercial space programs by people like Elon Musk (SpaceX), Jeff Bezos (Blue Origin), and Richard Branson (Virgin Galactic). Their public reasons for these efforts are to help humanity become a multi-planet species and to move manufacturing and other industry into space. While I certainly agree with these principles, a small part of me also wonders if at least some of the motivation behind private space programs is to provide their billionaire owners with an escape if everything goes to hell in a handbasket on Planet Earth. Even if that's the case, there is no Forest Hill on Mars or Mather Mansion on the moon. The space billionaires will need to build an entire off-planet infrastructure in orbit and on other worlds if they want a place to retreat. Both Musk and Bezos seem to understand this and are actively working toward achieving it. Good

---

[1] https://clevelandhistorical.org/items/show/83

for them, because building off-planet infrastructure and destinations in space is good for all of us.

Why? you may ask. Well, no child should remain dependent on their mother forever. At some point, it is important to break free and leave home. Humanity has been stuck on Mother Earth for too long, and we have very much outgrown the place. In fact, we have become like an unemployed, middle-aged child living in our mother's basement, wearing a bathrobe all day and playing video games while surrounded by empty pizza boxes and crushed beer cans. We have without question trashed her house, and we have also strained our poor old mother's ability to support us. Earth has been called the cradle of humanity, but as the writer Arthur C. Clarke famously stated, "One cannot remain in the cradle forever." It is time to give our mother a break and we need to leave. But in order to leave, we have to launch spaceships.

Moving into space and colonizing other worlds would help the environment and the climate crisis on Earth by removing a substantial number of people from the planet along with the most polluting energy and manufacturing industries. Elon Musk wants to put a million people on Mars this century. The moon could hold as many in underground warrens and pressure domes. Jeff Bezos wants to put people into orbiting space habitats.

So carry on, SpaceX, Blue Origin, and Virgin Galactic. Carry on NASA, JAXA, ESA, CNSA and Roscosmos. And carry on all you start-ups out there that are trying to develop new ways and improve old ways to get humanity into space. Because the alternatives are not great.

****

It should be patently obvious that a finite planet cannot support infinite growth. Yet many of our economic systems pretend that it can. They rely on the continuous growth of consumer spending, resource extraction, manufacturing capacity and population expansion to grow the economy. This endless growth leads to uncomfortable questions:

How much more GHG from fossil fuels can our atmosphere and oceans handle? Where are we getting all the exotic metals and rare earths needed to manufacture electric vehicles? How much more rainforest can we cut down for tropical hardwoods before the ecosystem collapses? How much more capacity do our landfills have for the disposal of increasingly less recycled plastic waste? What happens to the oceans and aquatic ecosystems if fish populations crash from overfishing, pollution, or the loss of plankton? How long can we sustain vegetable farming in places like the California desert with irrigation from a river that is drying up? What happens when we finally do cross Hubbert's peak oil and wells continue to decline?

At some point, this hits a wall and can go no further. We run out of room, we run out of resources, and we simply run out of planet. Using movie metaphors, I see three ways this can go. The first and worst is "Mad Max," where civilization collapses, billions die, and the remaining humans drive terrifying vehicles and fight fiercely with each other over whatever meager resources are left. The second is

"Soylent Green," which interestingly enough is supposed to take place in 2022, where civilization devolves into a highly-regulated, authoritarian society that doles out scarce resources following a strict system of rationing and also comes up with a creative solution for dealing with overpopulation (the "soylent" food is supposed to be soy and lentils, but the hero finds out the green one is people). The third movie is the "Star Trek" series, where we expand into space, explore new frontiers, and open up many new possibilities for the human race. I strongly favor the third option, even if we have to do battle with the Romulans and Klingons occasionally.

I don't think anyone wants to go with the crash of civilization as displayed in Mad Max. It looks kind of fun in the movie but would definitely not be fun in real life. Soylent Green is a slightly better option (except for the part about turning excess people into food) where the Earth's remaining resources are managed and rationed. This provides a possible alternative for extending the Anthropocene if we choose to turn away from space and remain confined to the Earth. The Soylent Green option is called "degrowth."

Degrowth is the opposite of current economic models. It sounds ominous, and it can be. In a broad sense, it recognizes the limits that the Earth places on resources and room to expand. Degrowth is a deliberate and planned slowdown in resource extraction and corresponding production and consumption. There is a risk that it could be instituted as a dystopian, heavy-handed government intervention and regulation, with Soylent Green-style rationing of vital products and services. However, proponents claim that degrowth is not necessarily synonymous with a reduction in the quality of life and say the actual goal is to reduce production of the most unsustainable and polluting products that disproportionately contribute to environmental degradation. Degrowth targets unnecessarily high levels of consumption by the world's very richest people. If you translated that as "Americans," you are probably correct.

The goal of degrowth is to lower economic dependence on processes of extraction and production that are not environmentally sustainable. These vary among different sectors of the economy and degrowth is focused on those with the greatest environmental impacts, not the economy as a whole. So for example, the energy and environmental costs of making steel from iron ore are considerably higher than the cost of making steel from recycled scrap iron. Scrap iron has already been through the reduction process described back in Chap. 6 to remove the oxygen from the iron ore. This is the most energy-intensive part of the steel making procedure and emits the greatest amount of GHG. Being able to avoid it reduces energy requirements and emissions significantly. Instead of mining and processing more iron ore, a more comprehensive scrap iron collection and recycling program could substantially reduce the environmental impacts of steel manufacturing. Old ships, demolished buildings, scrapped automobiles, abandoned railroad tracks, and even old vegetable cans are able to supply recycled scrap steel for re-processing. Every landfill and junkyard should have an electromagnet system for separating out steel. The same goes for recycled aluminum and other metals.

Degrowth also includes decarbonization of the energy sector, but it is more than just decarbonization. It supports expansion in other sectors of the economy that are not extractive or resource intensive, such as education and healthcare. Degrowth also seeks to decouple labor from economic growth and make it a political right instead. A reduction in worker productivity and notions like job-sharing are considered beneficial to society within a degrowth economy. "Sufficiency" is proposed as the organizing social principle for degrowth compared to the endless expansion required by current economic models.[2] One can have enough without going overboard.

The consumer economy is a relatively recent invention. People throughout history were happy to achieve sufficiency, and excess was seen as a moral failing, such as the ancient Greek story of King Midas and his greed for gold that ultimately led to tragedy. King Midas translates into modern times as the lifestyle of people with two SUVs, a pickup truck towing a large boat, twin jet skis, a giant camping trailer, and a snowmobile parked in front of their oversized, over-furnished, and rarely occupied McMansion. Such conspicuous consumption and accumulation of "toys" drives the capitalist economy and businesses love it, but it is not sustainable for most of the world.

Do we really need all this stuff? Of course not. We are surrounding ourselves with excess because we can, and because advertisers tell us we should. The more we buy, the more money businesses make and the stronger the economy becomes. However, as we pile up material goods, we are burning through resources, harming the environment, and damaging the climate. Once all these products end up in a landfill, as most eventually do, they cause even more harm. But if a business is not increasing its sales every year, it is considered problematic. Dividends fall and investors pull out. Workers lose jobs. The economy suffers. Our economic models that require endless growth can have severe consequences for those businesses who do not follow them.

From the perspective of someone sitting inside an endlessly expanding economy, degrowth proposals sound a bit insane. Yet, maybe there is something to them. Degrowth as an idea is not bad. Having a sufficiency instead of an excess is much better for the environment and the climate. Our culture is a long way from accepting such an idea, and before it can make any inroads a reassessment of our relationship with nature is required. Humanity must view itself as stewards of the Earth, not masters. Our economic models would have to shift from growth to maintenance. We need to develop an understanding that we are obliged to conserve and share resources and use them wisely. We ought to embrace the Native American notion that we have not inherited the Earth from our parents, but we are borrowing it from our children.

As a policy, degrowth has unsurprisingly gained zero traction in a world that is obsessed with growth. Any political leader who proposed degrowth right now as a way to deal with the climate and the environment would be committing political suicide. However, shifting our thinking away from the dogma of infinite growth can only help humanity as we move into the future.

****

---

[2] Sekulova, F., Rodríguez-Labajos, B., Kallis, G., and Schneider, F. (editors), 2013, Degrowth: From Theory to Practice: *Journal of Cleaner Production*, Special Issue, v. 38, p. 1–98.

Going into space is hard. I once had an engineer from the Jet Propulsion Laboratory (JPL) explain to me that a rocket engine is nothing more than a controlled explosion. The explosion is a bit on the slow side and channeled in a useful direction but combining a fuel and an oxidizer is a volatile mixture that burns violently when they come into contact. On the occasions when rockets do explode, this becomes obvious. Compared to the really dangerous early days of rocketry, rocket explosions are rare but still possible. The intent is to direct the controlled explosion downward, so the rocket flies upward, and the hope is that the explosion stays contained and lasts long enough to get the rocket and whatever cargo it is carrying all the way into space.

Rocket technology has come a long way since the early experiments of Robert Goddard with liquid fuels in Massachusetts and New Mexico, and the contemporaneous risky rocket engineering carried out by Jack Parsons in California. Parsons was one of the founders of JPL, and although he and Goddard met in 1935 and compared notes, they were never actual collaborators. Parsons and his Cal Tech colleagues were well known on campus as the "Suicide Squad" for their dangerous rocket launches, and he was killed in an explosion in 1952 at age 37.

Three of the most influential people who were models for space advocacy in the late twentieth century are Wernher von Braun, Carl Sagan, and Gerard K. O'Neill. Each of these men approached space advocacy from different directions, but each in their own way ignited public interest in space exploration.

Wernher von Braun (1912–1977) of NASA was responsible for spectacular projects like the moon landing where ordinary people could not participate but were proud of the accomplishments. Von Braun was born in Prussia (now part of Poland) and grew up in Berlin. He became fascinated with space travel at a young age and after graduating with a doctorate in physics from the University of Berlin in 1934, he joined the German Army Rocket Center at Peenemünde on the Baltic Sea, where he soon was named Technical Director. Because of an oversight in the Treaty of Versailles that ended World War I, rocketry for some reason was not included on the list of weapons forbidden to Germany.[3]

Von Braun was a card-carrying Nazi, although he always claimed that he never participated in any political activity. In an affidavit to the U.S. Army he stated that "my refusal to join the party would have meant that I would have to abandon the work of my life." Von Braun is credited with the design of the German A-4 rocket, later renamed the V-2 by Hitler that caused significant devastation in London late in the war. Von Braun's memoir was titled I Aim at the Stars, which commentator Mort Sahl said should be subtitled, "But sometimes I hit London."

Von Braun and his German rocketry team surrendered to the American Army at the end of the war and were transferred to the United States. (The bulk of the rocket specialists remaining at Peenemunde were captured by the Red Army and sent to the Soviet Union.) Von Braun worked on military rockets in Huntsville, Alabama for a

[3] Davies, N., 2006, Europe at War 1939–1945: No Simple Victory: London, Macmillan, 416 p. ISBN 9780333692851.

decade, including the Redstone, the first nuclear-capable ballistic missile. After the establishment of NASA in 1958, von Braun was named director of the Marshall Spaceflight Center in Huntsville. He is credited with the development of the Saturn V moon rocket and although many people worked hard to get men to the moon, Wernher von Braun gets a huge share of the credit. He popularized space travel in a series of specials he did for the Walt Disney television show in the 1960s, with animations of amazing conceptual spacecraft that would take humans to Mars and beyond. I remember watching these as a kid and being intrigued.

Another model for space advocacy in the late twentieth century was Carl Sagan (1934–1996), who explored the universe from a distance and got people to appreciate the grand scale of creation, leaving them awestruck. Sagan grew up in the New York City area and entered the University of Chicago as an undergraduate at age 16. After receiving his PhD from there in 1960, he taught briefly at Harvard but spent most of his career as an astronomy professor at Cornell University in New York.

Carl Sagan was a popularizer of astronomy and planetary science, publishing hundreds of articles, dozens of books, at least one science fiction novel about alien contact, and hosting a television series called "Cosmos" about exploring the universe. He advocated critical thinking, scientific skepticism, and from his work with planetary atmospheres he helped to develop the nuclear winter hypothesis for the catastrophic cooling of the Earth if a nuclear war incinerates cities and fills the air with smoke.[4] Interestingly, Sagan discovered the high surface temperature on Venus in 1962 by analyzing radio signals from the planet. He attributed this to greenhouse gas warming from the dense carbon dioxide atmosphere and recognized that similar greenhouse gases could affect the Earth's climate system. He testified as such before the U.S. Congress in 1985, and yet even for Carl Sagan they did nothing.

Sagan was an advisor to NASA almost from the beginning of the agency. As an example of his forward thinking, he was responsible for assembling the first physical messages sent into space on plaques attached to the Pioneer probes and as "golden records" on the Voyager spacecraft. These were the first human spacecraft to leave the solar system, and the messages are short, universal lessons about the nature of Earth and humanity that could potentially be understood by any extraterrestrial intelligence that might run across them. Long after the dawn of Weleftthescene, such records crossing the vastness of interstellar space might end up being the only trace of humanity left in the universe.

The third important person for space advocacy in the late twentieth century was Gerard K. O'Neill (1927–1992), who audaciously suggested that large numbers of humans could leave the Earth and live in space, making ordinary people feel like they could help settle the Solar System.

---

[4]Turco, R.P., Toon, O.B., Ackerman, T.P., Pollack, J.B., and Sagan, C., 1990, Climate and Smoke: An Appraisal of Nuclear Winter: *Science*, v. 247, no. 4939, p. 166–176, January 1990

Amazon founder Jeff Bezos wants to build orbiting space colonies as habitats for manufacturing and industry that sound very similar to the "high frontier" concepts developed by O'Neill at Princeton University in the 1970s.[5] I remember attending a lecture by Dr. O'Neill in graduate school when he was on tour promoting his book. He envisioned viable human space habitats inside large, long cylinders with the axis oriented toward the sun. I found the whole idea fascinating, and as I came to learn later about how O'Neill typically did things, all the details had been worked out and the questions were answered before they were even asked. He had very thoroughly covered every base.

This was not long after the energy crisis, and O'Neill's business model for the income to build his space colonies was to orbit a series of geostationary solar power satellites that would beam power down to receiver stations on Earth using microwaves and ending the energy shortages that seemed to plague nearly everyone back then. Had this actually come about, it would have displaced fossil fuels and possibly saved us from the climate crisis. The idea is not without some problems - water absorbs microwave frequencies, which is why food heats up in a microwave oven. Power beams from the satellites would have heated water vapor on their way through the atmosphere that might have caused local climate effects. However, this would have been less impactful than global warming of the entire atmosphere from elevated GHG levels.

The inside walls of O'Neill's habitat cylinders would be lined with soil and water for growing crops while also protecting the human inhabitants from radiation. A series of mirrors would allow sunlight to enter through long windows along the length of the cylinder to grow crops, and an axial spin would provide centripetal force for artificial gravity, making "up" the direction toward the center of the cylinder. The spin would also gyroscopically stabilize the cylinder and keep it pointed toward the sun. The mirrors could be angled to increase or reduce the amount of sunlight coming through the windows, making a more natural day cycle, and closed to block off light for darkness at "night."

O'Neill intended to place the cylinders at the L4 and L5 Lagrange points, stable locations on the Earth's orbit that lead and trail the planet around the sun. The habitats could be built at a relatively low cost if they were constructed in space using materials obtained from the moon or mined from the asteroids. To supply material to the habitats, O'Neill invented the "mass driver," an electromagnetic rail-gun type device that accelerates payloads at several hundred gravities using superconducting magnets. He planned to put one on the moon. Mass drivers on asteroids could be used to eject material from the asteroid into space as reaction mass and work like a rocket engine to move the whole thing over to L4 or L5. Some prototypes of mass drivers are in development, and recent modeling studies have shown that they can even be used for cargo launches from Earth if the acceleration is very high (a thousand gee, which would turn an astronaut into a pancake, so no passengers) using a

---

[5] O'Neill, G.K., 1977, The High Frontier: Human Colonies in Space: New York, NY: William Morrow and Company, 288 p.

short and straight trajectory through the atmosphere to get the payload into the vacuum of space before it burns up from air friction.

Jeff Bezos of Blue Origin may be familiar with the volumes of work that have already been done on orbital space colonies by Gerard O'Neill and those who came after him, but if not, I would urge him to look into it. O'Neill founded the Space Studies Institute in 1977, which is still active and promoting space colonization, mass drivers, new and exotic methods of spacecraft propulsion, and other cutting edge ideas.[6] O'Neill himself expected that the space habitats would be in place by the 1990s, and sadly he died in 1992 without seeing his vision realized. An excellent depiction of an "O'Neill cylinder" is shown near the end of the movie "Interstellar," when the Matthew McConaughey character returns to the Solar System.

Elon Musk is more interested in colonizing solid ground rather than floating around in orbit, and he has his sights set on Mars. Mars requires big rockets, and Musk's spacecraft company, Space Exploration Technologies Corporation or SpaceX has been launching and recovering spacecraft for about a decade (Fig. 11.1).

SpaceX has developed re-usable rocket boosters that gently touch down after launch and can be refueled and launched again. This saves an enormous amount of money over one-shot expendable boosters and has dropped the cost of spaceflight significantly. The SpaceX cost-to-orbit in the Dragon crew capsule on a Falcon-9 rocket is about $55 million per seat. In comparison, the per-seat cost on the retired Space Shuttle was $170 million, and the seats NASA has been contracting on Russian Soyuz launches cost $80 to $90 million.

The Falcon-9 rocket, the current SpaceX workhorse, has made about a hundred launches and landings at this writing. The Falcon Heavy, which is basically three

Fig. 11.1 A mannequin named "Starman" leaves Earth behind in a Tesla roadster after being launched by SpaceX. (*Source: Photograph by SpaceX in 2018, open access, public domain*).

---

[6] https://ssi.org/

Falcon-9 rockets strapped together, has been launched several times and can lift an amazing 64 metric tons into orbit. The so-called "Mars rocket" is named Starship. When assembled, it will be 30 feet (9 m) in diameter and stand nearly 400 feet (120 m) tall. This behemoth spacecraft will be capable of lifting 100 metric tons (some 220,000 pounds) of payload into low Earth orbit.[7]

Getting people to Mars is complicated and dangerous. NASA plans to return to the moon first under the Artemis program by using the new Space Launch System (SLS) for transit and a Starship upper stage for the actual landing from lunar orbit. This will give SpaceX some valuable performance data on Starship to optimize the craft for a flight to Mars. The plan at present is to send supplies on ahead to the Red Planet, and also to place supplies in Earth orbit. Starship will launch from the ground, refuel and take on supplies in low Earth orbit, and then head to Mars. The transit time will be between 80 and 150 days, with an average trip time to Mars of approximately 115 days. Once a colony is established, Musk plans to launch 100 people at a time on a Starship. To get a million people on Mars, SpaceX will need to make some 10,000 flights.

When discussing the future of space colonies, whether they be giant O'Neill cylinders in Earth orbit or habitats on the moon or Mars, the question arises as to whether or not folks would actually be willing to live in such places? I think the answer is yes, based on what people have done in the past to take advantage of jobs and economic opportunities. (I would sign up for the moon or an O'Neill habitat in a New York minute. I'd have to think a bit about Mars.) There were strong incentives to emigrate from Europe to America in the 18th and 19th centuries, such as economic opportunities, religious freedom, escape from persecution, and so forth. In the eighteenth century, it took weeks to cross the Atlantic in a sailing ship on an unpleasant and often dangerous sea voyage, yet people came. A space voyage of a few days to the moon or a few weeks to the L5 point would be quite comfortable by comparison. Even a three to five month trip to Mars would be tolerable (although this makes me hesitate).

By the late nineteenth century, ocean-crossing ships had become much safer, larger, and faster, and the immigrants came in droves. Three of my four grandparents were among them. My maternal grandparents left Bremen, Germany as newlyweds so my grandfather could get to a job in the steel mills of Youngstown, Ohio. My paternal grandfather left Bavaria, Germany and crossed the Atlantic as an 11-year old boy with his widower father to start a new life in Cleveland, where they had relatives. (My paternal grandmother was born in the U.S. and came from the large German community in St. Louis.) Human migrations have occurred throughout history because of overcrowding or a lack of opportunities. Relocating to a new land promised the ability to not only set one's own destiny, but to improve opportunities for future generations. These days we can add climate refugees to the list of potential colonizers.

---

[7] https://www.spacex.com/

The "New World" was not exactly paradise. Colonists braved arctic cold, desert heat, blizzards sweeping across the plains, malaria in mosquito-infested swamps, alligators, bears, jungle humidity, and the thin air of high mountains to settle in these places. Hostile natives were a constant threat. Subsurface lodgings on the moon or a pressure dome on Mars might not be perfect, but at least these are controlled environments without the dangers of unpredictable weather or wild animals.

The colonization of space also provides humanity with an opportunity to do this right and make up for some of the awful wrongs committed in the past. The worst voyages of all across the Atlantic were in the slave ships, where people were forced against their will to endure horrific, inhumane conditions. If they survived the voyage, they were sold into bondage. These people were not colonists and were not seeking opportunity. They would almost certainly not have been there if given the choice. Slavery was a shameful period in American history, and the repercussions are still with us today. Space should be open to everyone on Earth, and I mean everyone, no exceptions, full stop. On the other side of the coin, anyone who goes into space should go there of their own free will, not by coercion or force, no exceptions, full stop.

A complication with colonizing places on Earth is that the indigenous people who were already living in the so-called New World before the colonists showed up generally were not willing to give up their land without a fight. This resulted in many deaths and some very sad history in the Americas, Africa, Australia, India, New Zealand, and Hawaii, among other places. As far as we know, none of the other planets, moons, and asteroids in our own solar system have any existing inhabitants. There might be microbial life some places in the subsurface of Mars or perhaps in the oceans of Europa. We should let it be. There is plenty of other unoccupied real estate orbiting the sun. For once in our long and sordid history, humans can colonize a new land without decimating the natives.

Along with reducing environmental impacts on the Earth, there are other advantages to moving people and industry into outer space. Transferring the industrial base of humanity to the moon, for example, would allow manufacturers to produce goods without any concerns about damaging the environment because there is no lunar ecosystem to destroy. The sun provides free, clean, 24-hour energy. If nuclear power is needed, it can be built without fear of radioactive contamination. The dry vacuum on the moon might even be the ultimate answer to storing nuclear waste. Raw materials and resources are abundant on the moon, Mars, and in the asteroids. Moving operations into space that involve energy production, mining, minerals processing, materials refining, steelmaking, chemical manufacturing, smelting, metal plating, and other pollution-intensive heavy industries can keep our civilization expanding and moving forward technologically without sacrificing the environment.

It is much less energy intensive to move materials around from point to point in space than to lift them out of the Earth's deep gravity well. These economics suggest that it will be significantly more cost-effective to use materials that are already in space for the construction of various space habitats than to haul these up from Earth. Doing so would be as absurd as loading up a ship full of logs in Scotland to build a cabin in Nova Scotia.

I believe that some of our existing mining schools ought to establish space resource programs. We have an inkling of the resources that might be available in space, but no clear definition of what is located where. If we intend to use these materials, there are some very practical questions that need to be answered. Are the minerals on other planets concentrated in ores in the same manner as on Earth, in a different manner, or possibly not at all? What do these minerals look like? Are they oxides, sulfides, silicates, or do they exist as native metals not combined with anything? Metallic meteorites suggest the latter, but is this true everywhere, or just in certain bodies like asteroids? How do we mine and refine these minerals? Techniques for processing minerals on the Earth into usable materials like steel may or may not be applicable on other worlds, and new methods might be needed. For example, do we need a reductant like coke or hydrogen to remove oxygen from the iron or is this unnecessary? And how would you do this in the thin CO_2 atmosphere of Mars or in the absence of atmosphere on the moon? It would be good to have some answers to these questions before we actually get to these places and start trying to use the resources.

The moon is composed mostly of basalt and contains metals like iron, titanium and nickel. Mars has the largest volcano in the Solar System and also contains a variety of useful metals and materials. Asteroids tend to be rocky, metallic, or some combination of the two, and many approach the Earth closely enough to be easily reached on part of their orbits and mined.

Ores and other critical materials on Earth tend to be concentrated by geological activity such as sedimentation, evaporation, or alteration of rocks by hydrothermal fluids around volcanoes and deep fractures. Mars and the moon both have volcanic features, but we don't know if there was any hydrothermal mineralization activity involved. Some unusual "orange soil" discovered on the moon by astronaut (and geologist) Jack Schmitt during the Apollo 17 mission was initially thought to have a hydrothermal origin. It turned out to be a false alarm. After more detailed analysis, NASA concluded that the orange soil came from an ancient, 3.6 Ga fire-fountain eruption (these are a common type of basaltic lava eruption in Hawaii and Iceland). The lava cooled quickly into tiny, spherical particles with an unusual composition and orange color.

An asteroid called 16 Psyche with a diameter of about 140 miles (226 km) orbits between Mars and Jupiter. Psyche is a metallic rather than rocky body and it may be the remains of the core of a former planet. It is said to contain up to $700 quintillion in usable metals at current prices, which is far more than has ever been mined on Earth throughout the entirety of human history. This is equivalent to $93 billion for each human being now alive. NASA planned to launch a probe to 16 Psyche in 2022 to explore it and analyze the geochemistry but software testing problems have delayed this launch until at least 2023.

The future of humanity's expansion into space may include mass drivers and space elevators that can lift materials into Earth orbit at a fraction of the cost of rockets, and new propulsion systems using solar sails, nuclear rockets, and ion drives to move around the solar system. It is not inconceivable that human-made spacecraft could soon be voyaging to the nearest stars using sails pushed by microwave or infrared lasers on the Earth or moon. Humans may even go to the stars ourselves if we learn the secrets to suspending and reanimating life for ultra-long-distance travel, or how to get a spacecraft up near lightspeed.

As with most of the exploration and colonization efforts in human history, the government role will soon be superseded by private companies and entrepreneurs. We are already seeing this with SpaceX, Blue Origin, and their competitors. The human desire to innovate, to cash in on opportunities, and be the first one through the door can be exploited for the benefit of Earth to get people and industry off this planet. It will be good for society, it will be good for the environment, and it will be especially good for the climate.

Predicting the future is always a risk. After all, most predictions from the 1950s about the state of twenty-first century technology are wrong. Many things that were supposed to happen, like flying cars, personal robots, and a nuclear power plant on every corner did not come about, and some things that did happen, like the personal computer and the internet were pretty much completely missed. Surprisingly, one prediction that was uncannily spot-on was the cell phone. Mark Sullivan, the president of the Pacific Telephone and Telegraph Company predicted in the Tacoma News Tribune newspaper in 1953 that, "In its final development, the telephone will be carried about by the individual, perhaps as we carry a watch today. It probably will require no dial or equivalent, and I think the users will be able to see each other if they want as they talk." So not only did Mr. Sullivan nail the cell phone, but he also predicted speed dial, Zoom and Facetime.

Chris Impey, an astronomy professor at the University of Arizona has published a book where he bravely charts our future progress.[8] Dr. Impey's predictions suggest that in twenty years there will be a vibrant commercial space industry. In thirty years, humanity will have established small but successful colonies on the moon and Mars. In fifty years, mining technology will be gathering resources from the asteroids. And in a century there will be a group of humans born off-Earth who will never have set foot on humanity's home world. Impey claims these are not flights of fantasy but the logical extension of already available technologies.

I have my unpredicted laptop in one hand and my predicted cell phone in the other.

We shall see.

****

If we move heavy industry and manufacturing into space along with a sizable population of colonists to run these operations and their associated support functions, what happens to the Old Earth? Well, obviously we are not going to empty it out. Despite the millions who emigrated to America from Europe in the nineteenth century, there were still plenty of people left in the Old Country. Even if half the human

---

[8] Impey, C., 2015, <u>Beyond: Our Future in Space</u>: New York, NY, W.W. Norton & Company, 336 pages, ISBN: 978-0-393-35215-3

race goes off planet and into the space colonies, which is unlikely, some four billion people will remain on Earth.

What space manufacturing does is remove from Earth the most resource-intensive and environmentally destructive industries. These industries can operate more cheaply in space where there is unlimited free energy from the sun and no restrictions on what kind of contaminants and byproducts they can release. If materials can be easily obtained at a low energy cost from asteroids or the moon, space manufacturing will become the backbone of industry. There will be no need to mine, refine, smelt, or process materials on the Earth when they can be imported cheaply from space. Manufactured goods from the moon or the O'Neill habitats can be shipped to Earth at a low energy cost. The trip is mostly "downhill." The humans who remain on Earth will work in a service-based economy, and factories and mines will be replaced with office parks. It does occur to me, however, that the Law of Unintended Consequences may still be in force, even in space. Imagine the massive manufacturing colonies established at the Earth's Lagrange points freely dispose of contaminants by emitting them into space. Now imagine that the solar wind carries these into the outer solar system where they land on Jupiter's moon Europa, which may have life in the deep ocean under the ice cap. We could end up with a new environmental problem affecting another world, and maybe this is one of the things we ought to consider as we set up shop. I think expanding into space is the right answer for humanity, but we should try to do it without repeating the mistakes of the past.

The renowned late naturalist Edward O. Wilson (1929–2021) of Harvard University famously called for setting aside half the planet as a nature preserve. Wilson was a strong advocate for preserving biodiversity among plant and animal species and cautioned that avoiding catastrophic climate change was an integral part of this. His proposal was first outlined in 2016 and is called the Half-Earth Project. Wilson proposed that half of the planet's land and sea area should be protected to provide enough diverse habitat and well-connected ecosystems to slow and eventually reverse the rapid extinction of species that we are currently witnessing. There are about one million species at present that are threatened or endangered.

The Half Earth Project is probably not possible with crowded cities, burgeoning industries, deforestation, over-fishing, and extensive agricultural lands. About the best we seem to be able to accomplish is what the United Nations calls "30 by 30," a request that nations commit to conserving 30% of their land and water by 2030. Given that this is almost double the amount of land and water currently under some form of protection, it would represent some major progress even if it does fall well short of Wilson's goal.

However, if we transfer heavy industry, manufacturing, mining, and energy production to space, the Earth will have a fighting chance to recover from the ravages of humanity. Leaving the cradle will allow the Earth to heal, and taking our desire for unlimited growth out into the infinite universe will relieve the pressure on Earth's resources. Endangered species will have an opportunity to rebound, and the climate will have a chance to return to normal. It will take time, possibly centuries, but eventually the Earth will recover and return to a natural state. If humans are living on the planet in a sustainable manner with minimal impacts to air, water, landscapes, and climate, perhaps we will finally achieve the long-sought after "harmony with nature."

In reference to avoiding catastrophic climate change, Wilson said in an interview prior to the COP26 conference in 2021, "This is the most communal endeavor with a clear definable goal that humanity has ever had. If we don't get the kind of cooperation, ethical harmony and planning needed to make it work, the slope of human history will always be downward."

In other words, we need to stop damaging and start repairing the planet for our own survival and that of a million other species. And we need to do it now.

Glossary

Adsorption vs. absorption Adsorption is the attachment of gas molecules to an electrochemically active substrate such as carbon in coal; absorption is the movement of a liquid under capillary pressure into a porous material.

AMD Acid mine drainage produced when iron sulfide minerals associated with coal oxidize to sulfates, which then react with water to form sulfuric acid.

Amine A type of chemical that bonds to CO_2 and is used for carbon capture processes.

Angular unconformity A geological feature where horizontal sedimentary rocks overlie older strata that are tilted at a sharp angle.

Anion An ion in solution that has gained an electron and thus carries a negative electrical charge.

Anthropocene Unofficial term for the geologic time period (approximately 1950 to present) where human influence has dominated Earth processes.

Anthropogenic Human-produced

Arctic acceleration The more prominent manifestation of climate change effects in the polar regions of Earth compared to mid-latitudes or the tropics.

Base load Steady, minimum power levels retained in an electrical utility system to meet routine power demands.

Batholith a large body of intrusive igneous rock; smaller intrusions are called plutons.

BECCS Biofuel energy with carbon capture and storage: A method that uses photosynthesis to capture carbon dioxide from the air, convert it to a biofuel, and then capture and store the biofuel combustion products, effectively removing GHG from the atmosphere.

BCE Before the Common Era and Common Era (CE) are notations for the Gregorian calendar as religiously-neutral alternatives to Before Christ (BC) and Anno Domini (AD).

Biochar A type of charcoal derived from wood waste used to sequester carbon in soils.

Biogas Biological methane that can directly substitute for natural gas, produced from organic matter under anoxic conditions by a class of microbes known as methanogens.

Biogenic methane Natural gas generated by microbial activity on organic material in a sediment during the early stages of thermal maturation.

Biomineralization The conversion of CO_2 into solid carbonate (CO_3) minerals by organisms.

Biomining Using plants or microbes to concentrate critical minerals for harvesting.

Bitumen (asphalt) A heavy crude oil that is a common residual deposit in oil seeps after the lighter hydrocarbons have evaporated off.

BOP Blow-out preventer; a hydraulic ram designed to close off a wild well.

Brayton Cycle Open energy system where gas heated by combustion is compressed, flows across a turbine and produces power, exiting the system as exhaust; characterized by the absence of a phase change; a jet engine is an example.

BTEX Benzene, toluene, ethylbenzene, and xylenes; the main components of gasoline.

Btu British thermal unit; the quantity of heat required to raise the temperature of 1 pound of liquid water by 1 °F; the metric unit is a calorie: the energy required to raise the temperature of 1 gram of water by 1 °C (1 Btu equals about 250 calories).

CAPEX Capital expenses; the cost of building a facility or process including land, materials, housing, machinery, and anything else that's a one-time, permanent purchase.

Carbon cycle the balanced interchange of carbon among the atmosphere, oceans, soils, rocks, and biomass, unbalanced by fossil fuel combustion products.

Carbon tax Surcharge of fossil fuel to combat climate change

Carbonation A chemical reaction that binds carbon dioxide as solid carbonate minerals; for example: sodium hydroxide + CO_2 → sodium bicarbonate (baking soda).

Carbonic acid H_2CO_3, formed when carbon dioxide dissolves in water; responsible for ocean acidification.

Capacity factor The percentage of time an electric power source actually spends generating electricity.

Cation An ion in solution that has lost an electron and thus carries a positive electrical charge.

CBM Coal Bed Methane: natural gas associated with and produced from coal seams.

CC Combined Cycle; a type of super-efficient power plant that uses natural gas to fire a Brayton cycle gas turbine and then directs the hot exhaust to a boiler, where it creates steam for a Rankine cycle steam turbine.

CCS Carbon Capture and Storage technology; removal of CO_2 at a combustion source before it can be released into the atmosphere followed by sequestration.

CCUS Carbon Capture, Utilization, and Storage: An attempt by DOE to make CCS more economically viable by finding a use for the captured carbon that generates income.

CDR Carbon Dioxide Removal; a geoengineering technique to reduce carbon dioxide levels in the atmosphere by capturing emissions and by removing carbon dioxide from the air.

CFCs Chlorofluorocarbons, a class of inert, synthetic gases used for refrigeration and fire extinguishing that damage the protective ozone layer and are powerful greenhouse gases. Their use was banned internationally under the Montreal Protocol signed in 1987.

CH₄ Methane (a greenhouse gas)

Circular economy Atmospheric GHG is turned into useful products and chemicals like plastics or fuels where sequestration is durable, and any emissions are carbon-neutral.

CMG Compressed methane gas; biogas if generated by microbes.

CO₂ Carbon dioxide (a greenhouse gas)

COE Cost of electricity; comparison of CAPEX and OPEX of electricity from different sources.

Collective action theory The sociology of people in large groups to stand back and let others take responsibility for dealing with a crisis.

Comminution To reduce to small particles or pulverize

Condensate Short chain hydrocarbons such as butane, propane, ethane, etc. that occur in the vapor phase downhole, but condense to liquids at the surface.

Constantinus Africanus Tenth century Byzantine scholar who named petroleum after a Greek word for rock oil.

Critical minerals rare earth elements, cadmium, cobalt, nickel, and other scarce mineral commodities that are needed for electric vehicles, nuclear power, and other energy uses.

DAC Direct Air Capture; removal and sequestration of CO_2 directly from the atmosphere.

DAPL The Dakota Access Pipeline: opened in 2017, carries 31.5 million gallons of Bakken crude oil per day 1172 miles (1875 km) from North Dakota to Illinois.

Decline curve The rate at which oil and gas production drops during the life of a well.

Degrowth An economic philosophy focused on "sufficiency" rather than endless growth that seeks to reduce dependence on extraction and production processes that are not environmentally sustainable.

Diapir A sedimentary structure where less dense sediments rise up through overlying, denser sediments and deform them in the process.

Dirty bomb A terrorist weapon consisting of a conventional explosive loaded with radionuclides to widely spread radioactive contamination.

DOD The U.S. Department of Defense

DOE The U.S. Department of Energy

Dry gas Natural gas that occurs without any associated condensate such as butane or propane.

Durability A term related to the length of time a sequestration process will keep CO_2 isolated from the atmosphere. The minimum requirement is a century, and longer is better.

Ecofascism White supremacist doctrine that the climate crisis must be addressed to prevent non-white refugees from overrunning majority white nations.

EDBM Electro-dialysis bipolar membrane: a device that separates dissolved ions by electrical charge in a water solution.

Edwin Drake "Colonel" Drake drilled the first well specifically and deliberately in search of oil near Titusville, Pennsylvania in 1859.

EGS Enhanced (or engineered) geothermal systems: an engineering design to drill lateral wells deep into hot rock, hydraulically fracture between them to create flowpaths, and then extract geothermal heat by circulating fluid through the fractures.

EJ Environmental justice: the concern that industrial sites will be located in marginalized or minority neighborhoods forcing residents to put up with bad air, noise, and pollution.

EIA U.S. Energy Information Administration: a government agency that compiles annual data and trends on energy production and consumption in the U.S. and globally.

Embodied emissions Greenhouse gas emissions from the manufacturing process for materials like steel, concrete, or other manufactured products. Generally scope 1 and 2.

Enriched uranium Nuclear power plant fuel consisting mostly of $_{238}U$ enriched with a small percentage of $_{235}U$ to support a modest chain reaction and generate heat.

EOR Enhanced oil recovery: a process that uses carbon dioxide injection or water injection to push residual oil in the pore space of a nearly depleted oilfield to production wells.

EPA The U.S. Environmental Protection Agency

Errata Statements added to a scientific paper after publication to correct small errors.

ESG environmental, social and governance; so-called "woke" corporate concerns over quality of life issues that take priority over straight profits.

EV Electric vehicles operated purely on rechargeable battery power.

Externalized cost An environmental cost passed on to taxpayers.

Extremophiles Microbes that can survive at high temperatures, high pressures, high salinities, or other extreme environmental conditions.

FECM The DOE Office of Fossil Energy and Carbon Management, formerly just the Office of Fossil Energy.

Flambeau A gas flare maintained on early wells to prove to investors that the gas was flowing.

FORGE Frontier Observatory for Research in Geothermal Energy; Milford, Utah: DOE Geothermal Technology Office field experiments in enhanced geothermal systems.

Fracking Hydraulic fracturing; creating high permeability flowpaths into a rock by cracking the rock open using static water pressure.

Fugitive emissions leakage of natural gas, primarily methane, from surface equipment such as pipelines and compressors directly into the atmosphere.

Ga Giga-annum; geologic abbreviation for a billion years.

Geoengineering A process of planetary-wide engineering to adjust atmospheric composition, surface temperature, or other factors related to the climate.

Geothermal gradient Increase in temperature with depth in the Earth; varies from place to place but averages 25–30 °C/km (15 °F/1000 ft).

GDP Gross domestic product: The total value of all the finished goods and services produced within a nation over a specific time period, usually a year.

GCC Global Climate Coalition: A 1990s fossil fuel industry organization charged with casting doubt on the science behind anthropogenic climate change.

GHG Greenhouse gas; carbon dioxide, methane, water vapor, and other gases that absorb longwave IR radiated from the Earth and warm the atmosphere.

Greenwashing A company advertising itself as being environmentally-conscientious when they actually are not.

GT Gigaton (British spelling gigatonne) one billion metric tons

GTO The DOE Geothermal Technology Office under the Office of Energy Efficiency and Renewable Energy (EERE)

GW Gigawatt; one billion watts of electricity.

HV Hybrid vehicles; these have the option to operate electrically on battery power or use an internal combustion engine as driving conditions warrant. In standard HVs, the gasoline engine recharges the battery; plug-in HVs allow the battery to be recharged using line power and use the gasoline engine only when necessary.

Hydroelectric Electricity produced by the force of moving water, either flow through dams or along the gradient of the river. Often just called "hydro."

Hydrogen embrittlement The failure in metals, especially steel, when tiny hydrogen molecules enter the space between grains, reduce bonding and create fractures in the metal.

Hydrostatic pressure gradient The increase with depth of the fluid pressure in the pore systems of rocks; usually about 0.5 psi per foot.

Ibn Sina Ninth century Persian physician who wrote an encyclopedia of medicine that detailed the medicinal uses of petroleum.

IDF Israel Defense Forces; the Israeli military

Independents Industry term for medium-size oil companies

Induced seismicity Human-triggered earthquakes caused by injecting fluids into the ground, raising subsurface pore pressures and causing slip on existing faults.

Industrial Revolution Mid-eighteenth century transition in Great Britain from individual handcrafting of products to mass production manufacturing using machines.

IP Initial production; the start of oil and gas production from a new well, at generally the highest rate the well will ever see (see decline curve). The common non-petroleum use of IP is for Intellectual Property.

IPCC Intergovernmental Panel on Climate Change (United Nations)

IR Infrared radiation; wavelengths of electromagnetic radiation longer than red light that carry heat. Shortwave IR from the sun penetrates the atmosphere and warms the Earth, while longwave IR from the Earth is absorbed by GHG and heats the atmosphere.

ISL In-situ leaching; a solution mining method for extracting uranium by dissolving it underground and transporting it to the surface.

Ka Kilo-annum; a geologic abbreviation for a 1000 years.

Keeling Curve Data from the Mauna Loa Scripps/NOAA observatory showing a steady increase in atmospheric CO_2 concentrations since measurements began in 1957.

KT kiloton, 1000 tons

KW kilowatt; 1000 watts of electricity.

KWH Kilowatt-hour; a measure of electrical output. A 100 KWH battery will supply a KW of power for 100 hours.

LACE Levelized Avoided Cost of Electricity: the cost of electricity from not adding a specific generating technology but supplying the additional capacity using existing sources.

Lateral A horizontal or directional borehole drilled in shale to maximize contact with the rock.

LCA Life Cycle Analysis; a method of tallying up scope 1, 2, and 3 GHG emissions for a company or a facility from the beginning to the end of its existence.

LCOE Levelized Cost of Electricity; the cost per-kilowatt-hour to build and operate a generating plant.

Li Lithium-ion battery primarily used in EVs

Lithostatic pressure gradient increase with depth of the pressure on rocks from the weight of the overburden; usually about 1 psi per foot.

LNG Liquefied natural gas; a cryogenic liquid at $-260\ °F$ ($-126\ °C$) that occupies 600 times less volume than the gaseous phase.

LUST Leaking underground storage tank, primarily for gasoline, and responsible for significant amounts of groundwater contamination in the U.S.

Ma Mega-annum; a geologic abbreviation for a time period of one million years.

Majors Industry term for large, multinational oil companies

Montreal Protocol A 1987 agreement among all the nations of the world to ban the use of CFCs to protect the Earth's ozone layer.

MOX Mixed oxide fuel; reprocessed reactor fuel rods where the daughter products have been removed and the remaining fissionable materials have been remade into new fuel rods.

MT Megaton; one million tons.

MTBE Methyl-tertiary butyl ether; a compound added to gasoline in the late twentieth century to reduce air pollution. It severely contaminated groundwater when it leaked.

MTR Mountaintop Removal; a brutal coal mining process that strips overburden off the top of coal seams in highland areas and dumps it into surrounding ravines.

μm Micrometer: a unit of length one millionth of a meter, or one thousandth of a millimeter

MVA Monitoring, verification, and accounting: DOE procedures for keeping track of carbon dioxide gas sequestered in geologic formations in the subsurface.

MW megawatt; one million watts of electricity.

NDC Nationally-determined contributions: plans developed by countries that signed the 2015 Paris Agreement to meet the 1.5 degree C target for global warming.

NETL The DOE National Energy Technology Laboratory

NIMBY "Not in my back yard:" the desire to have infrastructure as long as it is located elsewhere out of sight and out of mind.

NiMH Nickel-metal hydride battery; primarily used in HVs

ODS Ozone-depleting substance; manufactured chemicals that break down high in the stratosphere and release chlorine or bromine atoms that destroy the ozone layer.

Oil Creek Association An oil producers group created in 1861 (2 years after the completion of the Drake well) to restrict output and maintain oil prices at $4 a barrel. Possibly the world's first oil cartel.

Oil window The zone of thermal maturity that will generate oil from the proper organic matter in sediment.

OPEX Operating expenses; the cost of operating a manufacturing or production process, including material, labor, and energy.

Oxy-combustion The process of burning coal or natural gas in an atmosphere of pure oxygen instead of air to increase the concentration of CO_2 in flue gases for easier capture.

P&A plug and abandon; the state-mandated process for properly sealing off an old oil or gas well. Involves setting a bridge plug above the production zone and filling the well with concrete. Many operators renew the lease annually as a much cheaper option.

Paris Agreement A legally binding, international treaty on climate change adopted by 196 parties in 2015 with the goal of limiting global warming to less than 2 °C.

Peak oil A concept developed by M. King Hubbert in 1956 describing the increase, apex, and decline of oil production as a conventional oil field is developed and produced.

Peak shaving supplying electrical power during periods of high demand or "peak loads."

Peer review A method for correcting scientific errors.

PETM The Paleocene-Eocene Thermal Maximum, a global temperature rise of about 8 °C that occurred around 56 Ma.

Phenocryst A larger crystal embedded in a finer-grained or glassy igneous rock, usually of a mineral that began to crystallize before the eruption.

Pluton a medium to small body of intrusive igneous rock; large intrusions are called batholiths.

Polar vortex A strong band of winter winds that blow around the North Pole at very high altitudes in the stratosphere, held in place by the polar jet stream.

Predatory pricing A method of undercutting competitors' prices to drive their business into bankruptcy, eliminating competition and building a monopoly.

ppm Parts per million; a chemical concentration

PV photovoltaic; solar cells that produce electricity directly from sunlight through the photoelectric effect first defined by Albert Einstein in 1905.

Rankine Cycle Closed system that uses a liquid-to-gas phase change, i.e., water into steam to drive a turbine; vapor phase is then condensed back to liquid and re-used.

REE Rare earth elements; important technology materials currently produced mostly in China.

Referees Recognized experts in the field who peer review the scientific work

"Rolling coal" An act of rebellion against climate by teenage boys that consists of spewing clouds of black diesel smoke from pickup trucks at bicyclists and electric vehicles.

SAGE Solar assisted geothermal energy; uses a shallow, porous aquifer and circulates groundwater through solar heating troughs at the surface, injecting the hot water back into the ground and using the aquifer to store heat.

Samuel Kier Mid-nineteenth century Pennsylvania entrepreneur who started selling petroleum from his saltwater wells as a medicine, then invented the kerosene lantern to displace whale oil as an illumination fuel.

Scope 1 emissions GHG emitted directly by company operations.

Scope 2 emissions GHG emissions from outside the company but are tied to company operations.

Scope 3 emissions Value chain GHG emissions from company products both upstream and downstream of manufacture.

Seneca oil Settler's term for petroleum collected from natural seeps by the Seneca tribe of the Iroquois Nation in North America for use as a salve, insect repellent, and cure-all tonic.

Sequestration The process by which captured carbon dioxide is kept isolated from the atmosphere for at least 100 years, and the longer the better.

Scientific hypothesis A conjecture based on data and observations

Scientific theory An attempt to construct a coherent explanation of the evidence by interpreting the data and experimental results used to test a hypothesis.

Shale gas Natural gas produced by fracking organic-rich, low permeability black shale.

Social cost of carbon The economic impact of de-carbonizing a segment of the economy.

SpaceX Space Exploration Technologies Corporation: a private company for launching satellites and manned spacecraft founded by Elon Musk.

Spent uranium The more abundant $_{238}U$ isotope after being separated from the more fissionable $_{235}U$ isotope.

Spindletop A low hill 3 miles south of Beaumont, Texas where the famous Lucas gusher occurred in January 1901.

SRM Solar radiation management; a geoengineering technology proposing to mitigate climate-induced higher temperatures by blocking some sunlight from reaching the ground.

Staged hydraulic fracturing A process of fracking for shale gas and tight oil that uses a series of individual fracks spaced along the length of a lateral wellbore.

Stray gas subsurface leakage of natural gas, primarily methane, from production wells into groundwater aquifers. It may or may not reach the atmosphere.

Supercritical fluid pressure and temperature conditions where compressed CO_2 gas suddenly condenses into a dense liquid.

TEA Techno-economic analysis for carbon capture and storage systems

Thermoelectric electric power produced by the force of steam or another working fluid pushing against a turbine driven by heat; the most common type of utility-scale power generation.

Thermogenic methane Methane gas generated from oil under high, late-stage thermal maturity.

Ticky-tacky suburban cookie-cutter tract homes built quickly after the war. A 1963 song by folk singer Pete Seeger called "Little Boxes" is about these homes.

Tight oil Typical light, low viscosity petroleum produced by fracking from low permeability rocks like shale.

TLA Three-letter acronym: Slang for companies located around the Washington D.C. beltway that thrive on Department of Defense, Department of Energy, and other government contracts. Also known as "beltway bandits."

Tonne British spelling for a metric ton, 1000 kg or 2205 pounds

Tonsteins Layers of volcanic ash deposited in a coal seam, considered a possible source of rare earth elements.

Town gas A manufactured fuel made by heating coal and water in the absence of oxygen to produce hydrogen and carbon monoxide gas for combustion.

Transuranics Heavy elements produced in a fission reactor such as plutonium that are heavier than uranium.

Trust An umbrella organization that controls a large number of subsidiary companies.

UIC Underground Injection Control: USEPA permitting system, largely implemented by state regulatory agencies, allowing the disposal of liquid wastes by injection into the ground.

USGS The United States Geological Survey.

Vesicles Frozen gas bubbles preserved in solidified lava.

Wet gas Natural gas that occurs in association with condensate.

Wokewashing A climate denial tactic that claims a transition away from fossil fuels will cause disproportionately serious harm to low income communities and countries.

Index

Printed in the United States
by Baker & Taylor Publisher Services